Matrix Theory - Classics and Advances

Edited by Mykhaylo Andriychuk

Published in London, United Kingdom

Matrix Theory - Classics and Advances
http://dx.doi.org/10.5772/intechopen.97927
Edited by Mykhaylo Andriychuk

Contributors
Mohammed Hadish, Elena A. Nikolaevskaya, Aleksandr N. Khimich, Igor A. Baranov, Amma Kazuo, Mykhaylo I. Andriychuk, Ioan R. Ciric, Sergey Zagorodnyuk, Huajun Huang, Ming-Cheng Tsai, Giovanni F. Crosta, Goong Chen, Gilles Burel, Hugo Pillin, Paul Baird, El-Houssaïn Baghious, Roland Gautier, Natalia Mihăilescu, Cristian N. Mihăilescu, Mihai Oane, Carmen Ristoscu, Muhammad Arif Mahmood, Ion N. Mihăilescu, Sung Kook Lee, Moon Ho Lee

First published in London, United Kingdom, 2023 by IntechOpen
IntechOpen is the global imprint of INTECHOPEN LIMITED, registered in England and Wales, registration number: 11086078, 5 Princes Gate Court, London, SW7 2QJ, United Kingdom

British Library Cataloguing-in-Publication Data
A catalogue record for this book is available from the British Library

Additional hard and PDF copies can be obtained from orders@intechopen.com

Matrix Theory - Classics and Advances
Edited by Mykhaylo Andriychuk
p. cm.
Print ISBN 978-1-80355-822-6
Online ISBN 978-1-80355-823-3
eBook (PDF) ISBN 978-1-80355-824-0

For EU product safety concerns:
IN TECH d.o.o., Prolaz Marije Krucifikse Kozulić 3, 51000 Rijeka, Croatia,
info@intechopen.com or visit our website at intechopen.com.

Meet the editor

Prof. Andriychuk received an MSc in Computational Mathematics from Lviv National University, Ukraine in 1976, a Ph.D. in Application of Computational Techniques from Kyiv National University, Ukraine in 1987, and a DSc in Mathematical Modelling from Lviv Polytechnic National University (LPNU), Ukraine in 2015. He has worked at the Pidstryhach Institute for Applied Problems of Mechanics and Mathematics (IAPMM), Ukraine for about 50 years. He is also a department head at the Pidstryhach Institute for Applied Problems of Mechanics and Mathematics (IAPMM), National Academy of Sciences of Ukraine and a professor at LPNU. He has published more than 180 papers in scientific journals and international conference proceedings on diffraction and antenna synthesis theory, optimization methods, and nonlinear integral equations. He is also the author of two books on antenna synthesis theory.

Contents

Preface

Matrix theory is a branch of mathematics that has been developing over many centuries and has been successfully used in both theoretical research and applied science.

In a theoretical sense, matrix theory is a powerful tool used to develop research in such areas of mathematics as algebra, combinatorics, graph theory, statistics, and so on. It is also used to solve many engineering problems in the fields of acoustics, fluid dynamics, electromagnetics, solid mechanics, build technology, and communications.

This book consists of two sections. Section 1 contains six chapters devoted to the development of such fields of matrix theory as pencils of matrices, semi-infinite matrices, matrices with perturbed elements, the specific product of matrices, homomorphisms of matrices, and extension for informatics.

In Chapter 1, H. Huang and M.-C. Tsai study the properties of matrices products related to the k-power preserver property. The authors introduce the subject and provide a literature review. The definitive theorem of Tsai is cited to emphasize the generality of results obtained in the author's subsequent research. The generalization of theorems on the k -power linear preservers starts for the case of a set of general matrices on the field of complex numbers, which is proven for both the case of positive and negative powers of k. The preliminary theorem, generalizing the known theorem of Chan and Lim, is applied to prove the result. The fundamental theorem is proved for a variety of sets (spaces) of matrices, such as complex Hermitian matrices, symmetric matrices, positive definite matrices, diagonal matrices, and triangular matrices. All the considered cases differ by assumptions that apply to the properties of operators in the initial condition of the theorems.

Chapter 2, by S. Zagorodnyuk, focuses on semi-infinite matrices, generalized eigenvalue problems, and orthogonal polynomials. The classical examples are Jacobi and Hessenberg matrices, which lead to orthogonal polynomials on the real line and to orthogonal polynomials on the unit circle. Pencils of semi-infinite matrices are related to various orthogonal systems of functions. The respective polynomials are defined as generalized eigenvectors of the pencil. The polynomials under investigation have a special orthogonality relation and they are useful for a series of physical and mathematical applications. The presented examples confirm that there is a certain relation to Sobolev orthogonal polynomials that is a challenge for further investigations.

In Chapter 3, A. Kazuo proposes the optimization matrix approach for the correct reordering of sentences in linguistics. The analysis starts with the usual methods based on the transformation of data characteristics that define the correctness of the transformed sentences. The chapter describes the maximal relative sequence evaluation and applies it to resolve the problem of correcting the arrangement of the words. To do this effectively, additional means of evaluation, such as the recovery distance, should be used. To use this approach in a PC environment, the authors use Excel. However, even though the problem data are not big in size, the number of necessary columns grows drastically. A more effective approach consists of introducing tools that decrease the required amount of memory. The illustrated application of the linearity matrix method confirms the effectiveness of the sentence reordering procedure in several examples. The evaluation of the obtained results demonstrates the possibility of using PC tools to check and correct big linguistic data.

In Chapter 4, A. N. Khimich et al. investigate the problems of weighted pseudoinverse matrices and weighted least squares (WLS). The first part of the chapter examines the sensitivity of the solution to the WLS problem with approximate initial data. The second part investigates the properties of a system of linear algebraic equations with approximate initial data and presents an algorithm for finding a weighted normal pseudosolution to the WLS problem with approximate initial data. The developed algorithm is extended for solving a WLS problem with symmetric positive semidefinite matrices and an approximate right side. The third part of the chapter analyses the exactness of the numerical solution to the WLS problem with approximate initial data, discusses the software-algorithmic approaches for improving the accuracy of computer solutions, and estimates the total error of the solution to the WLS problem.

Chapter 5, by M. Hadish is devoted to the evaluation of the errors of periodic functions by the Cesáro-Matrix product involving the conjugate Fourier series. The chapter presents an original approach related to the generalization of convergence of series that is not summarized in the classical sense. Evidence shows that the Cesáro-Matrix approach is a powerful tool for obtaining the sum of series when both the usual matrix approach and Cesáro means are not applicable. The authors prove two theorems that generalize the classical results related to slowly convergent series. Some important corollaries, which are perspective for extraction of convergence for the specific slowly convergent series, follow from the theorems. This technique has potential use in many engineering problems in which the computations lead to the calculation of slowly convergent series.

In Chapter 6, by Ivan I. Kyrchei the notions of the MPCEP inverse and CEPMP inverse are expanded to quaternion matrices and introduced new generalized inverses, the right and left MPCEPMP inverses. Direct method of their calculations, that is, their determinantal representations are obtained within the framework of theory of quaternion row-column determinants previously developed by the author. In consequence, these determinantal representations are derived in the case of complex matrices.

Section 2, consisting of six chapters, focuses on practical medicine, information theory, heat transfer, and antenna synthesis as related to the formation of COVID-19's genetic code, energy conversion processes, quantum information theory processing, solving differential and linear equations, and branching solutions to nonlinear integral equations.

In Chapter 7, S. K. Lee and M. H. Lee propose the analytical justification of the parameters of the Covid-19 genetic code predicted experimentally. This is realized by involving the information theory proof based on the doubly stochastic matrix. The genetic code model is considered in the framework of two symmetric probabilistic channels (DNA-RNA genetic code) with different parameters of input data, which differ from the classical ones proposed by E. Chargaff. Because the computational realization of the model was not implemented until now, the authors developed a simple solution using the information theory of doubly stochastic matrix over the Shannon symmetric channel. It was proved that DNA-RNA genetic code is some kind of block circulant jacket matrix. Moreover, the chapter explores the abnormal patterns by block circulant, upper-lower, and left-right schemes that cover the distorted signal as well as the Covid-19 evolution.

In Chapter 8, G. Burel et al. demonstrate the application of linear algebra fundamentals to quantum information processing. It is shown that in many practical cases a matrix representation of the quantum systems is a powerful tool because it allows the use of linear algebra to better understand their behaviours and to better implement simulation procedures. The authors focus on Joint EigenValue Decomposition (JEVD) for the development of quantum processing. The theoretical description of the method, which aims to find a common basis of eigenvectors of a set of matrices, is supported by the effective implementation of matrix-oriented programming languages (MATLAB or Octave). It is established how to determine the encoding matrix of a quantum code from a collection of Pauli errors that opens a perspective for future study related to the interception of quantum channels and identification of the quantum coder used by a non-cooperative transmitter. Using JEVD, the existence of a subspace of the whole Hilbert space, which captures the essence of the search process, is proved. In addition, an algorithm that allows us to check this result by simulation is given.

In Chapter 9, G. F. Crosta and G. Chen model three-phase, doubly fed induction (DFI) machines by the inductance matrices with related electric and magnetic quantities. It introduces the algebraic properties of the mutual (rotor-to-stator) inductance matrix, namely, its kernel, range, and left zero divisors. An exponential representation of an inductance matrix under suitable hypotheses is derived to obtain a simple recurrent formula for the powers of the corresponding infinitesimal generator. In addition, the transformation into an exponential form is derived axiomatically. The proof of the electric torque theorem is simplified owing to newly derived formula for the product of matrices that leads in relation to the Legendre transform. As a result, a simple realistic machine model with the broken three-fold rotor symmetry is discussed and some properties for the resulting mutual inductance matrix are obtained.

A new approach to solving a non-Fourier heat equation is developed by C. N. Mihăilescu et al. in Chapter 10. This leads to the necessity to check the validity/limits of the integral transform technique on finite domains. The proposed technique is based on the eigenvalues and eigenfunctions of the respective matrices, and its applicability to both the laser and electron beam processing problems is examined. The advantage of the method is its ability to obtain a solution with a small number of iterations and high accuracy for the like Fourier equation. However, additional efforts are needed to apply the approach proposed for the non-Fourier heat equation that is explained by the slow convergence. One such effort is applying the extra-boundary conditions. To avoid the problem with convergence, a new mixed approach is elaborated that provides the required characteristics of convergence.

In Chapter 11, M. Andriychuk focuses on the development of analytical-numerical methods for solving non-linear integral equations related to the generalized problem of phase optimization. The definitive property of such equations is that they are non-linear because of the specificity of the problem under consideration; therefore, the non-uniqueness of solutions appears. To extract a set of solutions, the respective homogeneous non-linear integral equation that results in a non-linear eigenvalue problem is used. Effective numerical algorithms are developed to find the respective eigenvalues and eigenfunctions. The study of the eigenvalues' peculiarities allows us to determine a set of points, in which the respective eigenvalues are equal to unity that determines the branching points of solutions. The total solution to the initial non-linear equation can be presented in terms of the obtained eigenfunctions. The data of calculations testify to the ability of the proposed approach to finding solutions to non-linear integral equations numerically without large computations.

Chapter 12, by I. R. Ciric, focuses on the application of matrix differential equations for solving systems of linear algebraic equations. The accurate solutions were derived in terms of a new kind of an infinite series of matrices, which are truncated and applied repeatedly to approximate the solution. Each new term in these matrix series is obtained by multiplication on a matrix, which becomes as conditioned tending to the identity matrix that results in the effective applying the computations based on the iterative procedure. The solution method is flexible to change the initial problem's parameters. Efficient computation of an approximate solution, applicable even to poorly conditioned systems, is demonstrated based on the alternate application of two different types of minimization of associated functionals. Large computation is not needed to obtain an approximate solution for large linear systems as compared to usual methods.

It is my great pleasure to thank all book authors for their enthusiasm, patience, and improvement of the chapters throughout the reviewing process. In addition, I express my sincere thanks to Ms. Jelena Vrdoljak for her professional support during the book's preparation.

Mykhaylo Andriychuk,
Pidstryhach Institute for Applied Problemsof Mechanics and Mathematics, NASU,
Lviv Polytechnic National University,
Lviv, Ukraine

Section 1

Theory and Progress

Chapter 1

Linear K-Power Preservers and Trace of Power-Product Preservers

Huajun Huang and Ming-Cheng Tsai

Abstract

Let V be the set of $n \times n$ complex or real general matrices, Hermitian matrices, symmetric matrices, positive definite (resp. semi-definite) matrices, diagonal matrices, or upper triangular matrices. Fix $k \in \mathbb{Z} \backslash \{0, 1\}$. We characterize linear maps $\psi : V \to V$ that satisfy $\psi(A^k) = \psi(A)^k$ on an open neighborhood S of I_n in V. The k-power preservers are necessarily k-potent preservers, and $k = 2$ corresponds to Jordan homomorphisms. Applying the results, we characterize maps $\phi, \psi : V \to V$ that satisfy "$tr\left(\phi(A)\psi(B)^k\right) = tr(AB^k)$ for all $A \in V$, $B \in S$, and ψ is linear" or "$tr\left(\phi(A)\psi(B)^k\right) = tr(AB^k)$ for all $A, B \in S$ and both ϕ and ψ are linear." The characterizations systematically extend existing results in literature, and they have many applications in areas like quantum information theory. Some structural theorems and power series over matrices are widely used in our characterizations.

Keywords: k-power, power preserver, trace preserver, power series of matrices

1. Introduction

Preserver problem is one of the most active research areas in matrix theory (e.g. [1–4]). Researchers would like to characterize the maps on a given space of matrices preserving certain subsets, functions or relations. One of the preserver problems concerns maps ψ on some sets V of matrices which preserves k-power for a fixed integer $k \geq 2$, that is, $\psi(A^k) = \psi(A)^k$ for any $A \in V$ (e.g. [3, 5, 6]). The k-power preservers form a special class of polynomial preservers. One important reason of this problem lies on the fact that the case $k = 2$ corresponds to Jordan homomorphisms. Moreover, every k-power preserver is also a k-potent preserver, that is, $A^k = A$ imply that $\psi(A)^k = \psi(A)$ for any $A \in V$. Some researches on k-potent preservers can be found in [6–8].

Given a field $\mathbb{F}$, let $\mathcal{M}_n(\mathbb{F})$, $\mathcal{S}_n(\mathbb{F})$, $\mathcal{D}_n(\mathbb{F})$, $\mathcal{N}_n(\mathbb{F})$, and $\mathcal{T}_n(\mathbb{F})$ denote the set of $n \times n$ general, symmetric, diagonal, strictly upper triangular, and upper triangular matrices over $\mathbb{F}$, respectively. When $\mathbb{F}$ is the complex field $\mathbb{C}$, we may write $\mathcal{M}_n$ instead of $\mathcal{M}_n(\mathbb{C})$, and so on. Let $\mathcal{H}_n$, $\mathcal{P}_n$, and $\overline{\mathcal{P}_n}$ denote the set of complex Hermitian, positive definite, and positive semidefinite matrices, and $\mathcal{H}_n(\mathbb{R}) = \mathcal{S}_n(\mathbb{R})$, $\mathcal{P}_n(\mathbb{R})$, and $\overline{\mathcal{P}_n(\mathbb{R})}$ the corresponding set of real matrices, respectively. A matrix space is a subspace of $\mathcal{M}_{m,n}(\mathbb{F})$ for certain $m, n \in \mathbb{Z}_+$. Let A^t (resp. A^*) denote the transpose (resp. conjugate transpose) of a matrix A.

In 1951, Kadison [9] showed that a Jordan $*$-isomorphism on $\mathcal{M}_n$, namely, a bijective linear map with $\psi(A^2) = \psi(A)^2$ and $\psi(A^*) = \psi(A)^*$ for all $A \in \mathcal{M}_n$, is the direct sum of a $*$-isomorphism and a $*$-anti-isomorphism. Hence $\psi(A) = UAU^*$ for all $A \in \mathcal{M}_n$ or $\psi(A) = UA^TU^*$ for all $A \in \mathcal{M}_n$ by [[3], Theorem A.8]. Let $k \geq 2$ be a fixed integer. In 1992, Chan and Lim ([5]) determined a nonzero linear operator $\psi : \mathcal{M}_n(\mathbb{F}) \to \mathcal{M}_n(\mathbb{F})$ (resp. $\psi : \mathcal{S}_n(\mathbb{F}) \to \mathcal{S}_n(\mathbb{F})$) such that $\psi(A^k) = \psi(A)^k$ for all $A \in \mathcal{M}_n(\mathbb{F})$ (resp. $\mathcal{S}_n(\mathbb{F})$) (See Theorems 3.1 and 5.1). In 1998, Brešar, Martindale, and Miers considered additive maps of general prime rings to solve an analogous problem by using the deep algebraic techniques ([10]). Monlár [[3], P6] described a particular case of their result which extends Theorem 3.1 to surjective linear operators on $\mathcal{B}(\mathcal{H})$. In 2004, Cao and Zhang determined additive k-power preserver on $\mathcal{M}_n(\mathbb{F})$ and $\mathcal{S}_n(\mathbb{F})$ ([11]). They also characterized injective additive k-power preserver on $\mathcal{T}_n(\mathbb{F})$ ([12] or [[6], Theorem 6.5.2]), which leads to injective linear k-power preserver on $\mathcal{T}_n(\mathbb{F})$ (see Theorem 8.1). In 2006, Cao and Zhang also characterized linear k-power preservers from $\mathcal{M}_n(\mathbb{F})$ to $\mathcal{M}_m(\mathbb{F})$ and from $\mathcal{S}_n(\mathbb{F})$ to $\mathcal{M}_m(\mathbb{F})$ (resp. $\mathcal{S}_m(\mathbb{F})$) [8].

In this article, given an integer $k \in \mathbb{Z}\backslash\{0, 1\}$, we show that a unital linear map $\psi : V \to W$ between matrix spaces preserving k-powers on a neighborhood of identity must preserve all integer powers (Theorem 2.1). Then we characterize, for $\mathbb{F} = \mathbb{C}$ and $\mathbb{R}$, linear operators on sets $V = \mathcal{M}_n(\mathbb{F})$, $\mathcal{H}_n$, $\mathcal{S}_n(\mathbb{F})$, $\mathcal{P}_n$, $\mathcal{P}_n(\mathbb{R})$, $\mathcal{D}_n(\mathbb{F})$, and $\mathcal{T}_n(\mathbb{F})$ that satisfy $\psi(A^k) = \psi(A)^k$ on an open neighborhood of I_n in V. In the following descriptions, $P \in \mathcal{M}_n(\mathbb{F})$ is invertible, $U \in \mathcal{M}_n(\mathbb{F})$ is unitary, $O \in \mathcal{M}_n(\mathbb{F})$ is orthogonal, and $\lambda \in \mathbb{F}$ satisfies that $\lambda^{k-1} = 1$.

1. $V = \mathcal{M}_n(\mathbb{F})$ (Theorem 3.4): $\psi(A) = \lambda PAP^{-1}$ or $\psi(A) = \lambda PA^tP^{-1}$.

2. $V = \mathcal{H}_n$ (Theorem 4.1): When k is even, $\psi(A) = U^*AU$ or $\psi(A) = U^*A^tU$. When k is odd, $\psi(A) = \pm U^*AU$ or $\psi(A) = \pm U^*A^tU$.

3. $V = \mathcal{S}_n(\mathbb{F})$ (Theorem 5.2): $\psi(A) = \lambda OAO^t$.

4. $V = \mathcal{P}_n$ or $\mathcal{P}_n(\mathbb{R})$ (Theorem 6.1): $\psi(A) = U^*AU$ or $\psi(A) = U^*A^tU$.

5. $V = \mathcal{D}_n(\mathbb{F})$ (Theorem 7.1): $\psi(A) = \psi(I_n)\mathrm{diag}\left(f_{p(1)}(A), \dots, f_{p(n)}(A)\right)$, in which $\psi(I_n)^k = \psi(I_n)$, $p : \{1, \dots, n\} \to \{0, 1, \dots, n\}$ is a function, and $f_i : \mathcal{D}_n(\mathbb{F}) \to \mathbb{F}$ $(i = 0, 1, \dots, n)$ satisfy that, for $A = \mathrm{diag}(a_1, \dots, a_n)$, $f_0(A) = 0$ and $f_i(A) = a_i$ for $i = 1, \dots, n$.

6. $V = \mathcal{T}_n(\mathbb{F})$ (Theorem 8.4 for $n \geq 3$): $\psi(A) = \lambda PAP^{-1}$ or $\psi(A) = \lambda PA^-P^{-1}$, in which $P \in \mathcal{T}_n(\mathbb{F})$ and $A^- = (a_{n+1-j,n+1-i})$ if $A = (a_{ij})$.

Our results on $\mathcal{M}_n(\mathbb{F})$ and $\mathcal{S}_n(\mathbb{F})$ extend Chan and Lim's results in Theorems 3.1 and 5.1, and result on $\mathcal{T}_n(\mathbb{F})$ extend Cao and Zhang's linear version result in [12].

Another topic is the study of a linear map ϕ from a matrix set S to another matrix set T preserving trace equation. In 1931, Wigner's unitary-antiunitary theorem [[3], p. 12] says that if ϕ is a bijective map defined on the set of all rank one projections on a Hilbert space H satisfying

$$\mathrm{tr}(\phi(A)\phi(B)) = \mathrm{tr}(AB), \tag{1}$$

then there is an either unitary or antiunitary operator U on H such that $\phi(P) = U^*PU$ or $\phi(P) = U^*P^tU$ for all rank one projections P. In 1963, Uhlhorn generalized Wigner's theorem to show that the same conclusion holds if the equality $\mathrm{tr}(\phi(P)\phi(Q)) = \mathrm{tr}(PQ)$ is replaced by $\mathrm{tr}(\phi(P)\phi(Q)) = 0 \Leftrightarrow \mathrm{tr}(PQ) = 0$ (see [13]).

In 2002, Molnár (in the proof of [[14], Theorem 1]) showed that maps ϕ on the space of all bounded linear operators on a Banach space $B(X)$ satisfying (1) for $A \in B(X)$, rank one operator $B \in B(X)$ are linear. In 2012, Li, Plevnik, and Šemrl [15] characterized bijective maps $\phi : S \to S$ satisfying $\mathrm{tr}(\phi(A)\phi(B)) = c \Leftrightarrow \mathrm{tr}(AB) = c$ for a given real number c, where S is $\mathcal{H}_n$, $\mathcal{S}_n(\mathbb{R})$, or the set of rank one projections.

In [[16], Lemma 3.6], Huang et al. showed that the following statements are equivalent for a unital map ϕ on $\mathcal{P}_n$:

1. $tr(\phi(A)\phi(B)) = tr(AB)$ for $A, B \in \mathcal{P}_n$;

2. $tr\left(\phi(A)\phi(B)^{-1}\right) = tr\left(AB^{-1}\right)$ for $A, B \in \mathcal{P}_n$;

3. $\phi(A) = U^*AU$ or U^*A^tU for a unitary matrix U.

The authors also determined the cases if ϕ is not assuming unital, the set $\mathcal{P}_n$ is replaced by another set like $\mathcal{M}_n$, $\mathcal{S}_n$, $\mathcal{T}_n$, or $\mathcal{D}_n$. In [[17], Theorem 3.8], Leung, Ng, and Wong considered the relation (1) on infinite dimensional space.

Let $\langle S \rangle$ denote the subspace spanned by a subset S of a vector space. Recently, Huang and Tsai studied two maps preserving trace of product [18]. Suppose two maps $\phi : V_1 \to W_1$ and $\psi : V_2 \to W_2$ between subsets of matrix spaces over a field $\mathbb{F}$ under some conditions satisfy

$$\mathrm{tr}(\phi(A)\psi(B)) = \mathrm{tr}(AB) \tag{2}$$

for all $A \in V_1$, $B \in V_2$. The authors showed that these two maps can be extended to bijective linear maps $\tilde{\phi} : \langle V_1 \rangle \to \langle W_1 \rangle$ and $\tilde{\psi} : \langle V_2 \rangle \to \langle W_2 \rangle$ that satisfy $\mathrm{tr}(\tilde{\phi}(A)\tilde{\psi}(B)) = \mathrm{tr}(AB)$ for all $A \in \langle V_1 \rangle$, $B \in \langle V_2 \rangle$ (see Theorem 2.2). Hence when a matrix space V is closed under conjugate transpose, every linear bijection $\phi : V \to V$ corresponds to a unique linear bijection $\psi : V \to V$ that makes (2) hold (see Corollary 2.3). Therefore, each of ϕ and ψ has no specific form.

One natural question is to ask when the following equality holds for a fixed $k \in \mathbb{Z}\backslash\{0, 1\}$:

$$\mathrm{tr}\left(\phi(A)\psi(B)^k\right) = \mathrm{tr}\left(AB^k\right). \tag{3}$$

The second major work of this paper is to use our descriptions of linear k-power preservers on an open neighborhood S of I_n in V to characterize maps $\phi, \psi : V \to V$ under one of the assumptions:

1. equality (3) holds for all $A \in V$, $B \in S$, and ψ is linear, or

2. equality (3) holds for all $A, B \in S$ and both ϕ and ψ are linear,

for the sets $V = \mathcal{M}_n, \mathcal{H}_n, \mathcal{P}_n, \mathcal{S}_n, \mathcal{D}_n, \mathcal{T}_n$, and their real counterparts. These results, together with Theorem 2.2 and the characterizations of maps $\phi_1, \cdots, \phi_m : V \to V$ $(m \geq 3)$ that satisfy $tr(\phi_1(A_1)\cdots\phi_m(A_m)) = tr(A_1\cdots A_m)$ in [18], make a comprehensive picture of the preservers of trace of matrix products in the related matrix spaces and sets.

In the following characterizations, $\mathbb{F} = \mathbb{C}$ or $\mathbb{R}$, $P, Q \in M_n(\mathbb{F})$ are invertible, $U \in M_n(\mathbb{F})$ is unitary, $O \in M_n(\mathbb{F})$ is orthogonal, and $c \in \mathbb{F}\backslash\{0\}$.

1. $V = \mathcal{M}_n(\mathbb{F})$ (Theorem 3.5):

   a. When $k = -1$, $\phi(A) = PAQ$ and $\psi(B) = PBQ$, or $\phi(A) = PA^tQ$ and $\psi(B) = PB^tQ$.

   b. When $k \in \mathbb{Z}\backslash\{-1, 0, 1\}$, $\phi(A) = c^{-k}PAP^{-1}$ and $\psi(B) = cPBP^{-1}$, or $\phi(A) = c^{-k}PA^tP^{-1}$ and $\psi(B) = cPB^tP^{-1}$.

2. $V = \mathcal{H}_n$ (Theorem 4.2):

   a. When $k = -1$, $\phi(A) = cP^*AP$ and $\psi(B) = cP^*BP$, or $\phi(A) = cP^*A^tP$ and $\psi(B) = cP^*B^tP$, for $c \in \{1, -1\}$.

   b. When $k \in \mathbb{Z}\backslash\{-1, 0, 1\}$, $\phi(A) = c^{-k}U^*AU$ and $\psi(B) = cU^*BU$, or $\phi(A) = c^{-k}U^*A^tU$ and $\psi(B) = cU^*B^tU$, for $c \in \mathbb{R}\backslash\{0\}$.

3. $V = \mathcal{S}_n(\mathbb{F})$ (Theorem 5.3):

   a. When $k = -1$, $\phi(A) = cPAP^t$ and $\psi(B) = cPBP^t$.

   b. When $k \in \mathbb{Z}\backslash\{-1, 0, 1\}$, $\phi(A) = c^{-k}OAO^t$ and $\psi(B) = cOBO^t$.

4. $V = \mathcal{P}_n$ and $\mathcal{P}_n(\mathbb{R})$ (Theorem 6.4): $\phi(A) = c^{-k}U^*AU$ and $\psi(B) = cU^*BU$, or $\phi(A) = c^{-k}U^*A^tU$ and $\psi(B) = cU^*B^tU$, in which $c \in \mathbb{R}^+$. Characterizations under some other assumptions are also given as special cases of Theorem 6.2 (Huang, Tsai [18]).

5. $V = \mathcal{D}_n(\mathbb{F})$ (Theorem 7.2): $\phi(A) = PC^{-k}AP^{-1}$, $\psi(B) = PCBP^{-1}$ where P is a permutation matrix and $C = \mathcal{D}_n(\mathbb{F})$ is diagonal and invertible.

6. $V = \mathcal{T}_n(\mathbb{F})$ (Theorem 8.5): ϕ and ψ send $\mathcal{N}_n(\mathbb{F})$ to $\mathcal{N}_n(\mathbb{F})$, $(D\circ\phi)|_{\mathcal{D}_n(\mathbb{F})}$ and $(D\circ\psi)|_{\mathcal{D}_n(\mathbb{F})}$ are characterized by Theorem 7.2, and $D\circ\phi = D\circ\phi\circ D$. Here D denotes the map that sends $A \in \mathcal{T}_n(\mathbb{F})$ to the diagonal matrix with the same diagonal as A.

The sets $\mathcal{M}_n, \mathcal{H}_n, \mathcal{P}_n, \mathcal{S}_n, \mathcal{D}_n$, and their real counterparts are closed under conjugate transpose. In these sets, $tr(AB) = \langle A^*, B\rangle$ for the standard inner product. Our trace of product preservers can also be interpreted as inner product preservers, which have wide applications in research areas like quantum information theory.

2. Preliminary

2.1 Linear operators preserving powers

We show below that: given $k \in \mathbb{Z} \backslash \{0, 1\}$, a unital linear map $\psi : V \to W$ between matrix spaces preserving k-powers on a neighborhood of identity in V must preserve all integer powers. Let $\mathbb{Z}_+$ (resp. $\mathbb{Z}_-$) denote the set of all positive (resp. negative) integers.

Theorem 2.1. *Let $\mathbb{F} = \mathbb{C}$ or $\mathbb{R}$. Let $V \subseteq \mathcal{M}_p(\mathbb{F})$ and $W \subseteq \mathcal{M}_q(\mathbb{F})$ be matrix spaces. Fix $k \in \mathbb{Z} \backslash \{0, 1\}$.*

1. *Suppose the identity matrix $I_p \in V$ and $A^k \in V$ for all matrices A in an open neighborhood $\mathcal{S}_V$ of I_p in V consisting of invertible matrices. Then*

$$\{AB + BA : A, B \in V\} \subseteq V, \tag{4}$$

$$\left\{A^{-1} : A \in V \text{ is invertible}\right\} \subseteq V. \tag{5}$$

In particular,

$$\{A^r : A \in V\} \subseteq V, \quad r \in \mathbb{Z}_+, \quad \text{and} \tag{6}$$

$$\{A^r : A \in V \text{ is invertible}\} \subseteq V, \quad r \in \mathbb{Z}_-. \tag{7}$$

2. *Suppose $I_p \in V$, $I_q \in W$, and $A^k \in V$ for all matrices A in an open neighborhood $\mathcal{S}_V$ of I_p in V consisting of invertible matrices. Suppose $\psi : V \to W$ is a linear map that satisfies the following conditions:*

$$\psi(I_p) = I_q, \tag{8}$$

$$\psi\left(A^k\right) = \psi(A)^k, \quad A \in \mathcal{S}_V. \tag{9}$$

Then

$$\psi(AB + BA) = \psi(A)\psi(B) + \psi(B)\psi(A), \quad A, B \in V, \tag{10}$$

$$\psi\left(A^{-1}\right) = \psi(A)^{-1}, \quad \text{invertible } A \in V. \tag{11}$$

In particular,

$$\psi(A^r) = \psi(A)^r, \quad A \in V, \ r \in \mathbb{Z}_+, \quad \text{and} \tag{12}$$

$$\psi(A^r) = \psi(A)^r, \quad \text{invertible } A \in V, \ r \in \mathbb{Z}_-. \tag{13}$$

Proof. We prove the complex case. The real case is done similarly.

1. For each $A \in V \backslash \{0\}$, there is $\epsilon > 0$ such that $I_p + xA \in \mathcal{S}_V$ for all $x \in \mathbb{C}$ with $|x| < \min\left\{\epsilon, \frac{1}{\|A\|}\right\}$. Thus

$$\left(I_p + xA\right)^k = I_p + xkA + x^2\frac{k(k-1)}{2}A^2 + \cdots \in V. \tag{14}$$

The second derivative

$$\frac{d^2}{dx^2}(I_p + xA)^k \bigg|_{x=0} = k(k-1)A^2 \in V. \tag{15}$$

Since $k \notin \{0, 1\}$, we have $A^2 \in V$ for all $A \in V$. Therefore, for $A, B \in V$,

$$AB + BA = (A + B)^2 - A^2 - B^2 \in V. \tag{16}$$

In particular, $A \in V$ implies that $A^r \in V$ for all $r \in \mathbb{Z}_+$.

Cayley-Hamilton theorem implies that every invertible matrix A satisfies that $A^{-1} = f(A)$ for a certain polynomial $f(x) \in \mathbb{F}[x]$. Therefore, $A^{-1} \in V$, so that $A^r \in V$ for all $r \in \mathbb{Z}_-$.

2. Now suppose (8) and (9) hold. The proof is proceeded similarly to the proof of part (1). For every $A \in V$, there is $\epsilon > 0$ such that for all $x \in \mathbb{C}$ with $|x| < \min\left\{\epsilon, \frac{1}{\|A\|}, \frac{1}{\|\psi(A)\|}\right\}$,

$$(\psi(I_p + xA))^k = I_q + xk\psi(A) + x^2\frac{k(k-1)}{2}\psi(A)^2 + \dots \in W, \tag{17}$$

$$\psi\left((I_p + xA)^k\right) = I_q + xk\psi(A) + x^2\frac{k(k-1)}{2}\psi(A^2) + \dots \in W. \tag{18}$$

since (17) and (18) equal, we have

$$\psi(A)^2 = \psi(A^2), \quad A \in V. \tag{19}$$

Therefore, for $A, B \in V$,

$$\psi\left((A + B)^2\right) = \psi(A + B)^2 \tag{20}$$

We get (10): $\psi(AB + BA) = \psi(A)\psi(B) + \psi(B)\psi(A)$. In particular $\psi(A^r) = \psi(A)^r$ for all $A \in V$ and $r \in \mathbb{Z}_+$.

Every invertible $A \in V$ can be expressed as $A^{-1} = f(A)$ for a certain polynomial $f(x) \in \mathbb{F}[x]$. Then $\psi(A^{-1}) = \psi(f(A)) = f(\psi(A))$ is commuting with $\psi(A)$. Hence

$$2\psi(A^{-1})\psi(A) = \psi(A^{-1})\psi(A) + \psi(A)\psi(A^{-1}) = \psi(A^{-1}A + AA^{-1}) = 2I_q. \tag{21}$$

We get $\psi(A^{-1}) = \psi(A)^{-1}$. Therefore, $\psi(A^r) = \psi(A)^r$ for all $r \in \mathbb{Z}_-$.

Theorem 2.1 is powerful in exploring k-power preservers in matrix spaces. Note that every k-power preserver is a k-potent preserver. Theorem 2.1 can also be used to investigate k-potent preservers in matrix spaces.

2.2 Two maps preserving trace of product

We recall two results about two maps preserving trace of product in [18]. They are handy in proving linear bijectivity of maps preserving trace of products. Recall that if S is a subset of a vector space, then $\langle S \rangle$ denotes the subspace spanned by S.

Theorem 2.2 (Huang, Tsai [18]). *Let $\phi : V_1 \to W_1$ and $\psi : V_2 \to W_2$ be two maps between subsets of matrix spaces over a field $\mathbb{F}$ such that:*

$$\dim\langle V_1\rangle = \dim\langle V_2\rangle \geq \max\{\dim\langle W_1\rangle, \dim\langle W_2\rangle\}.$$

1. *AB are well-defined square matrices for $(A,B) \in (V_1 \times V_2) \cup (W_1 \times W_2)$.*

2. *If $A \in \langle V_1\rangle$ satisfies that $\operatorname{tr}(AB) = 0$ for all $B \in \langle V_2\rangle$, then $A = 0$.*

3. *ϕ and ψ satisfy that*

$$\operatorname{tr}(\phi(A)\psi(B)) = \operatorname{tr}(AB), \quad A \in V_1, \ B \in V_2. \tag{22}$$

Then $\dim\langle V_1\rangle = \dim\langle V_2\rangle = \dim\langle W_1\rangle = \dim\langle W_2\rangle$ and ϕ and ψ can be extended to bijective linear map $\tilde{\phi} : \langle V_1\rangle \to \langle W_1\rangle$ and $\tilde{\psi} : \langle V_2\rangle \to \langle W_2\rangle$, respectively, such that

$$\operatorname{tr}\left(\tilde{\phi}(A)\tilde{\psi}(B)\right) = \operatorname{tr}(AB), \quad A \in \langle V_1\rangle, \ B \in \langle V_2\rangle. \tag{23}$$

A subset V of $\mathcal{M}_n$ is closed under conjugate transpose if $\{A^* : A \in V\} \subseteq V$. A real or complex matrix space V is closed under conjugate transpose if and only if V equals the direct sum of its subspace of Hermitian matrices and its subspace of skew-Hermitian matrices.

Corollary 2.3 (Huang, Tsai [18]). *Let V be a subset of $\mathcal{M}_n$ closed under conjugate transpose. Suppose two maps $\phi, \psi : V \to V$ satisfy that*

$$\operatorname{tr}(\phi(A)\psi(B)) = \operatorname{tr}(AB), \quad A, B \in V. \tag{24}$$

Then ϕ and ψ can be extended to linear bijections on $\langle V\rangle$. Moreover, when V is a vector space, every linear bijection $\phi : V \to V$ corresponds to a unique linear bijection $\psi : V \to V$ such that (24) holds. Explicitly, given an orthonormal basis $\{A_1, \dots, A_\ell\}$ of V with respect to the inner product $\langle A, B\rangle = \operatorname{tr}(A^*B)$, ψ is defined by $\psi(A_i) = B_i$ in which $\{B_1, \dots, B_\ell\}$ is a basis of V with $\operatorname{tr}\left(\phi\left(A_i^*\right)B_j\right) = \delta_{i,j}$ for all $i, j \in \{1, \dots, \ell\}$.

Corollary 2.3 shows that when a matrix space V is closed under conjugate transpose, every linear bijection $\phi : V \to V$ corresponds to a unique linear bijection $\psi : V \to V$ that makes (24) hold. The next natural thing is to determine ϕ and ψ that satisfy $tr\left(\phi(A)\psi(B)^k\right) = tr\left(AB^k\right)$ for a fixed $k \in \mathbb{Z}\backslash\{0,1\}$.

From now on, we focus on the fields $\mathbb{F} = \mathbb{C}$ or $\mathbb{R}$.

3. *k*-power linear preservers and trace of power-product preservers on $\mathcal{M}_n$ and $\mathcal{M}_n(\mathbb{R})$

3.1 *k*-power preservers on $\mathcal{M}_n$ and $\mathcal{M}_n(\mathbb{R})$

Chan and Lim described the linear k-power preservers on $\mathcal{M}_n$ and $\mathcal{M}_n(\mathbb{R})$ for $k \geq 2$ in [7, Theorem 1] as follows.

Theorem 3.1. *(Chan, Lim [5]) Let an integer $k \geq 2$. Let $\mathbb{F}$ be a field with $char(\mathbb{F}) = 0$ or $char(\mathbb{F}) > k$. Suppose that $\psi : \mathcal{M}_n(\mathbb{F}) \to \mathcal{M}_n(\mathbb{F})$ is a nonzero linear operator such that $\psi\left(A^k\right) = \psi(A)^k$ for all $A \in \mathcal{M}_n(\mathbb{F})$. Then there exist $\lambda \in \mathbb{F}$ with $\lambda^{k-1} = 1$ and an invertible matrix $P \in \mathcal{M}_n(\mathbb{F})$ such that*

$$\psi(A) = \lambda PAP^{-1}, \quad A \in \mathcal{M}_n(\mathbb{F}), \quad or \tag{25}$$

$$\psi(A) = \lambda PA^tP^{-1}, \quad A \in \mathcal{M}_n(\mathbb{F}). \tag{26}$$

(25) and (26) need not hold if ψ is zero or is a map on a subspace of $\mathcal{M}_n(\mathbb{F})$. The following are two examples. Another example can be found in maps on $\mathcal{D}_n(\mathbb{F})$ (Theorem 7.1).

Example 3.2. *The zero map $\psi(A) \equiv 0$ clearly satisfies $\psi(A^k) = \psi(A)^k$ for all $A \in \mathcal{M}_n$ but they are not of the form* (25) *or* (26).

Example 3.3. *Let $n = k + m$, $k, m \geq 2$, and consider the operator ψ on the subspace $W = \mathcal{M}_k \oplus \mathcal{M}_m$ of $\mathcal{M}_n$ defined by $\psi(A \oplus B) = A \oplus B^t$ for $A \in \mathcal{M}_k$ and $B \in \mathcal{M}_m$. Then $\psi(A^k) = \psi(A)^k$ for all $A \in W$ and $k \in \mathbb{Z}_+$, but ψ is not of the form* (25) *or* (26).

We now generalize Theorem 3.1 to include negative integers k and to assume the k-power preserving condition $\psi(A^k) = \psi(A)^k$ only on matrices nearby the identity.

Theorem 3.4. *Let $\mathbb{F} = \mathbb{C}$ or $\mathbb{R}$. Let an integer $k \in \mathbb{Z} \backslash \{0, 1\}$. Suppose that $\psi : \mathcal{M}_n(\mathbb{F}) \to \mathcal{M}_n(\mathbb{F})$ is a nonzero linear map such that $\psi(A^k) = \psi(A)^k$ for all A in an open neighborhood of I_n consisting of invertible matrices. Then there exist $\lambda \in \mathbb{F}$ with $\lambda^{k-1} = 1$ and an invertible matrix $P \in \mathcal{M}_n(\mathbb{F})$ such that*

$$\psi(A) = \lambda PAP^{-1}, \quad A \in \mathcal{M}_n(\mathbb{F}), \quad or \tag{27}$$

$$\psi(A) = \lambda PA^tP^{-1}, \quad A \in \mathcal{M}_n(\mathbb{F}). \tag{28}$$

Proof. We prove for the case $\mathbb{F} = \mathbb{C}$. The case $\mathbb{F} = \mathbb{R}$ can be done similarly. Obviously, $\psi(I_n) = \psi(I_n^k) = \psi(I_n)^k$.

1. First suppose $k \geq 2$. For each $A \in \mathcal{M}_n$, there exists $\epsilon > 0$ such that for all $x \in \mathbb{C}$ with $|x| < \epsilon$, the following two power series converge and equal:

$$\begin{aligned} (\psi(I_n + xA))^k &= \psi(I_n) + x\left(\sum_{i=0}^{k-1} \psi(I_n)^i \psi(A) \psi(I_n)^{k-1-i}\right) \\ &+ x^2\left(\sum_{i=0}^{k-2}\sum_{j=0}^{k-2-i} \psi(I_n)^i \psi(A) \psi(I_n)^j \psi(A) \psi(I_n)^{k-2-i-j}\right) + \cdots \end{aligned} \tag{29}$$

$$\psi\left((I_n + xA)^k\right) = \psi(I_n) + xk\psi(A) + x^2 \frac{k(k-1)}{2} \psi(A^2) + \cdots \tag{30}$$

Equating degree one terms above, we get

$$k\psi(A) = \sum_{i=0}^{k-1} \psi(I_n)^i \psi(A) \psi(I_n)^{k-1-i}. \tag{31}$$

Applying (31), we have

$$k\psi(I_n)\psi(A) - k\psi(A)\psi(I_n) = \psi(I_n)^k \psi(A) - \psi(A)\psi(I_n)^k = \psi(I_n)\psi(A) - \psi(A)\psi(I_n). \tag{32}$$

Hence $\psi(I_n)\psi(A) = \psi(A)\psi(I_n)$ for $A \in \mathcal{M}_n$, that is, $\psi(I_n)$ commutes with the range of ψ.

Now equating degree two terms of (29) and (30) and taking into account that $k \notin \{0, 1\}$, we have

$$\psi(I_n)^{k-2}\psi(A)^2 = \psi(A^2). \tag{33}$$

Define $\psi_1(A) = \psi(I_n)^{k-2}\psi(A)$ for $A \in \mathcal{M}_n$. Then $\psi_1(A^2) = (\psi_1(A))^2$ for all $A \in \mathcal{M}_n$. (31) and the assumption that ψ is nonzero imply that $\psi(I_n) \neq 0$. So $\psi_1(I_n)\psi(I_n) = \psi(I_n)^k = \psi(I_n) \neq 0$. Thus $\psi_1(I_n) \neq 0$ and ψ_1 is nonzero. By Theorem 3.1, there exists an invertible $P \in \mathcal{M}_n$ such that $\psi_1(A) = PAP^{-1}$ for $A \in \mathcal{M}_n$ or $\psi_1(A) = PA^tP^{-1}$ for $A \in \mathcal{M}_n$. Moreover, $\psi(I_n)$ commutes with all $\psi_1(A)$, so that $\psi(I_n) = \lambda I_n$ for a $\lambda \in \mathbb{C}$. By $I_n = \psi_1(I_n) = \psi(I_n)^{k-1}$, we get $\lambda^{k-1} = 1$. Therefore, $\psi(A) = \lambda\psi_1(A)$. We get (27) and (28).

1. Next Suppose $k < 0$. For every $A \in \mathcal{M}_n$, the power series expansions of $(\psi(I_n + xA))^{-k}$ and $\psi\left(\left((I_n + xA)^k\right)^{-1}\right)$ are equal when $|x|$ is sufficiently small:

$$(\psi(I_n + xA))^{-k} = \psi(I_n)^{-1} + x\left(\sum_{i=0}^{-k-1} \psi(I_n)^i\psi(A)\psi(I_n)^{-k-1-i}\right) + \cdots \tag{34}$$

$$\psi\left(\left((I_n + xA)^k\right)^{-1}\right) = \psi(I_n)^{-1} - xk\psi(I_n)^{-1}\psi(A)\psi(I_n)^{-1} + \cdots \tag{35}$$

Equating degree one terms of (34) and (35), we get

$$-k\psi(I_n)^{-1}\psi(A)\psi(I_n)^{-1} = \sum_{i=0}^{-k-1} \psi(I_n)^i\psi(A)\psi(I_n)^{-k-1-i}. \tag{36}$$

Therefore,

$$\begin{aligned} &-k\left(\psi(A)\psi(I_n)^{-1} - \psi(I_n)^{-1}\psi(A)\right) \\ &= \psi(I_n)\left(\sum_{i=0}^{-k-1} \psi(I_n)^i\psi(A)\psi(I_n)^{-k-1-i}\right) - \left(\sum_{i=0}^{-k-1} \psi(I_n)^i\psi(A)\psi(I_n)^{-k-1-i}\right)\psi(I_n) \\ &= \psi(I_n)^{-k}\psi(A) - \psi(A)\psi(I_n)^{-k} = \psi(I_n)^{-1}\psi(A) - \psi(A)\psi(I_n)^{-1}. \end{aligned} \tag{37}$$

We get $\psi(I_n)^{-1}\psi(A) = \psi(A)\psi(I_n)^{-1}$ for $A \in \mathcal{M}_n$. So $\psi(I_n)^{-1}$ and $\psi(I_n)$ commute with the range of ψ. The following power series are equal for every $A \in \mathcal{M}_n$ when $|x|$ is sufficiently small:

$$(\psi(I_n + xA))^k = \psi(I_n) + xk\psi(I_n)^{k-1}\psi(A) + x^2\frac{k(k-1)}{2}\psi(I_n)^{k-2}\psi(A)^2 + \cdots \tag{38}$$

$$\psi\left((I_n + xA)^k\right) = \psi(I_n) + xk\psi(A) + x^2\frac{k(k-1)}{2}\psi(A^2) + \cdots \tag{39}$$

Equating degree two terms of (38) and (39), we get $\psi(I_n)^{k-2}\psi(A)^2 = \psi(A^2)$. Let $\psi_1(A) := \psi(I_n)^{k-2}\psi(A) = \psi(I_n)^{-1}\psi(A)$. Then $\psi_1(A)^2 = \psi_1(A^2)$ and ψ_1 is nonzero. Using Theorem 3.1, we can get (27) and (39).

3.2 Trace of power-product preserers on $\mathcal{M}_n$ and $\mathcal{M}_n(\mathbb{R})$

Corollary 2.3 shows that every linear bijection $\phi : \mathcal{M}_n(\mathbb{F}) \to \mathcal{M}_n(\mathbb{F})$ corresponds to another linear bijection $\psi : \mathcal{M}_n(\mathbb{F}) \to \mathcal{M}_n(\mathbb{F})$ such that $tr(\phi(A)\psi(B)) = tr(AB)$ for all $A, B \in \mathcal{M}_n(\mathbb{F})$. When $m \geq 3$, maps $\phi_1, \cdots, \phi_m$ on $\mathcal{M}_n(\mathbb{F})$ that satisfy $tr(\phi_1(A_1)\cdots\phi_m(A_m)) = tr(A_1\cdots A_m)$ for $A_1, \ldots, A_m \in \mathcal{M}_n(\mathbb{F})$ are determined in [18].

If two maps on $\mathcal{M}_n(\mathbb{F})$ satisfy the following trace condition about k-powers, then they have specific forms.

Theorem 3.5. *Let $\mathbb{F} = \mathbb{C}$ or $\mathbb{R}$. Let $k \in \mathbb{Z}\backslash\{0, 1\}$. Let S be an open neighborhood of I_n consisting of invertible matrices. Then two maps $\phi, \psi : \mathcal{M}_n(\mathbb{F}) \to \mathcal{M}_n(\mathbb{F})$ satisfy that*

$$tr\left(\phi(A)\psi(B)^k\right) = tr\left(AB^k\right), \tag{40}$$

1. *for all $A \in \mathcal{M}_n(\mathbb{F})$, $B \in S$, and ψ is linear, or*

2. *for all $A, B \in S$ and both ϕ and ψ are linear,*

if and only if ϕ and ψ take the following forms:

a. *When $k = -1$, there exist invertible matrices $P, Q \in \mathcal{M}_n(\mathbb{F})$ such that*

$$\begin{pmatrix} \phi(A) = PAQ \\ \psi(B) = PBQ \end{pmatrix} \quad or \quad \begin{pmatrix} \phi(A) = PA^tQ \\ \psi(B) = PB^tQ \end{pmatrix} \quad A, B \in \mathcal{M}_n(\mathbb{F}). \tag{41}$$

b. *When $k \in \mathbb{Z}\backslash\{-1, 0, 1\}$, there exist $c \in \mathbb{F}\backslash\{0\}$ and an invertible matrix $P \in \mathcal{M}_n(\mathbb{F})$ such that*

$$\begin{pmatrix} \phi(A) = c^{-k}PAP^{-1} \\ \psi(B) = cPBP^{-1} \end{pmatrix} \quad or \begin{pmatrix} \phi(A) = c^{-k}PA^tP^{-1} \\ \psi(B) = cPB^tP^{-1} \end{pmatrix} \quad A, B \in \mathcal{M}_n(\mathbb{F}). \tag{42}$$

Proof. We prove the case $\mathbb{F} = \mathbb{C}$; the case $\mathbb{F} = \mathbb{R}$ can be done similarly.

Suppose assumption (2) holds. Then for every $A \in \mathcal{M}_n(\mathbb{F})$, there exists $c \in \mathbb{F}\backslash\{0\}$ such that $I_n - cA \in S$, so that for all $B \in S$:

$$\mathrm{tr}\left(B^k\right) = \mathrm{tr}\left((\phi(I_n - cA) + c\phi(A))\psi(B)^k\right) = \mathrm{tr}\left((I_n - cA)B^k\right) + c\mathrm{tr}\left(\phi(A)\psi(B)^k\right). \tag{43}$$

Thus $\mathrm{tr}\left(\phi(A)\psi(B)^k\right) = \mathrm{tr}\left(AB^k\right)$ for $A \in \mathcal{M}_n(\mathbb{F})$ and $B \in S$, which leads to assumption (1).

Now we prove the theorem under assumption (1), that is, (40) holds for all $A \in \mathcal{M}_n(\mathbb{F})$ and $B \in S$, and ψ is linear. Only the necessary part is needed to prove.

Let $S' = \left\{B \in \overline{\mathcal{P}_n} : B^{1/k} \in S\right\}$, which is an open neighborhood of I_n in $\overline{\mathcal{P}_n}$. Define $\tilde{\psi} : S' \to \mathcal{M}_n$ such that $\tilde{\psi}(B) = \psi\left(B^{1/k}\right)^k$. Then (40) implies that

$$\mathrm{tr}(\phi(A)\tilde{\psi}(B)) = \mathrm{tr}\left(\phi(A)\psi\left(B^{1/k}\right)^k\right) = \mathrm{tr}(AB), \quad A \in \mathcal{M}_n, \; B \in S'. \tag{44}$$

The complex span of S' is $\mathcal{M}_n$. By Theorem 2.2, ϕ is bijective linear, and $\tilde{\psi}$ can be extended to a linear bijection on $\mathcal{M}_n$.

The linearity of ψ and (40) imply that for every $B \in \mathcal{M}_n$, there exists $\epsilon > 0$ such that $I_n + xB \in S$ and the power series of $(I_n + xB)^k$ converges whenever $|x| < \epsilon$. Then

$$\mathrm{tr}\left(\phi(A)(\psi(I_n) + x\psi(B))^k\right) = \mathrm{tr}\left(A(I_n + xB)^k\right), \quad A \in \mathcal{M}_n, \; |x| < \epsilon. \tag{45}$$

1. First suppose $k \geq 2$. Equating degree one terms and degree $(k-1)$ terms on both sides of (45) respectively, we get the following identities for $A, B \in \mathcal{M}_n$:

$$\mathrm{tr}\left(\phi(A)\left(\sum_{i=0}^{k-1} \psi(I_n)^{k-1-i}\psi(B)\psi(I_n)^i\right)\right) = \mathrm{tr}(kAB), \tag{46}$$

$$\mathrm{tr}\left(\phi(A)\left(\sum_{i=0}^{k-1} \psi(B)^i\psi(I_n)\psi(B)^{k-1-i}\right)\right) = \mathrm{tr}\left(kAB^{k-1}\right). \tag{47}$$

Let $\{C_i : i = 1, \ldots, n^2\}$ be a basis of projection matrices (i.e. $C_i^2 = C_i$) in $\mathcal{M}_n$. For example, we may choose the following basis of rank 1 projections:

$$\{E_{ii} : 1 \leq i \leq n\} \cup \left\{\frac{1}{\sqrt{2}}\left(E_{ii} + E_{jj} + \delta E_{ij} + \bar{\delta} E_{ji}\right) : 1 \leq i < j \leq n, \delta \in \{1, \mathbf{i}\}\right\}. \tag{48}$$

By (40) and (47), for $A \in \mathcal{M}_n$ and $i = 1, \ldots, n^2$,

$$\mathrm{tr}\left(k\phi(A)\psi(C_i)^k\right) = \mathrm{tr}(kAC_i) = \mathrm{tr}\left(\phi(A)\left(\sum_{j=0}^{k-1} \psi(C_i)^j\psi(I_n)\psi(C_i)^{k-1-j}\right)\right). \tag{49}$$

By the bijectivity of ϕ,

$$k\psi(C_i)^k = \sum_{j=0}^{k-1} \psi(C_i)^j\psi(I)\psi(C_i)^{k-1-j}. \tag{50}$$

Therefore, for $i = 1, \ldots, n^2$,

$$\begin{aligned} 0 &= \psi(C_i)\left(\sum_{j=0}^{k-1} \psi(C_i)^j\psi(I_n)\psi(C_i)^{k-1-j}\right) - \left(\sum_{j=0}^{k-1} \psi(C_i)^j\psi(I_n)\psi(C_i)^{k-1-j}\right)\psi(C_i) \\ &= \psi(C_i)^k\psi(I_n) - \psi(I_n)\psi(C_i)^k. \end{aligned} \tag{51}$$

Since

$$\mathrm{tr}\left(A\psi(C_i)^k\right) = \mathrm{tr}\left(\phi^{-1}(A)C_i^k\right) = \mathrm{tr}\left(\phi^{-1}(A)C_i\right), \quad A \in \mathcal{M}_n, \; i = 1, \ldots, n^2, \tag{52}$$

the only matrix $A \in \mathcal{M}_n$ such that $\text{tr}\left(A\psi(C_i)^k\right) = 0$ for all $i \in \{1, \ldots, n^2\}$ is the zero matrix. So $\left\{\psi(C_i)^k : i = 1, \ldots, n^2\right\}$ is a basis of $\mathcal{M}_n$. (51) implies that $\psi(I_n) = cI_n$ for certain $c \in \mathbb{C}\backslash\{0\}$.

(46) shows that

$$c^{k-1}\text{tr}(\phi(A)\psi(B)) = \text{tr}(AB), \quad A, B \in \mathcal{M}_n. \tag{53}$$

Therefore,

$$c^{k-1}\text{tr}\left(\phi(A)\psi\left(B^k\right)\right) = \text{tr}\left(AB^k\right) = \text{tr}\left(\phi(A)\psi(B)^k\right), \quad A \in \mathcal{M}_n, \ B \in S. \tag{54}$$

The bijectivity of ϕ shows that $c^{k-1}\psi\left(B^k\right) = \psi(B)^k$ for $B \in S$, that is,

$$c^{-1}\psi\left(B^k\right) = \left[c^{-1}\psi(B)\right]^k, \quad B \in S. \tag{55}$$

Notice that $c^{-1}\psi(I_n) = I_n$. By Theorem 3.4, there is an invertible $P \in \mathcal{M}_n$ such that ψ is of the form $\psi(B) = cPBP^{-1}$ or $\psi(B) = cPB^tP^{-1}$ for $B \in \mathcal{M}_n$. Consequently, we get (42).

2. Now suppose $k < 0$. Then $\psi(I_n)$ is invertible. For every $B \in \mathcal{M}_n$ and sufficiently small x, we have the power series expansion:

$$\begin{aligned}
&(\psi(I_n) + x\psi(B))^k \\
&= \left[\left(I_n + x\psi(I_n)^{-1}\psi(B)\right)^{-1}\psi(I_n)^{-1}\right]^{|k|} \\
&= \left[\left(I_n - x\psi(I_n)^{-1}\psi(B) + x^2\psi(I_n)^{-1}\psi(B)\psi(I_n)^{-1}\psi(B) + \cdots\right)\psi(I_n)^{-1}\right]^{|k|} \\
&= \psi(I_n)^k - x\left(\sum_{i=1}^{|k|}\psi(I_n)^{-i}\psi(B)\psi(I_n)^{k-1+i}\right) \\
&\quad + x^2\left(\sum_{i=1}^{|k|}\sum_{j=1}^{|k|+1-i}\psi(I_n)^{-i}\psi(B)\psi(I_n)^{-j}\psi(B)\psi(I_n)^{k-2+i+j}\right) + \cdots
\end{aligned} \tag{56}$$

Equating degree one terms and degree two terms of (45) respectively and using (56), we get the following identities for $A, B \in \mathcal{M}_n$:

$$\text{tr}\left(\phi(A)\left(\sum_{i=1}^{|k|}\psi(I_n)^{-i}\psi(B)\psi(I_n)^{k-1+i}\right)\right) = \text{tr}(|k|AB), \tag{57}$$

$$\text{tr}\left(\phi(A)\left(\sum_{i=1}^{|k|}\sum_{j=1}^{|k|+1-i}\psi(I_n)^{-i}\psi(B)\psi(I_n)^{-j}\psi(B)\psi(I_n)^{k-2+i+j}\right)\right) = \text{tr}\left(\frac{k(k-1)}{2}AB^2\right). \tag{58}$$

(57) and (40) imply that

$$\sum_{i=1}^{|k|}\psi(I_n)^{-i}\psi\left(B^k\right)\psi(I_n)^{k-1+i} = |k|\psi(B)^k, \quad B \in S. \tag{59}$$

Let $F_r(B)$ denote the degree r coefficient in the power series of $(\psi(I_n) + x\psi(B))^k$. Then (57) and (58) show that:

$$\frac{k-1}{2}F_1(B^2) = F_2(B), \quad B \in \mathcal{M}_n. \tag{60}$$

Denote $\psi_1(B) := \psi(B)\psi(I_n)^{-1}$. We discuss the cases $k = -1$ and $k \neq -1$.

a. When $k = -1$, (60) leads to

$$\psi(B^2) = \psi(B)\psi(I_n)^{-1}\psi(B), \quad B \in \mathcal{M}_n. \tag{61}$$

So $\psi_1(B^2) = \psi_1(B)^2$ for $B \in \mathcal{M}_n$. Note that $\psi_1(I_n) = I_n$. By Theorem 3.4, there exists an invertible $P \in \mathcal{M}_n$ such that $\psi_1(B) = PBP^{-1}$ or $\psi_1(B) = PB^tP^{-1}$ for $B \in \mathcal{M}_n$. Let $Q := P^{-1}\psi(I_n)$. Then Q is invertible, and $\psi(B) = PBQ$ or $\psi(B) = PB^tQ$ for $B \in \mathcal{M}_n$. Using (40), we get (41).

b. Suppose the integer $k < -1$. Then (60) implies that

$$\frac{k-1}{2}\left(\psi(I_n)^{-1}F_1(B^2) - F_2(B^2)\psi(I_n)^{-1}\right) = \psi(I_n)^{-1}F_2(B) - F_2(B)\psi(I_n)^{-1}, \tag{62}$$

which gives

$$\begin{aligned} &\frac{1-k}{2}\left(\psi(I_n)^k\psi(B^2) - \psi(B^2)\psi(I_n)^k\right) \\ &= \psi(I_n)^k\psi(B)\psi(I_n)^{-1}\psi(B) - \psi(B)\psi(I_n)^{-1}\psi(B)\psi(I_n)^k. \end{aligned} \tag{63}$$

In other words, for $B \in \mathcal{M}_n$:

$$\psi(I_n)^k\left(\frac{1-k}{2}\psi_1(B^2) - \psi_1(B)^2\right) = \left(\frac{1-k}{2}\psi_1(B^2) - \psi_1(B)^2\right)\psi(I_n)^k. \tag{64}$$

Let $B = I_n + xE$ for an arbitrary matrix $E \in \mathcal{M}_n$. Then (64) becomes

$$\begin{aligned} &\psi(I_n)^k\left[x(-1-k)\psi_1(E) + x^2\left(\frac{1-k}{2}\psi_1(E^2) - \psi_1(E)^2\right)\right] \\ &= \left[x(-1-k)\psi_1(E) + x^2\left(\frac{1-k}{2}\psi_1(E^2) - \psi_1(E)^2\right)\right]\psi(I_n)^k. \end{aligned} \tag{65}$$

The equality on degree one terms shows that $\psi(I_n)^k$ commutes with all $\psi_1(E)$. Hence $\psi(I_n)^k$ commutes with the range of ψ. (??) can be rewritten as

$$\operatorname{tr}\left(\left(\sum_{i=1}^{|k|}\psi(I_n)^{k-1+i}\phi(A)\psi(I_n)^{-i}\right)\psi(B)\right) = \operatorname{tr}(|k|AB), \quad A, B \in \mathcal{M}_n. \tag{66}$$

By Theorem 2.2, ψ is a linear bijection and its range is $\mathcal{M}_n$. So $\psi(I_n)^k = \mu I_n$ for certain $\mu \in \mathbb{C}$.

Now by (59), for $B \in S$:

$$\begin{aligned} &|k|\psi(I_n)\psi(B)^k - |k|\psi(B)^k\psi(I_n) \\ &= \psi(I_n)\left(\sum_{i=1}^{|k|}\psi(I_n)^{-i}\psi(B^k)\psi(I_n)^{k-1+i}\right) - \left(\sum_{i=1}^{|k|}\psi(I_n)^{-i}\psi(B^k)\psi(I_n)^{k-1+i}\right)\psi(I_n) \\ &= \psi(B^k)\psi(I_n)^k - \psi(I_n)^k\psi(B^k) = 0 \end{aligned} \tag{67}$$

So $\psi(I_n)$ commutes with $\psi(B)^k$ for $B \in S$. In particular, $\psi(I_n)$ commutes with $\tilde{\psi}(B) = \psi(B^{1/k})^k$ for $B \in S'$. The complex span of S' is $\mathcal{M}_n$, and $\tilde{\psi}$ can be extended to a linear bijection on $\mathcal{M}_n$. Hence $\psi(I_n) = cI_n$ for certain $c \in \mathbb{C}\backslash\{0\}$. By (59), we get $\psi_1(B^k) = \psi_1(B)^k$ for $B \in S$. Note that $\psi_1(I_n) = I_n$. By Theorem 3.4, there is an invertible $P \in \mathcal{M}_n$ such that $\psi_1(B) = PBP^{-1}$ or $\psi_1(B) = PB^tP^{-1}$. Then $\psi(B) = cPBP^{-1}$ or $\psi(B) = cPB^tP^{-1}$. Using (40), we get (42).

Remark 3.6 *The following modifications could be applied to the proof of Theorem 3.5 for* $\mathbb{F} = \mathbb{R}$*:*

1. *Let S' be the collection of matrices $A = QDQ^{-1}$, in which D is nonnegative diagonal and $Q \in \mathcal{M}_n(\mathbb{R})$ is invertible, such that $A^{1/k} = QD^{1/k}Q^{-1} \in S$.*

2. *We may choose the following basis of rank 1 projections of $\mathcal{M}_n(\mathbb{R})$ to substitute* (48)*:*

$$\begin{aligned} &\{E_{ii} : 1 \le i \le n\} \cup \left\{\frac{1}{\sqrt{2}}(E_{ii} + E_{jj} + E_{ij} + E_{ji}) : 1 \le i < j \le n\right\} \\ &\cup \{\omega_1 E_{ii} + \omega_2 E_{jj} + E_{ij} - E_{ji} : 1 \le i < j \le n\}, \end{aligned} \tag{68}$$

in which ω_1, ω_2 are distinct roots of $x^2 - x - 1 = 0$..

The arguments in the above proof will be applied analogously to maps on the other sets discussed in this paper.

4. *k*-power linear preservers and trace of power-product preservers on $\mathcal{H}_n$

4.1 *k*-power linear preservers on $\mathcal{H}_n$

We give a result that determine linear operators on $\mathcal{H}_n$ that satisfy $\psi(A^k) = \psi(A)^k$ on a neighborhood of I_n in $\mathcal{H}_n$ for certain $k \in \mathbb{Z}\backslash\{0, 1\}$

Theorem 4.1. *Fix $k \in \mathbb{Z}\backslash\{0, 1\}$. A nonzero linear map $\psi : \mathcal{H}_n \to \mathcal{H}_n$ satisfies that*

$$\psi(A^k) = \psi(A)^k \tag{69}$$

on an open neighborhood of I_n consisting of invertible matrices if and only if ψ is of the following forms for certain unitary matrix $U \in \mathcal{M}_n$:

1. *When k is even,*

$$\psi(A) = U^*AU, \quad A \in \mathcal{H}_n; \quad or \quad \psi(A) = U^*A^tU, \quad A \in \mathcal{H}_n. \tag{70}$$

2. *When k is odd,*

$$\psi(A) = \pm U^*AU, \quad A \in \mathcal{H}_n; \quad or \quad \psi(A) = \pm U^*A^tU, \quad A \in \mathcal{H}_n. \tag{71}$$

Proof. It suffices to prove the necessary part. Suppose (69) holds on an open neighborhood S of I_n in $\mathcal{H}_n$.

1. First assume $k \geq 2$. Replacing $\mathcal{M}_n$ by $\mathcal{H}_n$ in part (1) of the proof of Theorem 3.4 up to (33), we can prove that $\psi(I_n)$ commutes with the range of ψ, and $\psi_1(A) := \psi(I_n)^{k-2}\psi(A)$ is a nonzero linear map that satisfies $\psi_1(A^2) = \psi_1(A)^2$ for $A \in \mathcal{H}_n$.

Every matrix in $\mathcal{M}_n$ can be uniquely expressed as $A + \mathbf{i}B$ for $A, B \in \mathcal{H}_n$. Extend ψ_1 to a map $\tilde{\psi} : \mathcal{M}_n \to \mathcal{M}_n$ such that

$$\tilde{\psi}(A + \mathbf{i}B) = \psi_1(A) + \mathbf{i}\psi_1(B), \quad A, B \in \mathcal{H}_n. \tag{72}$$

It is straightforward to check that $\tilde{\psi}$ is a complex linear bijection. Moreover, for $A, B \in \mathcal{H}_n$,

$$\begin{aligned} \psi_1(AB + BA) &= \psi_1\left((A + B)^2\right) - \psi_1(A^2) - \psi_1(B^2) \\ &= \psi_1(A + B)^2 - \psi_1(A)^2 - \psi_1(B)^2 \\ &= \psi_1(A)\psi_1(B) + \psi_1(B)\psi_1(A). \end{aligned}$$

It implies that

$$\tilde{\psi}\left((A + \mathbf{i}B)^2\right) = \tilde{\psi}(A + \mathbf{i}B)^2, \quad A, B \in \mathcal{H}_n.$$

By Theorem 3.1, there is an invertible matrix $U \in \mathcal{M}_n$ such that

a. $\tilde{\psi}(A) = UAU^{-1}$ for all $A \in \mathcal{M}_n$, or

b. $\tilde{\psi}(A) = UA^tU^{-1}$ for all $A \in \mathcal{M}_n$.

First suppose $\tilde{\psi}(A) = UAU^{-1}$. The restriction of $\tilde{\psi}$ on $\mathcal{H}_n$ is $\psi_1 : \mathcal{H}_n \to \mathcal{H}_n$. Hence for $A \in \mathcal{H}_n$, we have $UAU^{-1} = \left(UAU^{-1}\right)^* = U^{-*}AU^*$; then $U^*UA = AU^*U$ for all $A \in \mathcal{H}_n$, which shows that $U^*U = cI_n$ for certain $c \in \mathbb{R}^+$. By adjusting a scalar if necessary, we may assume that U is unitary. So $\psi(I_n)^{k-2}\psi(A) = UAU^*$. Then $\psi(I_n)^{k-1} = I_n$, so that $\psi(I_n) = I_n$ when k is even and $\psi(I_n) \in \{I_n, -I_n\}$ when k is odd. Thus $\psi(A) = UAU^*$ when k is even and $\psi(A) = \pm UAU^*$ when k is odd. Similarly for the case $\tilde{\psi}(A) = UA^tU^{-1}$. Therefore, (70) or (71) holds.

2. Now assume that $k < 0$. Replacing $\mathcal{M}_n$ by $\mathcal{H}_n$ in part (2) of the proof of Theorem 3.4, we can show that $\psi(I_n)$ commutes with the range of ψ, and furthermore the nonzero linear map $\psi_1(A) := \psi(I_n)^{-1}\psi(A)$ satisfies that $\psi_1(A^2) = \psi_1(A)^2$. By arguments in the preceding paragraphs, we get (70) or (71).

4.2 Trace of power-product preservers on $\mathcal{H}_n$

By Corollary 2.3, every linear bijection $\phi : \mathcal{H}_n \to \mathcal{H}_n$ corresponds to another linear bijection $\psi : \mathcal{H}_n \to \mathcal{H}_n$ such that $\operatorname{tr}(\phi(A)\psi(B)) = \operatorname{tr}(AB)$ for all $A, B \in \mathcal{H}_n$. When $m \geq 3$, linear maps $\phi_1, \cdots, \phi_m : \mathcal{H}_n \to \mathcal{H}_n$ that satisfy $\operatorname{tr}(\phi_1(A_1)\cdots\phi_m(A_m)) = \operatorname{tr}(A_1\cdots A_m)$ are characterized in [18].

Theorem 4.2. *Let $k \in \mathbb{Z}\backslash\{0, 1\}$. Let S be an open neighborhood of I_n in $\mathcal{H}_n$ consisting of invertible Hermitian matrices. Then two maps $\phi, \psi : \mathcal{H}_n \to \mathcal{H}_n$ satisfy that*

$$\operatorname{tr}\left(\phi(A)\psi(B)^k\right) = \operatorname{tr}\left(AB^k\right), \tag{73}$$

1.*for all $A \in \mathcal{H}_n$, $B \in S$, and ψ is linear, or*

2.*for all $A, B \in S$ and both ϕ and ψ are linear,*

if and only if ϕ and ψ take the following forms:

a. *When $k = -1$, there exist an invertible matrix $P \in \mathcal{M}_n$ and $c \in \{1, -1\}$ such that*

$$\begin{pmatrix} \phi(A) = cP^*AP \\ \psi(B) = cP^*BP \end{pmatrix} \quad A, B \in \mathcal{H}_n; \quad or \quad \begin{pmatrix} \phi(A) = cP^*A^tP \\ \psi(B) = cP^*B^tP \end{pmatrix} \quad A, B \in \mathcal{H}_n. \tag{74}$$

b. *When $k \in \mathbb{Z}\backslash\{-1, 0, 1\}$, there exist a unitary matrix $U \in \mathcal{M}_n$ and $c \in \mathbb{R}\backslash\{0\}$ such that*

$$\begin{pmatrix} \phi(A) = c^{-k}U^*AU \\ \psi(B) = cU^*BU \end{pmatrix} \quad A, B \in \mathcal{H}_n; \quad or \quad \begin{pmatrix} \phi(A) = c^{-k}U^*A^tU \\ \psi(B) = cU^*B^tU \end{pmatrix} \quad A, B \in \mathcal{H}_n. \tag{75}$$

Proof. Assumption (2) leads to assumption (1) (cf. the proof of Theorem 3.5). We prove the theorem under assumption (1). It suffices to prove the necessary part.

1. When $k \geq 2$, in the part (1) of proof of Theorem 3.5, through replacing $\mathcal{M}_n$ by $\mathcal{H}_n$, complex numbers by real numbers, and Theorem 3.1 or Theorem 3.4 by Theorem 4.1, we can prove that (ϕ, ψ) has the forms in (75).

2. When $k < 0$, in the part (2) of proof of Theorem 3.5, through replacing $\mathcal{M}_n$ by $\mathcal{H}_n$ and complex numbers by real numbers, we can get the corresponding equalities of (56) $\sim$ (60) on $\mathcal{H}_n$. The case $k < -1$ can be proved completely analogously with the help of Theorem 4.1.

For the case $k = -1$, the equality corresponding to (60) can be simplified as

$$\psi\left(B^2\right) = \psi(B)\psi(I_n)^{-1}\psi(B), \quad B \in \mathcal{H}_n. \tag{76}$$

Let $\psi_1(B) := \psi(I_n)^{-1}\psi(B)$. Then $\psi_1 : \mathcal{H}_n \to \mathcal{M}_n$ is a nonzero real linear map that satisfies $\psi_1\left(B^2\right) = \psi_1(B)^2$ for $B \in \mathcal{H}_n$. Extend ψ_1 to a complex linear map $\tilde{\psi} : \mathcal{M}_n \to \mathcal{M}_n$ such that

$$\tilde{\psi}(A + \mathbf{i}B) := \psi_1(A) + \mathbf{i}\psi_1(B), \quad A, B \in \mathcal{H}_n. \tag{77}$$

Similarly to the arguments in part (1) of the proof of Theorem 4.1, we have $\tilde{\psi}\left((A + \mathbf{i}B)^2\right) = (\tilde{\psi}(A + \mathbf{i}B))^2$ for all $A, B \in \mathcal{H}_n$. Using Theorem 3.4 and the fact that $\tilde{\psi}(I_n) = \psi_1(I_n) = I_n$, we can prove that there is an invertible $P \in \mathcal{M}_n$ such that for all $B \in \mathcal{H}_n$, either $\psi_1(B) = P^{-1}BP$ or $\psi_1(B) = P^{-1}B^tP$. So

$$\psi(B) = \psi(I_n)P^{-1}BP, \quad B \in \mathcal{H}_n; \quad \text{or} \tag{78}$$

$$\psi(B) = \psi(I_n)P^{-1}B^tP, \quad B \in \mathcal{H}_n. \tag{79}$$

If $\psi(B) = \psi(I_n)P^{-1}BP$ for $B \in \mathcal{H}_n$, then $\psi(I_n)P^{-1}BP = \left(\psi(I_n)P^{-1}BP\right)^* = P^*BP^{-*}\psi(I_n)$, which gives

$$\left(P^{-*}\psi(I_n)P^{-1}\right)B = B\left(P^{-*}\psi(I_n)P^{-1}\right), \quad B \in \mathcal{H}_n. \tag{80}$$

Hence $P^{-*}\psi(I_n)P^{-1} = cI_n$ for certain $c \in \mathbb{R}\backslash\{0\}$. We have $\psi(I_n) = cP^*P$ so that $\psi(B) = cP^*BP$ for $B \in \mathcal{H}_n$. Similarly for the case $\psi(B) = \psi(I_n)P^{-1}B^tP$. Adjusting c and P by scalar factors simultaneously, we may assume that $c \in \{1, -1\}$. It implies (74).

Remark 4.3. *Theorem 4.2 does not hold if ψ is not assumed to be linear. Let k be a positive even integer. Let $\tilde{\psi} : \mathcal{H}_n \to \mathcal{H}_n$ be any bijective linear map such that $\tilde{\psi}\left(\overline{\mathcal{P}_n}\right) \subseteq \overline{\mathcal{P}_n}$. For example, $\tilde{\psi}$ may be a completely positive map of the form $\tilde{\psi}(B) = \sum_{i=1}^r N_i^* BN_i$ for $r \geq 2$, $N_1, \ldots, N_r \in \mathcal{M}_n$ linearly independent, and at least one of $N_1, \ldots, N_r$ is invertible. By Corollary 2.3, there is a linear bijection $\phi : \mathcal{H}_n \to \mathcal{H}_n$ such that $\operatorname{tr}(\phi(A)\tilde{\psi}(B)) = \operatorname{tr}(AB)$ for all $A, B \in \mathcal{H}_n$. Let $\psi : \mathcal{H}_n \to \mathcal{H}_n$ be defined by $\psi(B) = \tilde{\psi}\left(B^k\right)^{1/k}$. Then*

$$\operatorname{tr}\left(\phi(A)\psi(B)^k\right) = \operatorname{tr}\left(\phi(A)\tilde{\psi}\left(B^k\right)\right) = \operatorname{tr}\left(AB^k\right), \quad A, B \in \mathcal{H}_n.$$

Obviously, ψ may be non-linear, and the choices of pairs (ϕ, ψ) are much more than those in (74) and (75).

5. *k*-power linear preservers and trace of power-product preservers on $\mathcal{S}_n$ and $\mathcal{S}_n(\mathbb{R})$

5.1 *k*-power linear preservers on $\mathcal{S}_n$ and $\mathcal{S}_n(\mathbb{R})$

Chan and Lim described the linear k-power preservers on $\mathcal{S}_n(\mathbb{F})$ for $k \geq 2$ in [7, Theorem 2] as follows.

Theorem 5.1. *(Chan, Lim [5]) Let an integer $k \geq 2$. Let $\mathbb{F}$ be an algebraic closed field with char$(\mathbb{F}) = 0$ or char$(\mathbb{F}) > k$. Suppose that $\psi : \mathcal{S}_n(\mathbb{F}) \to \mathcal{S}_n(\mathbb{F})$ is a nonzero linear operator such that $\psi\left(A^k\right) = \psi(A)^k$ for all $A \in \mathcal{S}_n(\mathbb{F})$. Then there exist $\lambda \in \mathbb{F}$ with $\lambda^{k-1} = 1$ and an orthogonal matrix $O \in \mathcal{M}_n(\mathbb{F})$ such that*

$$\psi(A) = \lambda OAO^t, \quad A \in \mathcal{S}_n. \tag{81}$$

We generalize Theorem 5.1 to include the case $\mathcal{S}_n(\mathbb{R})$, to include negative integers k, and to assume the k-power preserving condition only on matrices nearby the identity.

Theorem 5.2. *Let $k \in \mathbb{Z} \backslash \{0,1\}$. Let $\mathbb{F} = \mathbb{C}$ or $\mathbb{R}$. Suppose that $\psi : \mathcal{S}_n(\mathbb{F}) \to \mathcal{S}_n(\mathbb{F})$ is a nonzero linear map such that $\psi(A^k) = \psi(A)^k$ for all A in an open neighborhood of I_n in $\mathcal{S}_n(\mathbb{F})$ consisting of invertible matrices. Then there exist $\lambda \in \mathbb{F}$ with $\lambda^{k-1} = 1$ and an orthogonal matrix $O \in \mathcal{M}_n(\mathbb{F})$ such that*

$$\psi(A) = \lambda OAO^t, \quad A \in \mathcal{S}_n(\mathbb{F}). \tag{82}$$

Proof. It suffices to prove the necessary part. In both $k \geq 2$ and $k < 0$ cases, using analogous arguments as parts (1) and (2) of the proof of Theorem 3.4, we get that $\psi(I_n)$ commutes with the range of ψ, and the nonzero map $\psi_1(A) := \psi(I_n)^{k-2}\psi(A)$ satisfies that $\psi_1(A^2) = \psi_1(A)^2$ for $A \in \mathcal{S}_n(\mathbb{F})$. Then

$$\begin{aligned} \psi_1(A)\psi_1(B) + \psi_1(B)\psi_1(A) &= \psi_1(A+B)^2 - \psi_1(A)^2 - \psi_1(B)^2 \\ &= \psi_1\left((A+B)^2\right) - \psi_1(A^2) - \psi_1(B^2) \\ &= \psi_1(AB + BA). \end{aligned} \tag{83}$$

In particular, $\psi_1(A)\psi_1(A^r) + \psi_1(A^r)\psi_1(A) = 2\psi_1(A^{r+1})$ for $r \in \mathbb{Z}_+$. Using induction, we get $\psi_1(A^\ell) = \psi_1(A)^\ell$ for all $A \in \mathcal{S}_n(\mathbb{F})$ and $\ell \in \mathbb{Z}_+$. By [26, Corollary 6.5.4], there is an orthogonal matrix $O \in \mathcal{M}_n(\mathbb{F})$ such that $\psi_1(A) = OAO^t$. Since $\psi(I_n)$ commutes with the range of ψ_1, we have $\psi(I_n) = \lambda I_n$ for certain $\lambda \in \mathbb{F}$ in which $\lambda^{k-1} = 1$. So $\psi(A) = \lambda OAO^t$ as in (82).

Obviously, in $\mathbb{F} = \mathbb{R}$ case, (82) has $\lambda = 1$ when k is even and $\lambda \in \{1, -1\}$ when k is odd.

5.2 Trace of power-product preservers on $\mathcal{S}_n$ and $\mathcal{S}_n(\mathbb{R})$

Corollary 2.3 shows that every linear bijection $\phi : \mathcal{S}_n(\mathbb{F}) \to \mathcal{S}_n(\mathbb{F})$ corresponds to another linear bijection $\psi : \mathcal{S}_n(\mathbb{F}) \to \mathcal{S}_n(\mathbb{F})$ such that $\operatorname{tr}(\phi(A)\psi(B)) = \operatorname{tr}(AB)$ for all $A, B \in \mathcal{S}_n(\mathbb{F})$. When $m \geq 3$, maps $\phi_1, \cdots, \phi_m : \mathcal{S}_n(\mathbb{F}) \to \mathcal{S}_n(\mathbb{F})$ that satisfy $\operatorname{tr}(\phi_1(A_1)\cdots\phi_m(A_m)) = \operatorname{tr}(A_1\cdots A_m)$ are determined in [18].

We characterize the trace of power-product preserver for $\mathcal{S}_n(\mathbb{F})$ here.

Theorem 5.3. *Let $\mathbb{F} = \mathbb{C}$ or $\mathbb{R}$. Let $k \in \mathbb{Z} \backslash \{0,1\}$. Let S be an open neighborhood of I_n in $\mathcal{S}_n(\mathbb{F})$ consisting of invertible matrices. Then two maps $\phi, \psi : \mathcal{S}_n(\mathbb{F}) \to \mathcal{S}_n(\mathbb{F})$ satisfy that*

$$\operatorname{tr}\left(\phi(A)\psi(B)^k\right) = \operatorname{tr}(AB^k), \tag{84}$$

1. *for all $A \in \mathcal{S}_n(\mathbb{F})$, $B \in S$, and ψ is linear, or*

2. *for all $A, B \in S$ and both ϕ and ψ are linear,*

if and only if ϕ and ψ take the following forms:

a. *When $k = -1$, there exist an invertible matrix $P \in \mathcal{M}_n(\mathbb{F})$ and $c \in \mathbb{F} \backslash \{0\}$ such that*

$$\phi(A) = cPAP^t, \quad \psi(B) = cPBP^t, \quad A, B \in \mathcal{S}_n(\mathbb{F}). \tag{85}$$

We may choose $c = 1$ for $\mathbb{F} = \mathbb{C}$ and $c \in \{1, -1\}$ for $\mathbb{F} = \mathbb{R}$.

b. *When* $k \in \mathbb{Z}\backslash\{-1, 0, 1\}$, *there exist* $c \in \mathbb{F}\backslash\{0\}$ *and an orthogonal matrix* $O \in \mathcal{M}_n(\mathbb{F})$ *such that*

$$\phi(A) = c^{-k} OAO^t, \quad \psi(B) = cOBO^t, \quad A, B \in \mathcal{S}_n(\mathbb{F}). \tag{86}$$

Proof. Assumption (2) leads to assumption (1) (cf. the proof of Theorem 3.5). We prove the theorem under assumption (1). It suffices to prove the necessary part.

Obviously, $\mathcal{S}_n \cap \mathcal{H}_n = \mathcal{S}_n(\mathbb{R})$ and $\mathcal{S}_n \cap \mathcal{P}_n = \mathcal{P}_n(\mathbb{R})$. Let $S' := \{B \in \mathcal{P}_n(\mathbb{R}) : B^{1/k} \in S\}$, which is an open neighborhood of I_n in $\mathcal{P}_n(\mathbb{R})$ and whose real (resp. complex) span is $\mathcal{S}_n(\mathbb{R})$ (resp. $\mathcal{S}_n$). Using an analogous argument of the proof of Theorem 3.5, and replacing $\mathcal{M}_n$ by $\mathcal{S}_n(\mathbb{F})$, replacing the basis (48) of $\mathcal{M}_n$ by the following basis of rank 1 projections in $\mathcal{S}_n(\mathbb{F})$:

$$\{E_{ii} : 1 \le i \le n\} \cup \left\{ \frac{1}{\sqrt{2}} (E_{ii} + E_{jj} + E_{ij} + E_{ji}) : 1 \le i < j \le n \right\}, \tag{87}$$

and replacing the usage of Theorem 3.4 by that of Theorem 5.2, we can prove the case $k \ge 2$, and for $k < 0$, we can get the corresponding equalities up to (60).

Define a linear map $\psi_1 : \mathcal{S}_n(\mathbb{F}) \to \mathcal{M}_n(\mathbb{F})$ by $\psi_1(B) := \psi(B)\psi(I_n)^{-1}$.

When $k = -1$, we get the corresponding equality of (61), so that $\psi_1(B^2) = \psi_1(B)^2$ for $B \in \mathcal{S}_n(\mathbb{F})$. Similar to the proof of Theorem 5.2, we get $\psi_1(B^r) = \psi_1(B)^r$ for all $r \in \mathbb{Z}_+$. By [26, Theorem 6.5.3], there is an invertible matrix $P \in \mathcal{M}_n(\mathbb{F})$ such that $\psi_1(B) = PBP^{-1}$, so that $\psi(B) = PBP^{-1}\psi(I_n)$ for $B \in \mathcal{S}_n(\mathbb{F})$. Since $\psi(B) = \psi(B)^t$, we get

$$(P^{-1}\psi(I_n)P^{-t})B = B(P^{-1}\psi(I_n)P^{-t}), \quad B \in \mathcal{S}_n(\mathbb{F}). \tag{88}$$

Therefore, $P^{-1}\psi(I_n)P^{-t} = cI_n$ for certain $c \in \mathbb{F}\backslash\{0\}$, so that $\psi(B) = cPBP^t$ for all $B \in \mathcal{S}_n(\mathbb{F})$. Consequently, we get (85). The remaining claims are obvious.

When $k < -1$, using analogous argument as in the proof of $k < -1$ case of Theorem 3.5 and applying Theorem 5.2, we can get (86).

6. *k*-power linear preservers and trace of power-product preservers on $\mathcal{P}_n$ and $\mathcal{P}_n(\mathbb{R})$

In this section, we will determine k-power linear preservers and trace of power-product preservers on maps $\mathcal{P}_n \to \overline{\mathcal{P}_n}$ (resp. $\mathcal{P}_n(\mathbb{R}) \to \overline{\mathcal{P}_n(\mathbb{R})}$). Properties of such maps can be applied to maps $\mathcal{P}_n \to \mathcal{P}_n$ and $\overline{\mathcal{P}_n} \to \overline{\mathcal{P}_n}$ (resp. $\mathcal{P}_n(\mathbb{R}) \to \mathcal{P}_n(\mathbb{R})$ and $\overline{\mathcal{P}_n(\mathbb{R})} \to \overline{\mathcal{P}_n(\mathbb{R})}$).

6.1 *k*-power linear preservers on $\mathcal{P}_n$ and $\mathcal{P}_n(\mathbb{R})$

Theorem 6.1. *Fix* $k \in \mathbb{Z}\backslash\{0, 1\}$. *A nonzero linear map* $\psi : \mathcal{P}_n \to \overline{\mathcal{P}_n}$ *(resp.* $\psi : \mathcal{P}_n(\mathbb{R}) \to \overline{\mathcal{P}_n(\mathbb{R})}$*) satisfies that*

$$\psi(A^k) = \psi(A)^k \tag{89}$$

on an open neighborhood of I_n in $\mathcal{P}_n$ (resp. $\mathcal{P}_n(\mathbb{R})$) if and only if there is a unitary (resp. real orthogonal) matrix $U \in M_n$ such that

$$\psi(A) = U^*AU, \quad A \in \mathcal{P}_n; \quad or \quad \psi(A) = U^*A^tU, \quad A \in \mathcal{P}_n. \tag{90}$$

Proof. We prove the case $\psi : \mathcal{P}_n \to \overline{\mathcal{P}_n}$. The sufficient part is obvious. About the necessary part, the nonzero linear map $\psi : \mathcal{P}_n \to \overline{\mathcal{P}_n}$ can be easily extended to a linear map $\tilde{\psi} : \mathcal{H}_n \to \mathcal{H}_n$ that satisfies $\tilde{\psi}(A^k) = \tilde{\psi}(A)^k$ on an open neighborhood of I_n. By Theorem 4.1, we immediately get (90).

The case $\psi : \mathcal{P}_n(\mathbb{R}) \to \overline{\mathcal{P}_n(\mathbb{R})}$ can be similarly proved using Theorem 5.2.

6.2 Trace of powered product preservers on $\mathcal{P}_n$ and $\mathcal{P}_n(\mathbb{R})$

Now consider the maps $\mathcal{P}_n \to \overline{\mathcal{P}_n}$ (resp. $\mathcal{P}_n(\mathbb{R}) \to \overline{\mathcal{P}_n(\mathbb{R})}$) that preserve trace of powered products. Unlike $\mathcal{M}_n$ and $\mathcal{H}_n$, the set $\mathcal{P}_n$ (resp. $\mathcal{P}_n(\mathbb{R})$) is not a vector space. The trace of powered product preservers of two maps have the following forms.

Theorem 6.2 (Huang, Tsai [18]). *Let $a, b, c, d \in \mathbb{R}\backslash\{0\}$. Two maps $\phi, \psi : \mathcal{P}_n \to \overline{\mathcal{P}_n}$ satisfy*

$$\mathrm{tr}\left(\phi(A)^a\psi(B)^b\right) = \mathrm{tr}(A^cB^d), \quad A, B \in \mathcal{P}_n, \tag{91}$$

if and only if there exists an invertible $P \in \mathcal{M}_n$ such that

$$\begin{cases} \phi(A) = (P^*A^cP)^{1/a} \\ \psi(B) = (P^{-1}B^dP^{-*})^{1/b} \end{cases} or \begin{cases} \phi(A) = \left[P^*(A^t)^cP\right]^{1/a} \\ \psi(B) = \left[P^{-1}(B^t)^dP^{-*}\right]^{1/b} \end{cases} A, B \in \mathcal{P}_n. \tag{92}$$

Theorem 6.3 (Huang, Tsai [18]). *Given an integer $m \geq 3$ and real numbers $\alpha_1, \ldots, \alpha_m, \beta_1, \ldots, \beta_m \in \mathbb{R}\backslash\{0\}$, maps $\phi_i : \mathcal{P}_n \to \overline{\mathcal{P}_n}$ $(i = 1, \ldots, m)$ satisfy that*

$$\mathrm{tr}(\phi_1(A_1)^{\alpha_1} \cdots \phi_m(A_m)^{\alpha_m}) = \mathrm{tr}\left(A_1^{\beta_1} \cdots A_m^{\beta_1}\right), \quad A_1, \ldots, A_m \in \mathcal{P}_n, \tag{93}$$

if and only if they have the following forms for certain $c_1, \ldots, c_m \in \mathbb{R}_+$ with $c_1 \cdots c_m = 1$:

1. *When m is odd, there exists a unitary matrix $U \in \mathcal{M}_n$ such that for $i = 1, \ldots, m$:*

$$\phi_i(A) = c_i^{1/\alpha_i}U^*A^{\beta_i/\alpha_i}U, \quad A \in \mathcal{P}_n. \tag{94}$$

2. *When m is even, there exists an invertible $M \in \mathcal{M}_n$ such that for $i = 1, \ldots, m$:*

$$\phi_i(A) = \begin{cases} c_i^{1/\alpha_i}(M^*A^{\beta_i}M)^{1/\alpha_i}, & i \text{ is odd}, \\ c_i^{1/\alpha_i}(M^{-1}A^{\beta_i}M^{-*})^{1/\alpha_i}, & i \text{ is even}, \end{cases} \quad A \in \mathcal{P}_n. \tag{95}$$

Both Theorems 6.2 and 6.3 can be analogously extended to maps $\mathcal{P}_n(\mathbb{R}) \to \overline{\mathcal{P}_n(\mathbb{R})}$ without difficulties.

Theorem 6.2 determines maps $\phi, \psi : \mathcal{P}_n \to \overline{\mathcal{P}_n}$ that satisfy (91) throughout their domain. If we only assume the equality (91) for (A, B) in certain subset of $\mathcal{P}_n \times \mathcal{P}_n$ and assume certain linearity of ϕ and ψ, then ϕ and ψ may have slightly different forms. We determine the case $a = c = 1$ and $b = d = k \in \mathbb{Z}\backslash\{0\}$ here.

Theorem 6.4. *Let* $k \in \mathbb{Z}\backslash\{0\}$. *Let* S *be an open neighborhood of* I_n *in* $\mathcal{P}_n$. *Two maps* $\phi, \psi : \mathcal{P}_n \to \overline{\mathcal{P}_n}$ *satisfy*

$$\operatorname{tr}\left(\phi(A)\psi(B)^k\right) = \operatorname{tr}\left(AB^k\right), \tag{96}$$

3.*for all* $A, B \in \mathcal{P}_n$, *or*

4.*for all* $A \in S$, $B \in \mathcal{P}_n$, *and* ϕ *is linear,*

if and only if there exists an invertible $P \in \mathcal{M}_n$ such that

$$\begin{cases} \phi(A) = P^* A P \\ \psi(B) = \left(P^{-1} B^k P^{-*}\right)^{1/k} \end{cases} \quad or \quad \begin{cases} \phi(A) = P^* A^t P \\ \psi(B) = \left[P^{-1} (B^t)^k P^{-*}\right]^{1/k} \end{cases} \quad A, B \in \mathcal{P}_n. \tag{97}$$

The maps ϕ *and* ψ *satisfy* (96)

1.*for all* $A \in \mathcal{P}_n$, $B \in S$, *and* ψ *is linear, or*

2.*for all* $A, B \in S$ *and both* ϕ *and* ψ *are linear,*

if and only if when $k \in \{-1, 1\}$, ϕ *and* ψ *take the form* (97), *and when* $k \in \mathbb{Z}\backslash\{-1, 0, 1\}$, *there exist a unitary matrix* $U \in \mathcal{M}_n$ *and* $c \in \mathbb{R}^+$ *such that*

$$\begin{cases} \phi(A) = c^{-k} U^* A U \\ \psi(B) = c U^* B U \end{cases} \quad or \quad \begin{cases} \phi(A) = c^{-k} U^* A^t U \\ \psi(B) = c U^* B^t U \end{cases} \quad A, B \in \mathcal{P}_n. \tag{98}$$

Proof. It suffices to prove the necessary part.

The case of assumption (1) has been proved by Theorem 6.2.

Similar to the proof of Theorem 3.5, assumption (2) implies assumption (1); assumption (4) implies assumption (3). It remains to prove the case with assumption (3).

When $k = 1$, assumption (3) is analogous to assumption (2), and we get (97).

Suppose $k \in \mathbb{Z}\backslash\{1, 0\}$. Let $\psi_1 : \mathcal{P}_n \to \overline{\mathcal{P}_n}$ be defined by $\psi_1(B) := \psi\left(B^{1/k}\right)^k$. Let $\mathcal{S}_1 := \left\{B \in \mathcal{P}_n : B^{1/k} \in S\right\}$. Then (97) with assumption (2) becomes

$$\operatorname{tr}(\phi(A)\psi_1(B)) = \operatorname{tr}(AB), \quad A \in \mathcal{P}_n, \ B \in \mathcal{S}_1. \tag{99}$$

Let $\tilde{\psi} : \mathcal{H}_n \to \mathcal{H}_n$ be the linear extension of ψ. By Theorem 2.2, ϕ can be extended to a linear bijection $\tilde{\phi} : \mathcal{H}_n \to \mathcal{H}_n$ such that

$$\operatorname{tr}\left(\tilde{\phi}(A)\tilde{\psi}(B)^k\right) = \operatorname{tr}\left(\tilde{\phi}(A)\psi_1\left(B^k\right)\right) = \operatorname{tr}\left(AB^k\right), \quad A \in \mathcal{H}_n, \ B \in S. \tag{100}$$

By Theorem 4.2 and taking into account the ranges of ϕ and ψ, we see that when $k = -1$, ϕ and ψ take the form of (97), and when $k \in \mathbb{Z}\backslash\{-1, 0, 1\}$, ϕ and ψ take the form of (98).

Theorem 6.4 has counterpart results for $\phi, \psi : \mathcal{P}_n(\mathbb{R}) \to \overline{\mathcal{P}_n(\mathbb{R})}$ and the proof is analogous using Theorem 5.3 instead of Theorem 4.2.

7. *k*-power linear preservers and trace of power-product preservers on $\mathcal{D}_n$ and $\mathcal{D}_n(\mathbb{R})$

Let $\mathbb{F} = \mathbb{C}$ or $\mathbb{R}$. Define the function diag : $\mathbb{F}^n \to \mathcal{D}_n(\mathbb{F})$ to be the linear bijection that sends each $(c_1, \cdots, c_n)^t$ to the diagonal matrix with $c_1, \ldots, c_n$ (in order) as the diagonal entries. Define $\text{diag}^{-1} : \mathcal{D}_n(\mathbb{F}) \to \mathbb{F}^n$ the inverse map of diag.

With the settings, every linear map $\psi : \mathcal{D}_n(\mathbb{F}) \to \mathcal{D}_n(\mathbb{F})$ uniquely corresponds to a matrix $L_\psi \in \mathcal{M}_n(\mathbb{F})$ such that

$$\psi(A) = \text{diag}\left(L_\psi \text{diag}^{-1}(A)\right), \quad A \in \mathcal{D}_n(\mathbb{F}). \tag{101}$$

7.1 *k*-power linear preservers on $\mathcal{D}_n$ and $\mathcal{D}_n(\mathbb{R})$

We define the linear functionals $f_i : \mathcal{D}_n(\mathbb{F}) \to \mathbb{F}$ $(i = 0, 1, \ldots, n)$, such that for each $A = \text{diag}(a_1, \ldots, a_n) \in \mathcal{D}_n(\mathbb{F})$,

$$f_0(A) = 0; \quad f_i(A) = a_i, \quad i = 1, \ldots, n. \tag{102}$$

Theorem 7.1. *Let $\mathbb{F} = \mathbb{C}$ or $\mathbb{R}$. Let $k \in \mathbb{Z}\backslash\{0, 1\}$. Let S be an open neighborhood of I_n in $\mathcal{D}_n(\mathbb{F})$. A linear map $\psi : \mathcal{D}_n(\mathbb{F}) \to \mathcal{D}_n(\mathbb{F})$ satisfies that*

$$\psi(A^k) = \psi(A)^k, \quad A \in S, \tag{103}$$

if and only if

$$\psi(A) = \psi(I_n)\text{diag}\left(f_{p(1)}(A), \ldots, f_{p(n)}(A)\right), \quad A \in \mathcal{D}_n(\mathbb{F}), \tag{104}$$

in which $\psi(I_n)^k = \psi(I_n)$ and $p : \{1, \ldots, n\} \to \{0, 1, \ldots, n\}$ is a function such that $p(i) \neq 0$ when $k < 0$ for $i = 1, \ldots, n$. In particular, a linear bijection $\psi : \mathcal{D}_n(\mathbb{F}) \to \mathcal{D}_n(\mathbb{F})$ satisfies (103) *if and only if there is a diagonal matrix $C \in \mathcal{M}_n(\mathbb{F})$ with $C^{k-1} = I_n$ and a permutation matrix $P \in \mathcal{M}_n(\mathbb{F})$ such that*

$$\psi(A) = PCAP^{-1}, \quad A \in \mathcal{D}_n(\mathbb{F}). \tag{105}$$

Proof. For every $A = \text{diag}(a_1, \ldots, a_n) \in \mathcal{D}_n(\mathbb{F})$, when $x \in \mathbb{F}$ is sufficiently close to 0, we have $I_n + xA \in S$ and the power series of $(I_n + xA)^k$ converges, so that $\psi\left((I_n + xA)^k\right) = \psi(I_n + xA)^k$.

$$\psi\left((I_n + xA)^k\right) = \psi(I_n) + xk\psi(A) + x^2\frac{k(k-1)}{2}\psi(A^2) + \cdots \tag{106}$$

$$\psi(I_n + xA)^k = \psi(I_n) + xk\psi(I_n)^{k-1}\psi(A) + x^2\frac{k(k-1)}{2}\psi(I_n)^{k-2}\psi(A)^2 + \cdots \tag{107}$$

So for all $A \in \mathcal{D}_n(\mathbb{F})$:

$$\psi(A) = \psi(I_n)^{k-1}\psi(A), \tag{108}$$

$$\psi(A^2) = \psi(I_n)^{k-2}\psi(A)^2. \tag{109}$$

The linear map $\psi_1(A) := \psi(I_n)^{k-2}\psi(A)$ satisfies that

$$\psi_1(A^2) = \psi_1(A)^2, \quad A \in \mathcal{D}_n(\mathbb{F}). \tag{110}$$

By (101), let $L_{\psi_1} = (\ell_{ij}) \in \mathcal{M}_n(\mathbb{F})$ such that $\text{diag}^{-1}(\psi_1(A)) = L_{\psi_1}(\text{diag}^{-1}(A))$ for $A \in \mathcal{D}_n(\mathbb{F})$. Then (110) implies that for all $A = \text{diag}(a_1, \dots, a_n) \in \mathcal{D}_n(\mathbb{F})$:

$$\sum_{j=1}^{n} \ell_{ij}a_j^2 = \left(\sum_{j=1}^{n} \ell_{ij}a_j\right)^2, \quad i = 1, 2, \dots, n. \tag{111}$$

Therefore, each row of L_{ψ_1} has at most one nonzero entry and each nonzero entry must be 1. We get

$$\psi_1(A) = \text{diag}\left(f_{p(1)}(A), \dots, f_{p(n)}(A)\right) \tag{112}$$

in which $p : \{1, \dots, n\} \to \{0, 1, \dots, n\}$ is a function. Suppose $\psi(I_n) = \text{diag}(\lambda_1, \dots, \lambda_n)$. Then (108) implies that $\psi(A) = \psi(I_n)\psi_1(A)$ has the form (104). Obviously, $\psi(I_n)^k = \psi(I_n)$ and when $k < 0$, each $p(i) \neq 0$ for $i = 1, \dots, n$. Moreover, when ψ is a linear bijection, (112) shows that $\psi_1(A) = PAP^{-1}$ for a permutation matrix P. (105) can be easily derived.

7.2 Trace of power-product preservers on $\mathcal{D}_n$ and $\mathcal{D}_n(\mathbb{R})$

In [18], we show that two maps $\phi, \psi : \mathcal{D}_n(\mathbb{F}) \to \mathcal{D}_n(\mathbb{F})$ satisfy $\text{tr}(\phi(A)\psi(B)) = \text{tr}(AB)$ for $A, B \in \mathcal{D}_n(\mathbb{F})$ if and only if there exists an invertible $N \in \mathcal{M}_n(\mathbb{F})$ such that

$$\phi(A) = \text{diag}(N\text{diag}^{-1}(A)), \ \psi(B) = \text{diag}(N^{-t}\text{diag}^{-1}(B)), \quad A, B \in \mathcal{D}_n(\mathbb{F}). \tag{113}$$

When $m \geq 3$, the maps $\phi_1, \dots, \phi_m : \mathcal{D}_n(\mathbb{F}) \to \mathcal{D}_n(\mathbb{F})$ satisfying $tr(\phi_1(A_1)\cdots\phi_m(A_m)) = tr(A_1\cdots A_m)$ for $A_1, \dots, A_m \in \mathcal{D}_n(\mathbb{F})$ are also determined in [18].

Next we consider the trace of power-product preserver on $\mathcal{D}_n(\mathbb{F})$.

Theorem 7.2. *Let $\mathbb{F} = \mathbb{C}$ or $\mathbb{R}$. Let $k \in \mathbb{Z}\backslash\{0, 1\}$. Let S be an open neighborhood of I_n in $\mathcal{D}_n(\mathbb{F})$. Two maps $\phi, \psi : \mathcal{D}_n(\mathbb{F}) \to \mathcal{D}_n(\mathbb{F})$ satisfy that*

$$\text{tr}\left(\phi(A)\psi(B)^k\right) = \text{tr}(AB^k), \tag{114}$$

1. *for all $A \in \mathcal{D}_n(\mathbb{F}), \ B \in S$, and ψ is linear, or*

2. *for all $A, B \in S$ and both ϕ and ψ are linear,*

if and only if there exist an invertible diagonal matrix $C \in \mathcal{D}_n(\mathbb{F})$ and a permutation matrix $P \in \mathcal{M}_n(F)$ such that

$$\phi(A) = PC^{-k}AP^{-1}, \quad \psi(B) = PCBP^{-1}, \quad A, B \in \mathcal{D}_n(\mathbb{F}). \tag{115}$$

Proof. Assumption (2) leads to assumption (1) (cf. the proof of Theorem 3.5). We prove the theorem under assumption (1).

For every $B \in \mathcal{D}_n(\mathbb{F})$, $I_n + xB \in S$ and the power series of $(I_n + xB)^k$ converges when $x \in \mathbb{F}$ is sufficiently close to 0, so that

$$\mathrm{tr}\left(\phi(A)\psi(I_n + xB)^k\right) = \mathrm{tr}\left(A(I_n + xB)^k\right) \tag{116}$$

Comparing degree one terms and degree two terms in the power series of the above equality, respectively, we get the following equalities for $A, B \in \mathcal{D}_n(\mathbb{F})$:

$$\mathrm{tr}\left(\phi(A)\psi(B)\psi(I_n)^{k-1}\right) = \mathrm{tr}(AB), \tag{117}$$

$$\mathrm{tr}\left(\phi(A)\psi(B)^2\psi(I_n)^{k-2}\right) = \mathrm{tr}\left(AB^2\right). \tag{118}$$

Applying Theorem 2.2 to (117), $\psi(I_n)$ is invertible and both ϕ and ψ are linear bijections. (117) and (119) imply that $\psi(B^2)\psi(I_n)^{k-1} = \psi(B)^2\psi(I_n)^{k-2}$. Let $\psi_1(B) := \psi(B)\psi(I_n)^{-1}$. Then $\psi_1(B^2) = \psi_1(B)^2$ for $B \in \mathcal{D}_n(\mathbb{F})$. By Theorem 7.1 and $\psi_1(I_n) = I_n$, there exists a permutation matrix $P \in \mathcal{M}_n(F)$ such that $\psi_1(B) = PBP^{-1}$ for $B \in \mathcal{D}_n(\mathbb{F})$. So $\psi(B) = \psi(I_n)PBP^{-1} = PCBP^{-1}$ for $C := P^{-1}\psi(I_n)P \in \mathcal{D}_n(\mathbb{F})$. Then (114) implies (115).

8. *k*-power injective linear preservers and trace of power-product preservers on $\mathcal{T}_n$ and $\mathcal{T}_n(\mathbb{R})$

8.1 *k*-power preservers on $\mathcal{T}_n(\mathbb{F})$

The characterization of injective linear k-power preserver on $\mathcal{T}_n(\mathbb{F})$ can be derived from Cao and Zhang's characterization of injective additive k-power preserver on $\mathcal{T}_n(\mathbb{F})$ ([12] or [[6], Theorem 6.5.2]).

Theorem 8.1 (Cao and Zhang [12]). *Let $k \geq 2$ and $n \geq 3$. Let $\mathbb{F}$ be a field with char$(\mathbb{F}) = 0$ or char$(\mathbb{F}) > k$. Then $\psi : \mathcal{T}_n(\mathbb{F}) \mapsto \mathcal{T}_n(\mathbb{F})$ is an injective linear map such that $\psi(A^k) = \psi(A)^k$ for all $A \in \mathcal{T}_n(\mathbb{F})$ if and only if there exists a $(k-1)$th root of unity λ and an invertible matrix $P \in \mathcal{T}_n(\mathbb{F})$ such that*

$$\psi(A) = \lambda PAP^{-1}, \quad A \in \mathcal{T}_n(\mathbb{F}), \quad or \tag{119}$$

$$\psi(A) = \lambda PA^{-}P^{-1}, \quad A \in \mathcal{T}_n(\mathbb{F}), \tag{120}$$

where $A^{-} = \left(a_{n+1-j,n+1-i}\right)$ if $A = \left(a_{ij}\right)$.

Example 8.2. *When $n = 2$, the injective linear maps that satisfy $\psi(A^k) = \psi(A)^k$ for $A \in \mathcal{T}_2(\mathbb{F})$ send $A = \begin{pmatrix} a_{11} & a_{12} \\ 0 & a_{22} \end{pmatrix}$ to the following $\psi(A)$:*

$$\lambda\begin{pmatrix} a_{11} & ca_{12} \\ 0 & a_{22} \end{pmatrix}, \quad \lambda\begin{pmatrix} a_{22} & ca_{12} \\ 0 & a_{11} \end{pmatrix}, \tag{121}$$

in which $\lambda^{k-1} = 1$ and $c \in \mathbb{F} \backslash \{0\}$..

Example 8.3. *Theorem 8.1 does not hold if ψ is not assumed to be injective. Let $n = 3$ and suppose $\psi : \mathcal{T}_3(\mathbb{F}) \to \mathcal{T}_3(\mathbb{F})$ is a linear map that sends $A = (a_{ij})_{3\times 3} \in \mathcal{T}_3(\mathbb{F})$ to one of the following $\psi(A)$ $(c, d \in \mathbb{F})$:*

$$\begin{pmatrix} a_{11} & ca_{12} & 0 \\ 0 & a_{22} & da_{23} \\ 0 & 0 & a_{33} \end{pmatrix}, \begin{pmatrix} a_{33} & 0 & 0 \\ 0 & a_{11} & 0 \\ 0 & 0 & a_{22} \end{pmatrix}, \begin{pmatrix} a_{22} & 0 & ca_{12} \\ 0 & 0 & 0 \\ 0 & 0 & a_{11} \end{pmatrix}. \tag{122}$$

Then each ψ satisfies that $\psi(A^k) = \psi(A)^k$ for every positive integer k but it is not of the forms in Theorem 8.1.

We extend Theorem 8.1 to the following result that includes negative k-powers and that only assumes k-power preserving in a neighborhood of I_n.

Theorem 8.4. *Let $\mathbb{F} = \mathbb{C}$ or $\mathbb{R}$. Let integers $k \neq 0, 1$ and $n \geq 3$. Suppose that $\psi : \mathcal{T}_n(\mathbb{F}) \to \mathcal{T}_n(\mathbb{F})$ is an injective linear map such that $\psi(A^k) = \psi(A)^k$ for all A in an open neighborhood of I_n in $\mathcal{T}_n(\mathbb{F})$ consisting of invertible matrices. Then there exist $\lambda \in \mathbb{F}$ with $\lambda^{k-1} = 1$ and an invertible matrix $P \in \mathcal{T}_n(\mathbb{F})$ such that*

$$\psi(A) = \lambda PAP^{-1}, \quad A \in \mathcal{T}_n(\mathbb{F}), \quad or \tag{123}$$

$$\psi(A) = \lambda PA^{-}P^{-1}, \quad A \in \mathcal{T}_n(\mathbb{F}). \tag{124}$$

where $A^{-} = (a_{n+1-j,n+1-i}) = J_n A^t J_n$ if $A = (a_{ij})$, J_n is the anti-diagonal identity.

Proof. Obviously ψ is a linear bijection. Follow the same process in the proof of Theorem 3.4. In both $k \geq 2$ and $k < 0$ cases we have $\psi(I_n)$ commutes with the range of ψ, so that $\psi(I_n) = \lambda I_n$ for $\lambda \in \mathbb{F}$ and $\lambda^{k-1} = 1$. Moreover, let $\psi_1(A) := \psi(I_n)^{-1}\psi(A)$, then ψ_1 is injective linear and $\psi_1(A^2) = \psi_1(A)^2$ for $A \in \mathcal{T}_n(\mathbb{F})$. Theorem 8.1 shows that $\psi_1(A) = PAP^{-1}$ or $\psi_1(A) = PA^{-}P^{-1}$ for certain invertible $P \in \mathcal{T}_n(\mathbb{F})$. It leads to (123) and (124).

8.2 Trace of power-product preservers on $\mathcal{T}_n$ and $\mathcal{T}_n(\mathbb{R})$

Theorem 2.2 or Corollary 2.3 does not work for maps on $\mathcal{T}_n(\mathbb{F})$. However, the following trace preserving result can be easily derived from Theorem 7.2. We have $\mathcal{T}_n(\mathbb{F}) = \mathcal{D}_n(\mathbb{F}) \oplus \mathcal{N}_n(\mathbb{F})$. Let $D(A)$ denote the diagonal matrix that takes the diagonal of $A \in \mathcal{T}_n(\mathbb{F})$.

Theorem 8.5. *Let $\mathbb{F} = \mathbb{C}$ or $\mathbb{R}$. Let $k \in \mathbb{Z} \backslash \{0, 1\}$. Let S be an open neighborhood of I_n in $\mathcal{T}_n(\mathbb{F})$ consisting of invertible matrices. Then two maps $\phi, \psi : \mathcal{T}_n(\mathbb{F}) \to \mathcal{T}_n(\mathbb{F})$ satisfy that*

$$\mathrm{tr}\left(\phi(A)\psi(B)^k\right) = \mathrm{tr}(AB^k), \tag{125}$$

1. *for all $A \in \mathcal{T}_n(\mathbb{F})$, $B \in S$, and ψ is linear, or*

2. *for all $A, B \in S$ and both ϕ and ψ are linear,*

if and only if ϕ *and* ψ *send* $\mathcal{N}_n(\mathbb{F})$ *to* $\mathcal{N}_n(\mathbb{F})$, $(D\circ\phi)|_{\mathcal{D}_n(\mathbb{F})}$ *and* $(D\circ\psi)|_{\mathcal{D}_n(\mathbb{F})}$ *are linear bijections characterized by* (115) *in Theorem 7.2, and* $D\circ\phi = D\circ\phi\circ D$.

Proof. The sufficient part is easy to verify. We prove the necessary part here. Let $\phi' := (D\circ\phi)|_{\mathcal{D}_n(\mathbb{F})}$ and $\psi' := (D\circ\psi)|_{\mathcal{D}_n(\mathbb{F})}$. Then $\phi', \psi' : \mathcal{D}_n(\mathbb{F}) \to \mathcal{D}_n(\mathbb{F})$ satisfy $tr\left(\phi'(A)\psi'(B)^k\right) = tr(AB^k)$ for $A, B \in \mathcal{D}_n(\mathbb{F})$. So they are characterized by (115). The bijectivity of ϕ' and ψ' implies that ϕ and ψ must send $\mathcal{N}_n(\mathbb{F})$ to $\mathcal{N}_n(\mathbb{F})$ in order to satisfy (125). Moreover, ϕ should send matrices with same diagonal to matrices with same diagonal, which implies that $D\circ\phi = D\circ\phi\circ D$.

9. Conclusion

We characterize linear maps $\psi : V \to V$ that satisfy $\psi(A^k) = \psi(A)^k$ on an open neighborhood S of I_n in V, where $k \in \mathbb{Z}\backslash\{0, 1\}$ and V is the set of $n \times n$ general matrices, Hermitian matrices, symmetric matrices, positive definite (resp. semi-definite) matrices, diagonal matrices, or upper triangular matrices, over the complex or real field. The characterizations extend the existing results of linear k-power preservers on the spaces of general matrices, symmetric matrices, and upper triangular matrices.

Applying the above results, we determine the maps $\phi, \psi : V \to V$ on the preceding sets V that satisfy $tr\left(\phi(A)\psi(B)^k\right) = tr(AB^k)$

1. for all $A \in V$, $B \in S$, and ψ is linear, or

2. for all $A, B \in S$ and both ϕ and ψ are linear.

These results, together with Theorem 2.2 about maps satisfying $tr(\phi(A)\psi(B)) = tr(AB)$ and the characterizations of maps $\phi_1, \cdots, \phi_m : V \to V$ $(m \geq 3)$ satisfying $tr(\phi_1(A_1)\cdots\phi_m(A_m)) = tr(A_1\cdots A_m)$ in [18], make a comprehensive picture of the preservers of trace of matrix products in the related matrix spaces and sets. Our results can be interpreted as inner product preservers when V is close under conjugate transpose, in which wide applications are found.

There are a few prospective directions to further the researches.

First, for a polynomial or an analytic function $f(x)$ and a matrix set V, we can consider "local" linear f-preservers, that is, linear operators $\psi : V \to V$ that satisfy $\psi(f(A)) = f(\psi(A))$ on an open subset S of V. A linear f-preserver ψ on S also preserves matrices annihilated by f on S, that is, $f(A) = 0$ $(A \in S)$ implies $f(\psi(A)) = 0$. When $S = V$ is M_n, $B(H)$, or some operator algebras, extensive studies have been done on operators preserving elements annihilated by a polynomial f; for examples, the results on M_n by R. Howard in [19], by P. Botta, S. Pierce, and W. Watkins in [20], and by C.-K. Li and S. Pierce in [21], on $B(H)$ by P. Šemrl [22], on linear maps $\psi : B(H) \to B(K)$ by Z. Bai and J. Hou in [23], and on some operator algebras by J. Hou and S. Hou in [24]. We may further explore linear f-preservers for a multivariable function $f(x_1, \ldots, x_r)$, that is, operator ψ satisfying $\psi(f(A_1, \ldots, A_r)) = f(\psi(A_1), \ldots, \psi(A_r))$. The corresponding annihilator preserver problem has been studied in some special cases, for example, on M_n for homogeneous multilinear polynomials by A. E. Guterman and B. Kuzma in [25].

Second, it is interesting to further investigate maps $\phi, \psi : V \to V$ that satisfy $tr(f(\phi(A))g(\psi(B))) = tr(f(A)g(B))$ for some polynomials or analytic functions $f(x)$ and $g(x)$. This is equivalent to the inner product preserver problem $\langle f(\phi(A))^*, g(\psi(B))\rangle = \langle f(A)^*, g(B)\rangle$ when V is close under conjugate transpose. More generally, given a multivariable function $h(x_1, \dots, x_m)$, we can ask what combinations of linear operators $\phi_1, \dots, \phi_m : V \to V$ satisfy that $tr(h(\phi_1(A_1), \dots, \phi_m(A_m)) = tr(h(A_1, \dots, A_m))$. The research on this area seems pretty new. No much has been discovered by the authors.

Author details

Huajun Huang[1*] and Ming-Cheng Tsai[2]

1 Department of Mathematics and Statistics, Auburn University, AL, USA

2 General Education Center, Taipei University of Technology, Taipei, Taiwan

*Address all correspondence to: huanghu@auburn.edu

References

[1] Li CK, Pierce S. Linear preserver problems. American Mathematical Monthly. 2001;**108**:591-605

[2] Li CK, Tsing NK. Linear preserver problems: A brief introduction and some special techniques. Linear Algebra and its Applications. 1992;**162–164**: 217-235

[3] Molnár L. Selected Preserver Problems on Algebraic Structures of Linear Operators and on Function Spaces, Lecture Notes in Mathematics. Vol. 1895. Berlin: Springer-Verlag; 2007

[4] Pierce S et al. A survey of linear preserver problems. Linear and Multilinear Algebra. 1992;**33**:1-129

[5] Chan GH, Lim MH. Linear preservers on powers of matrices. Linear Algebra and its Applications. 1992;**162–164**: 615-626

[6] Zhang X, Tang X, Cao C. Preserver Problems on Spaces of Matrices. Beijing: Science Press; 2006

[7] Brešar M, Šemrl P. Linear transformations preserving potent matrices. Proceedings of the American Mathematical Society. 1993;**119**:81-86

[8] Zhang X, Cao CG. Linear k-power/k-potent preservers between matrix spaces. Linear Algebra and its Applications. 2006;**412**:373-379

[9] Kadison RV. Isometries of operator algebras. Annals of Mathematics. 1951; **54**:325-338

[10] Brešar M, Martindale WS, Miers CR. Maps preserving nth powers. Communications in Algebra. 1998;**26**: 117-138

[11] Cao C, Zhang X. Power preserving additive maps (in Chinese). Advances in Mathematics. 2004;**33**:103-109

[12] Cao C, Zhang X. Power preserving additive operators on triangular matrix algebra (in Chinese). Journal of Mathematics. 2005;**25**:111-114

[13] Uhlhorn U. Representation of symmetry transformations in quantum mechanics. Arkiv för Matematik. 1963; **23**:307-340

[14] Molnár L. Some characterizations of the automorphisms of $\mathscr{B}\mathscr{H}$ and CX. Proceedings of the American Mathematical Society. 2002;**130**(1):111-120

[15] Li CK, Plevnik L, Šemrl P. Preservers of matrix pairs with a fixed inner product value. Operators and Matrices. 2012;**6**:433-464

[16] Huang H, Liu CN, Tsai MC, Szokol P, Zhang J. Trace and determinant preserving maps of matrices. Linear Algebra and its Applications. 2016;**507**: 373-388

[17] Leung CW, Ng CK, Wong NC. Transition probabilities of normal states determine the Jordan structure of a quantum system. Journal of Mathematical Physics. 2016;**57**:015212

[18] Huang H, Tsai MC. Maps Preserving Trace of Products of Matrices. Available from: https://arxiv.org/abs/2103.12552 [Preprint]

[19] Howard R. Linear maps that preserve matrices annihilated by a polynomial. Linear Algebra and its Applications. 1980;**30**:167-176

[20] Botta P, Pierce S, Watkins W. Linear transformations that preserve the

nilpotent matrices. Pacific Journal of Mathematics. 1983;**104**(1):39-46

[21] Li C-K, Pierce S. Linear operators preserving similarity classes and related results. Canadian Mathematical Bulletin. 1994;**37**(3):374-383

[22] Šemrl P. Linear mappings that preserve operators annihilated by a polynomial. Journal of Operator Theory. 1996;**36**(1):45-58

[23] Bai Z, Hou J. Linear maps and additive maps that preserve operators annihilated by a polynomial. Journal of Mathematical Analysis and Applications. 2002;**271**(1):139-154

[24] Hou J, Hou S. Linear maps on operator algebras that preserve elements annihilated by a polynomial. Proceedings of the American Mathematical Society. 2002;**130**(8):2383-2395

[25] Guterman AE, Kuzma B. Preserving zeros of a polynomial. Communications in Algebra. 2009;**37**(11):4038-4064

Chapter 2

Pencils of Semi-Infinite Matrices and Orthogonal Polynomials

Sergey Zagorodnyuk

Abstract

Semi-infinite matrices, generalized eigenvalue problems, and orthogonal polynomials are closely related subjects. They connect different domains in mathematics—matrix theory, operator theory, analysis, differential equations, etc. The classical examples are Jacobi and Hessenberg matrices, which lead to orthogonal polynomials on the real line (OPRL) and orthogonal polynomials on the unit circle (OPUC). Recently there turned out that pencils (i.e., operator polynomials) of semi-infinite matrices are related to various orthogonal systems of functions. Our aim here is to survey this increasing subject. We are mostly interested in pencils of symmetric semi-infinite matrices. The corresponding polynomials are defined as generalized eigenvectors of the pencil. These polynomials possess special orthogonality relations. They have physical and mathematical applications that will be discussed. Examples show that there is an unclarified relation to Sobolev orthogonal polynomials. This intriguing connection is a challenge for further investigations.

Keywords: semi-infinite matrix, pencil, orthogonal polynomials, Sobolev orthogonality, difference equation

1. Introduction

In this section, we will introduce the main objects of this chapter along with some brief historical notes.

By operator pencils or operator polynomials one means polynomials of a complex variable λ whose coefficients are linear bounded operators acting in a Banach space X:

$$L(\lambda) = \sum_{j=0}^{m} \lambda^j A_j, \tag{1}$$

where $A_j : X \to X$ $(j = 0, \dots, m)$, see, for example, [1, 2]. Parlett in ref. [3, p. 339] stated that the term *pencil* was introduced by Gantmacher in ref. [4] for matrix expressions, and Parlett explained how this term came from optics and geometry. In this chapter, we shall be mainly interested in pencils of banded semi-infinite matrices that are related to different kinds of scalar orthogonal polynomials. The classical example of such a relation is the case of orthogonal polynomials on the real line

(OPRL) and Jacobi matrices, see, for example, refs. [5, 6]. If $\{p_n(x)\}_{n=0}^{\infty}$ is a set of orthonormal OPRL and J is the corresponding Jacobi matrix, then the following relation holds:

$$(J - xE)\vec{p}(x) = 0, \tag{2}$$

where $\vec{p}(x) = (p_0(x), p_1(x), \dots)^T$, is a vector of polynomials (here the superscript T means the transposition), and E is the identity matrix (having units on the main diagonal and zeros elsewhere). In other words, $\vec{p}$ is an eigenfunction of the pencil $J - xE$. It is surprising that mathematicians rarely talked about the relation (2) in such a manner. The next classical example is the case of orthogonal polynomials on the unit circle (OPUC) and the corresponding three-term recurrence relation, see ref. [7, p. 159]. More recently there appeared CMV matrices, which are also related to OPUC, see, for example, ref. [8]. We should notice that besides orthogonal polynomials, there are other systems of functions that are closely related to semi-infinite matrices. Here we can mention biorthogonal polynomials and rational functions, see, for example, [9, 10] and references therein.

A natural generalization of OPRL is matrix orthogonal polynomials on the real line (MOPRL). MOPRL was introduced by Krein in 1949 [11]. They satisfy the relation of type (2), with J replaced by a block Jacobi matrix, and with $\vec{p}$ replaced by a vector of matrix polynomials. It turned out that MOPRL is closely related to orthogonal polynomials on the radial rays in the complex plane, see refs. [12, 13]. We shall discuss this case in Section 2.

Another possible generalization of relation (2) is the following one:

$$(J_5 - xJ_3)\vec{p}(x) = 0, \tag{3}$$

where J_3 is a Jacobi matrix, and J_5 is a real symmetric semi-infinite five-diagonal matrix with positive numbers on the second subdiagonal, see ref. [14]. These polynomials contain OPRL as a proper subclass. In general, they possess some special orthogonality relations. These polynomials will be discussed in Section 3.

Another natural generalization of OPRL is Sobolev orthogonal polynomials, see a recent survey in ref. [15]. During last years there appeared several examples of Sobolev polynomials, which are eigenfunctions of pencils of differential or difference operators. This subject will be discussed in Section 4.

Notations. As usual, we denote by $\mathbb{R}, \mathbb{C}, \mathbb{Z}, \mathbb{Z}_+$, the sets of real numbers, complex numbers, positive integers, integers, and nonnegative integers, respectively. By $\mathbb{C}_{m,n}$ we mean a set of all complex matrices of size $(m \times n)$. By $\mathbb{P}$ we denote the set of all polynomials with complex coefficients. The superscript T means the transposition of a matrix.

By l_2 we denote the usual Hilbert space of all complex sequences $c = (c_n)_{n=0}^{\infty} = (c_0, c_1, c_2, \dots)^T$ with the finite norm $\|c\|_{l_2} = \sqrt{\sum_{n=0}^{\infty}|c_n|^2}$. The scalar product of two sequences $c = (c_n)_{n=0}^{\infty}, d = (d_n)_{n=0}^{\infty} \in l_2$ is given by $(c, d)_{l_2} = \sum_{n=0}^{\infty} c_n \overline{d_n}$. We denote $\vec{e}_m = (\delta_{n,m})_{n=0}^{\infty} \in l_2$, $m \in \mathbb{Z}_+$. By $l_{2,fin}$ we denote the set of all finite vectors from l_2, that is, vectors with all, but finite number, components being zeros.

By $\mathfrak{B}(\mathbb{R})$ we denote the set of all Borel subsets of $\mathbb{R}$. If σ is a (non-negative) bounded measure on $\mathfrak{B}(\mathbb{R})$ then by L^2_σ we denote a Hilbert space of all (classes of equivalences of) complex-valued functions f on $\mathbb{R}$ with a finite norm $\|f\|_{L^2_\sigma} = \sqrt{\int_{\mathbb{R}} |f(x)|^2 d\sigma}$. The scalar product of two functions $f,g \in L^2_\sigma$ is given by $(f,g)_{L^2_\sigma} = \int_{\mathbb{R}} f(x)\overline{g(x)} d\sigma$. By $[f]$ we denote the class of equivalence in L^2_σ, which contains the representative f. By $\mathcal{P}$ we denote a set of all (classes of equivalence which contain) polynomials in L^2_σ. As usual, we sometimes use the representatives instead of their classes in formulas. Let B be an arbitrary linear operator in L^2_σ with the domain $\mathcal{P}$. Let $f(\lambda) \in \mathbb{P}$ be nonzero and of degree $d \in \mathbb{Z}_+, f(\lambda) = \sum_{k=0}^{d} d_k \lambda^k$, $d_k \in \mathbb{C}$. We set

$$f(B) = \sum_{k=0}^{d} d_k B^k; \quad B^0 := E\Big|_{\mathcal{P}}.$$

If $f \equiv 0$, then $f(B) := 0|_{\mathcal{P}}$.

If H is a Hilbert space then $(\cdot,\cdot)_H$ and $\|\cdot\|_H$ mean the scalar product and the norm in H, respectively. Indices may be omitted in obvious cases. For a linear operator A in H, we denote by $D(A)$ its domain, by $R(A)$ its range, by Ker A its null subspace (kernel), and A^* means the adjoint operator if it exists. If A is invertible then A^{-1} means its inverse. $\overline{A}$ means the closure of the operator, if the operator is closable. If A is bounded then $\|A\|$ denotes its norm.

2. Pencils $J_{2N+1} - \lambda^N E$ and orthogonal polynomials on radial rays in the complex plane

Throughout this section N will denote a fixed natural number. Let J_{2N+1} be a complex Hermitian semi-infinite $(2N+1)$-diagonal matrix. Let $\{p_n(\lambda)\}_{n=0}^{\infty}$, $\deg p_n = n$ be a set of complex polynomials, which satisfy the following relation:

$$(J_{2N+1} - \lambda^n E)\vec{p}(\lambda) = 0, \tag{4}$$

where $\vec{p}(\lambda) = (p_0(\lambda), p_1(\lambda), \ldots)^T$, is a vector of polynomials, and E is the identity matrix. Polynomials, which satisfy (4) with real coefficients, were first studied by Durán in ref. [16], following a suggestion of Marcellán. As it was already noticed in the Introduction, these polynomials are related to MOPRL. Namely, the following polynomials:

$$P_n(x) = \begin{pmatrix} R_{N,0}(p_{nN})(x) & R_{N,1}(p_{nN})(x) & \cdots & R_{N,N-1}(p_{nN})(x) \\ R_{N,0}(p_{nN+1})(x) & R_{N,1}(p_{nN+1})(x) & \cdots & R_{N,N-1}(p_{nN+1})(x) \\ \vdots & \vdots & \ddots & \vdots \\ R_{N,0}(p_{nN+N-1})(x) & R_{N,1}(p_{nN+N-1})(x) & \cdots & R_{N,N-1}(p_{nN+N-1})(x) \end{pmatrix} \tag{5}$$

are orthonormal MOPRL [12, Theorem]. Here

$$R_{N,m}(p)(t) = \sum_n \frac{p^{(nN+m)}(0)}{(nN+m)!} t^n, \qquad p \in \mathbb{P}, \ 0 \le m \le N-1. \tag{6}$$

Conversely, from a given set $\left\{P_n(x) = \left(P_{n,m,j}\right)_{m,j=0}^{N-1}\right\}_{n=0}^{\infty}$ of orthonormal MOPRL (suitably normed) one can construct scalar polynomials:

$$p_{nN+m}(x) = \sum_{j=0}^{N-1} x^j P_{n,m,j}\left(x^N\right), \qquad n \in \mathbb{N}, \ 0 \le m \le N-1, \tag{7}$$

which satisfy relation (4) [12]. Writing the corresponding matrix orthonormality conditions for P_n and equating the entries on both sides, one immediately gets orthogonality conditions for p_n:

$$\int_{\mathbb{R}} \left(R_{N,0}(p_n)(x), R_{N,1}(p_n)(x), \dots, R_{N,N-1}(p_n)(x)\right) d\mu \overline{\begin{pmatrix} R_{N,0}(p_m)(x) \\ R_{N,1}(p_m)(x) \\ \vdots \\ R_{N,N-1}(p_m)(x) \end{pmatrix}} = \tag{8}$$
$$= \delta_{n,m}, \qquad n,m \in \mathbb{Z}_+,$$

where μ is a $(N \times N)$ matrix measure. In the case of real coefficients in (4), this property was obtained by Durán in ref. [17].

Polynomials $\{p_n(\lambda)\}_{n=0}^{\infty}$ also satisfy the following orthogonality relations on radial rays in the complex plane [13]:

$$\int_{L_N} \left(p_n(\lambda), p_n(\lambda\varepsilon), \dots, p_n\left(\lambda\varepsilon^{N-1}\right)\right) dW(\lambda) \overline{\begin{pmatrix} p_m(\lambda) \\ p_m(\lambda\varepsilon) \\ \vdots \\ p_m\left(\lambda\varepsilon^{N-1}\right) \end{pmatrix}} +$$
$$+ \left(p_n(0), p_n'(0), \dots, p_n^{(N-1)}(0)\right) M \overline{\begin{pmatrix} p_m(0) \\ p_m'(0) \\ \vdots \\ p_m^{(N-1)}(0) \end{pmatrix}} = \delta_{n,m}, \qquad n,m \in \mathbb{Z}_+, \tag{9}$$

where $W(\lambda)$ is a non-decreasing matrix-valued function on $L_N \backslash \{0\}$; $M \in \mathbb{R}_{N,N}$, $M \ge 0$; $L_N = \left\{\lambda \in \mathbb{R} : \ \lambda^N \in \mathbb{R}\right\}$; ε is a primitive N-th root of unity. At $\lambda = 0$ the integral is understood as improper. Relation (9) can be derived from a Favard-type theorem in ref. [12, Theorem], but in ref. [13] we proceeded in another way. Relation (9) easily shows that the following classes of polynomials are included in the class of polynomials from (4):

A. OPRL;

B. orthogonal polynomials with respect to a scalar measure on radial rays L_N;

C. discrete Sobolev orthogonal polynomials on $\mathbb{R}$, having one discrete Sobolev term.

A detailed investigation of polynomials in the case (B) was done by Milovanovi'c, see ref. [18] and references therein. In particular, interesting examples of orthogonal polynomials were constructed and zero distribution of polynomials was studied. Discrete Sobolev polynomials from the case (C) may possess higher-order differential equations. This subject has a long history, see historical remarks in recent papers [19, 20]. For polynomials (9) some simple general properties of zeros were studied in ref. [21], while a Christoffel type formula was constructed in ref. [22]. In ref. [12] there was studied a more general case of relation (4), with a polynomial $h(\lambda)$ instead of λ^N.

3. Pencils $J_5 - xJ_3$ and orthogonal polynomials

Let J_3 be a Jacobi matrix and J_5 be a semi-infinite real symmetric five-diagonal matrix with positive numbers on the second subdiagonal. A set $\Theta = (J_3, J_5, \alpha, \beta)$, where $\alpha > 0$, $\beta \in \mathbb{R}$, is said to be *a Jacobi-type pencil (of matrices)* [14]. With a Jacobi-type pencil of matrices Θ one associates a system of polynomials $\{p_n(\lambda)\}_{n=0}^{\infty}$, which satisfies the following relations:

$$p_0(\lambda) = 1, \quad p_1(\lambda) = \alpha\lambda + \beta, \tag{10}$$

and

$$(J_5 - \lambda J_3)\vec{p}(\lambda) = 0, \tag{11}$$

where $\vec{p}(\lambda) = (p_0(\lambda), p_1(\lambda), p_2(\lambda), \cdots)^T$. Polynomials $\{p_n(\lambda)\}_{n=0}^{\infty}$ are said to be *associated with the Jacobi-type pencil of matrices* Θ.

Observe that for each system of OPRL with $p_0 = 1$ one can take J_3 to be the corresponding Jacobi matrix, $J_5 = J_3^2$, and α, β being the coefficients of p_1 ($p_1(\lambda) = \alpha\lambda + \beta$). Then, this system is associated with $\Theta = (J_3, J_5, \alpha, \beta)$. Let us mention two other circumstances where Jacobi-type pencils arise in a natural way.

1. **Discretization of a 4-th order differential operator.** Ben Amara, Vladimirov, and Shkalikov investigated the following linear pencil of differential operators [23]:

$$(py'') - \lambda(-y'' + cry) = 0. \tag{12}$$

The initial conditions are: $y(0) = y'(0) = y(1) = y'(1) = 0$, or $y(0) = y'(0) = y'(1) = (py'')'(1) + \lambda a y(1) = 0$. Here $p, r \in C[0, 1]$ are uniformly positive, while the parameters c and a are real. Eq. (12) has several physical applications, which include a motion of a partially fixed bar with additional constraints in the elasticity theory [23]. The discretization of this equation leads to a Jacobi-type pencil, see ref. [24].

2. **Partial sums of series of OPRL.** Let $\{g_n(x)\}_{n=0}^{\infty}$ ($\deg g_n = n$) be orthonormal OPRL with positive leading coefficients. Let $\{c_k\}_{k=0}^{\infty}$ be a set of arbitrary positive numbers. Then polynomials

$$p_n(x) := \frac{1}{c_0 g_0} \sum_{j=0}^{n} c_j g_j(x), \qquad n \in \mathbb{Z}_+, \tag{13}$$

are associated with a Jacobi-type pencil [25, Theorem 1]. Polynomials p_n are normed partial sums of the following formal power series:

$$\sum_{j=0}^{\infty} c_j g_j(x).$$

We shall return to such sums below.

From the definition of a Jacobi type pencil we see that matrices J_3 and J_5 have the following form:

$$J_3 = \begin{pmatrix} b_0 & a_0 & 0 & 0 & 0 & \cdots \\ a_0 & b_1 & a_1 & 0 & 0 & \cdots \\ 0 & a_1 & b_2 & a_2 & 0 & \cdots \\ \vdots & \vdots & \vdots & \ddots & & \end{pmatrix}, \qquad a_k > 0,\ b_k \in \mathbb{R},\ k \in \mathbb{Z}_+; \tag{14}$$

$$J_5 = \begin{pmatrix} \alpha_0 & \beta_0 & \gamma_0 & 0 & 0 & 0 & \cdots \\ \beta_0 & \alpha_1 & \beta_1 & \gamma_1 & 0 & 0 & \cdots \\ \gamma_0 & \beta_1 & \alpha_2 & \beta_2 & \gamma_2 & 0 & \cdots \\ 0 & \gamma_1 & \beta_2 & \alpha_3 & \beta_3 & \gamma_3 & \cdots \\ \vdots & \vdots & \vdots & \vdots & \ddots & & \end{pmatrix}, \quad \alpha_n, \beta_n \in \mathbb{R},\ \gamma_n > 0,\ n \in \mathbb{Z}_+. \tag{15}$$

Set

$$u_n := J_3 \vec{e}_n = a_{n-1} \vec{e}_{n-1} + b_n \vec{e}_n + a_n \vec{e}_{n+1}, \tag{16}$$

$$w_n := J_5 \vec{e}_n = \gamma_{n-2} \vec{e}_{n-2} + \beta_{n-1} \vec{e}_{n-1} + \alpha_n \vec{e}_n + \beta_n \vec{e}_{n+1} + \gamma_n \vec{e}_{n+2}, \qquad n \in \mathbb{Z}_+. \tag{17}$$

Here and in what follows by $\vec{e}_k$ with negative k we mean (vector) zero. The following operator:

$$Af = \frac{\zeta}{\alpha}\left(\vec{e}_1 - \beta \vec{e}_0\right) + \sum_{n=0}^{\infty} \xi_n w_n,$$

$$f = \zeta \vec{e}_0 + \sum_{n=0}^{\infty} \xi_n u_n \in l_{2,fin}, \quad \zeta, \xi_n \in \mathbb{C}, \tag{18}$$

with $D(A) = l_{2,fin}$ is called *the associated operator for the Jacobi-type pencil* Θ. In the sums in (18), only a finite number of ξ_n are nonzero. In what follows we shall always assume this in the case of elements from the linear span. In particular, the following relation holds:

$$AJ_3 \vec{e}_n = J_5 \vec{e}_n, \qquad n \in \mathbb{Z}_+.$$

Then

$$AJ_3 = J_5. \tag{19}$$

The matrices J_3 and J_5 define linear operators with the domain $l_{2,fin}$, which we denote by the same letters.

For an arbitrary nonzero polynomial $f(\lambda) \in \mathbb{P}$ of degree $d \in \mathbb{Z}_+, f(\lambda) = \sum_{k=0}^{d} d_k \lambda^k$, $d_k \in \mathbb{C}$, we set $f(A) = \sum_{k=0}^{d} d_k A^k$. Here $A^0 := E|_{l_{2,fin}}$. For $f(\lambda) \equiv 0$, we set $f(A) = 0|_{l_{2,fin}}$. The following relations hold [14]:

$$\vec{e}_n = p_n(A)\vec{e}_0, \qquad n \in \mathbb{Z}_+; \tag{20}$$

$$\left(p_n(A)\vec{e}_0, p_m(A)\vec{e}_0\right)_{l_2} = \delta_{n,m}, \qquad n, m \in \mathbb{Z}_+. \tag{21}$$

Denote by $\{r_n(\lambda)\}_{n=0}^{\infty}$, $r_0(\lambda) = 1$, the system of polynomials satisfying

$$J_3\vec{r}(\lambda) = \lambda\vec{r}(\lambda), \quad \vec{r}(\lambda) = (r_0(\lambda), r_1(\lambda), r_2(\lambda), \dots)^T. \tag{22}$$

These polynomials are orthonormal on the real line with respect to a (possibly nonunique) nonnegative finite measure σ on the Borel subsets of $\mathbb{R}$ (Favard's theorem). Consider the following operator:

$$U\sum_{n=0}^{\infty} \xi_n \vec{e}_n = \left[\sum_{n=0}^{\infty} \xi_n r_n(x)\right], \qquad \xi_n \in \mathbb{R}, \tag{23}$$

which maps $l_{2,fin}$ onto $\mathcal{P}$. Here, by $\mathcal{P}$ we denote a set of all (classes of equivalence which contain) polynomials in L^2_σ. Denote

$$\mathcal{A} = \mathcal{A}_\sigma = UAU^{-1}. \tag{24}$$

The operator $\mathcal{A} = \mathcal{A}_\sigma$ is said to be *the model representation in* L^2_σ *of the associated operator* A.

Theorem 1.1 ([14]) Let $\Theta = (J_3, J_5, \alpha, \beta)$ be a Jacobi-type pencil. Let $\{r_n(\lambda)\}_{n=0}^{\infty}$, $r_0(\lambda) = 1$, be a system of polynomials satisfying (22) and σ be their (arbitrary) orthogonality measure on $\mathfrak{B}(\mathbb{R})$. The associated polynomials $\{p_n(\lambda)\}_{n=0}^{\infty}$ satisfy the following relations:

$$\int_{\mathbb{R}} p_n(\mathcal{A})(1)\overline{p_m(\mathcal{A})(1)}d\sigma = \delta_{n,m}, \qquad n, m \in \mathbb{Z}_+, \tag{25}$$

where $\mathcal{A}$ is the model representation in L^2_σ of the associated operator A.

There appears a natural question: *what are the characteristic properties of the operator* $\mathcal{A}$*?* The answer is given by the following theorem.

Theorem 1.2 ([24, Corollary 1]) Let σ be a nonnegative measure on $\mathfrak{B}(\mathbb{R})$ with all finite power moments, $\int_{\mathbb{R}} d\sigma = 1$, $\int_{\mathbb{R}} |g(x)|^2 d\sigma > 0$, for any nonzero complex polynomial g. A linear operator $\mathcal{A}$ in L^2_σ is a model representation in L^2_σ of the associated operator of a Jacobi-type pencil if and only if the following properties hold:

i. $D(\mathcal{A}) = \mathcal{P}$;

ii. For each $k \in \mathbb{Z}_+$ it holds:

$$\mathcal{A}x^k = \xi_{k,k+1}x^{k+1} + \sum_{j=0}^{k} \xi_{k,j}x^j, \tag{26}$$

where $\xi_{k,k+1} > 0$, $\xi_{k,j} \in \mathbb{R}$ $(0 \le j \le k)$;

iii. The operator $\mathcal{A}\Lambda_0$ is symmetric. Here, by Λ_0 we denote the operator of the multiplication by an independent variable in L^2_σ restricted to $\mathcal{P}$.

There is a general subclass of Jacobi-type pencils, for which elements much more can be said about their associated operators and models [24]. Here we used some ideas from the general theory of operator pencils, see ref. [1, Chapter IV, p. 163].

Let $\Theta = (J_3, J_5, \alpha, \beta)$ be a Jacobi-type pencil and $\mathcal{A}$ be a model representation in L^2_σ of the associated operator of Θ. By Theorem 1.2 we see that $\mathcal{A}\Lambda_0$ is symmetric:

$$(\mathcal{A}\Lambda_0[u(\lambda)], [v(\lambda)])_{L^2_\sigma} = ([u(\lambda)], \mathcal{A}\Lambda_0[v(\lambda)])_{L^2_\sigma}, \qquad u, v \in \mathcal{P}. \tag{27}$$

Suppose that the measure σ is supported inside a finite real segment $[a, b]$, $0 < a < b < +\infty$, that is, $\sigma(\mathbb{R}\backslash[a, b]) = 0$. In this case, the operator Λ of the multiplication by an independent variable has a bounded inverse on the whole L^2_σ. Using (27) we may write:

$$\left(\Lambda^{-1}\mathcal{A}[\lambda u(\lambda)], [\lambda v(\lambda)]\right)_{L^2_\sigma} = \left(\Lambda^{-1}[\lambda u(\lambda)], \mathcal{A}[\lambda v(\lambda)]\right)_{L^2_\sigma}, \qquad u, v \in \mathcal{P}. \tag{28}$$

Denote $\mathcal{P}_0 = \Lambda\mathcal{P}$ and $\mathcal{A}_0 = \mathcal{A}|_{\mathcal{P}_0}$. Then

$$\left(\Lambda^{-1}\mathcal{A}_0 f, g\right)_{L^2_\sigma} = \left(\Lambda^{-1} f, \mathcal{A}_0 g\right)_{L^2_\sigma}, \qquad f, g \in \mathcal{P}_0. \tag{29}$$

Then $\mathcal{A}_0$ is symmetric with respect to the form $\left(\Lambda^{-1}\cdot, \cdot\right)_{L^2_\sigma}$. Thus, *in this case, the operator $\mathcal{A}$ is a perturbation of a symmetric operator*.

Consider two examples of Jacobi-type pencils which show that Sobolev orthogonality is close to them.

Example 3.1. ([26]). Let σ be a nonnegative measure on $\mathfrak{B}(\mathbb{R})$ with all finite power moments, $\int_\mathbb{R} d\sigma = 1$, $\int_\mathbb{R} |g(x)|^2 d\sigma > 0$, for any nonzero complex polynomial g. The following operator:

$$\mathcal{A}[p(\lambda)] = \Lambda_0[p(\lambda)] + p(0)[c\lambda + d], \qquad p \in \mathbb{P}, \tag{30}$$

where $c > -1$ and $d \in \mathbb{R}$, satisfies properties (*i*)-(*iii*) of Theorem 1.2. Let J_3 be the Jacobi matrix, corresponding to the measure σ, and $J_5 = J_3^2$. Define α, β in the following way:

$$\alpha = \frac{1}{\xi_{0,1}\sqrt{\Delta_1}}, \qquad \beta = -\frac{\xi_{0,1}s_1 + \xi_{0,0}}{\xi_{0,1}\sqrt{\Delta_1}}. \tag{31}$$

Here s_j are the power moments of σ, while $\Delta_n := \det(s_{k+l})_{k,l=0}^n$, $n \in \mathbb{Z}_+$, $\Delta_{-1} := 1$ are the Hankel determinants. The coefficients $\xi_{k,j}$ are taken from property (ii) of Theorem 1.2.

Let $\Theta = (J_3, J_5, \alpha, \beta)$. Denote by $\{p_n(\lambda)\}_{n=0}^{\infty}$ the associated polynomials to the pencil Θ, and denote by $\{r_n(\lambda)\}_{n=0}^{\infty}$ the orthonormal polynomials (with positive leading coefficients) with respect to the measure σ. Then

$$p_n(\lambda) = \frac{1}{c+1} r_n(\lambda) - \frac{d}{c+1} \frac{r_n(\lambda) - r_n(0)}{\lambda} + \frac{c}{c+1} r_n(0), \qquad n \in \mathbb{Z}_+; \tag{32}$$

$$r_n(\lambda) = (c+1)p_n(\lambda) + (c+1)d \frac{p_n(\lambda) - p_n(d)}{\lambda - d} - cp_n(d), \qquad n \in \mathbb{Z}_+. \tag{33}$$

In (32), (33) we mean the limit values at $\lambda = 0$ and $\lambda = d$, respectively. The following recurrence relation, involving three subsequent associated polynomials, holds:

$$\lambda p_n(\lambda) = \frac{p_n(d)}{c+1}(c\lambda + d) + a_{n-1}p_{n-1}(\lambda) + b_n p_n(\lambda) + a_n p_{n+1}(\lambda), \; n \in \mathbb{Z}_+, \; (\lambda \in \mathbb{C}). \tag{34}$$

The following orthogonality relations hold:

$$\begin{aligned} &\int_{\mathbb{R}\setminus\{d\}} (p_n(\lambda), p_n(d)) \begin{pmatrix} (c+1)^2 \left(\frac{\lambda}{\lambda-d}\right)^2 & (-c-1)\dfrac{\lambda(c\lambda+d)}{(\lambda-d)^2} \\ (-c-1)\dfrac{\lambda(c\lambda+d)}{(\lambda-d)^2} & \left(\frac{c\lambda+d}{\lambda-d}\right)^2 \end{pmatrix} \begin{pmatrix} p_m(\lambda) \\ p_m(d) \end{pmatrix} d\sigma + \\ &+ (p_n(d), p_n'(d)) \begin{pmatrix} 1 & (c+1)d \\ (c+1)d & (c+1)^2 d^2 \end{pmatrix} \begin{pmatrix} p_m(d) \\ p_m'(d) \end{pmatrix} \sigma(\{d\}) = \delta_{n,m}, \; n, m \in \mathbb{Z}_+. \end{aligned} \tag{35}$$

Polynomials $p_n(\lambda)$ can have multiple or complex roots.

Suppose additionally that σ and J_3 correspond to orthonormal Jacobi polynomials $r_n(\lambda) = P_n(\lambda; a, b)$ $(a, b > -1)$ and $c = 0$; $d = 1$. In this case, the associated polynomial p_n $(n \in \mathbb{Z}_+)$:

$$p_n(\lambda) = r_n(\lambda) - \frac{r_n(\lambda) - r_n(0)}{\lambda}, \tag{36}$$

is a unique, up to a constant multiple, real n-th degree polynomial solution of the following 4-th order differential equation:

$$\begin{aligned} &-(t+1)t(t-1)^2 y^{(4)}(t) + (t-1)\left(-(a+b+10)t^2 + (b-a)t + 4\right)y^{(3)}(t) \\ &+ + \left(-3(2a+2b+8)t^2 + (a+9b+22)t + 3a - 3b\right)y''(t) \\ &+ (-6(a+b+2)t + 2a + 6b + 8)y'(t) \\ &+ + \lambda_n \left(t(t-1)y''(t) + 2(2t-1)y'(t) + 2y(t)\right) \\ &= 0, \end{aligned} \tag{37}$$

where $\lambda_n = n(n+a+b+1)$.

Moreover, there exists a unique $\lambda_n \in \mathbb{R}$, such that differential Eq. (37) has a real n-th degree polynomial solution.

Example 3.2. ([25]). Recall that Jacobi polynomials $P_n^{(\alpha,\beta)}(x)$:

$$P_n^{(\alpha,\beta)}(x) = \binom{n+\alpha}{n} {}_2F_1\left(-n, n+\alpha+\beta+1; \alpha+1; \frac{1-x}{2}\right), \quad n \in \mathbb{Z}_+,$$

are orthogonal on $[-1,1]$ with respect to the weight $w(x) = (1-x)^\alpha(1+x)^\beta$, $\alpha, \beta > -1$. Orthonormal polynomials have the following form:

$$\widehat{P}_0^{(\alpha,\beta)}(x) = \frac{1}{\sqrt{2^{\alpha+\beta+1}\mathrm{B}(\alpha+1,\beta+1)}},$$

$$\widehat{P}_n^{(\alpha,\beta)}(x) = \sqrt{\frac{(2n+\alpha+\beta+1)n!\Gamma(n+\alpha+\beta+1)}{2^{\alpha+\beta+1}\Gamma(n+\alpha+1)\Gamma(n+\beta+1)}} P_n^{(\alpha,\beta)}(x), \quad n \in \mathbb{N}.$$

Let $c > 0$ be an arbitrary positive number. Set

$$D_{\alpha,\beta,c} := (x^2-1)\frac{d^2}{dx^2} + [(\alpha+\beta+2)x + \alpha - \beta]\frac{d}{dx} + c, \tag{38}$$

$$l_{n,c} := c + n(n+\alpha+\beta+1). \tag{39}$$

Define the following polynomials:

$$P_n(\alpha,\beta,c,t_0;x) := \sum_{k=0}^{n} \frac{1}{l_{k,c}} \widehat{P}_k^{(\alpha,\beta)}(t_0)\widehat{P}_k^{(\alpha,\beta)}(x), \qquad n \in \mathbb{Z}_+, \tag{40}$$

where $t_0 \geq 1$ is an arbitrary parameter. Notice that normed by eigenvalues polynomial kernels of some Sobolev orthogonal polynomials appeared earlier in literature, see ref. [27].

Theorem 1.3. Let $\alpha, \beta > -1$; $c > 0$, and $t_0 \geq 1$, be arbitrary parameters. Polynomials $P_n(x) = P_n(\alpha,\beta,c,t_0;x)$, from (40), are Sobolev orthogonal polynomials on $\mathbb{R}$:

$$\int_{-1}^{1} (P_n(x), P_{n'}(x), P_{n''}(x)) M_{\alpha,\beta,c}(x) \begin{pmatrix} P_m(x) \\ P'_m(x) \\ P''_m(x) \end{pmatrix} (t_0 - x)(1-x)^\alpha(1+x)^\beta dx = \tag{41}$$

$$= A_n \delta_{n.m}, \qquad n, m \in \mathbb{Z}_+,$$

where A_n are some positive numbers and

$$M_{\alpha,\beta,c} == \begin{pmatrix} c \\ (\alpha+\beta+2)x + \alpha - \beta \\ x^2 - 1 \end{pmatrix} \left(c, (\alpha+\beta+2)x + \alpha - \beta, x^2 - 1\right). \tag{42}$$

For $P_n(\alpha,\beta,c,1;x)$ the following differential equation holds:

$$D_{\alpha+1,\beta,0} D_{\alpha,\beta,c} P_n(\alpha,\beta,c,1;x) = l_{n,0} D_{\alpha,\beta,c} P_n(\alpha,\beta,c,1;x), \quad n \in \mathbb{Z}_+, \tag{43}$$

where $D_{\alpha,\beta,c}$, $l_{n,c}$ are defined by (38), (39).

4. Pencils of banded matrices and Sobolev orthogonality

Let $\mathcal{K}$ denote the real line or the unit circle. The following problem was stated in ref. [28], see also ref. [29]:

Problem 1. *To describe all Sobolev orthogonal polynomials* $\{y_n(z)\}_{n=0}^{\infty}$ *on* $\mathcal{K}$, *satisfying the following two properties:*

a. *Polynomials* $y_n(z)$ *satisfy the following differential equation:*

$$Ry_n(z) = \lambda_n S y_n(z), \qquad n = 0, 1, 2, \dots, \tag{44}$$

where R, S are linear differential operators of finite orders, having complex polynomial coefficients not depending on n; $\lambda_n \in \mathbb{C}$;

b. *Polynomials* $y_n(z)$ *obey the following difference equation:*

$$L\vec{y}(z) = zM\vec{y}(z), \quad \vec{y}(z) = (y_0(z), y_1(z), \dots)^T, \tag{45}$$

where L, M are semi-infinite complex banded (i.e., having a finite number of non-zero diagonals) matrices.

Relation (44) shows that y_n is an eigenfunction of the operator pencil $R - \lambda S$, while relation (45) means that vectors of $y_n(z)$ are eigenfunctions of the operator pencil $L - zM$. We emphasize that in Problem 1 we do not exclude OPRL or OPUC. They are formally considered as Sobolev orthogonal polynomials with the derivatives of order 0. In this way, we may view systems from Problem 1 as generalizations of systems of classical orthogonal polynomials (see, e.g., the book [30], and papers [20, 31, 32] for more recent developments on this subject, as well as references therein). Related topics are also studied for systems of biorthogonal rational functions, see, for example, ref. [33]. Conditions (a), (b) of Problem 1 are close to bispectral problems, and in particular, to the Bochner-Krall problem (see refs. [31, 34–36] and papers cited therein).

One example of Sobolev orthogonal polynomials, which satisfy conditions of Problem 1, we have already met in Example 3.2. In ref. [37] there was proposed a way to construct such systems of polynomials. Let $\{p_n(x)\}_{n=0}^{\infty}$ (p_n has degree n and real coefficients) be orthogonal polynomials on $[\mathbf{a}, \mathbf{b}] \subseteq \mathbb{R}$ with respect to a weight function $w(x)$:

$$\int_{\mathbf{a}}^{\mathbf{b}} p_n(x) p_m(x) w(x) dx = A_n \delta_{n,m}, \qquad A_n > 0, \quad n, m \in \mathbb{Z}_+. \tag{46}$$

The weight w is supposed to be continuous on $(\mathbf{a}, \mathbf{b})$. Denote

$$D_\xi y(x) = \sum_{k=0}^{\xi} d_k(x) y^{(k)}(x), \tag{47}$$

where d_k and y are real polynomials of x: $d_\xi \neq 0$. Let us fix a positive integer ξ, and consider the following differential equation:

$$D_\xi y(x) = p_n(x), \tag{48}$$

where D_ξ is defined as in Eq. (47), and $n \in \mathbb{Z}_+$. The following assumption plays a key role here.

Condition 1. *Suppose that for each $n \in \mathbb{Z}_+$, the differential Eq. (48) has a real n-th degree polynomial solution $y(x) = y_n(x)$.*

If Condition 1 is satisfied, by relations (46),(48) we immediately obtain that $\{y_n(x)\}_{n=0}^{\infty}$ are Sobolev orthogonal polynomials:

$$\int_{\mathbf{a}}^{\mathbf{b}} \left(y_n(x), y_n'(x), \dots, y_n^{(\xi)}(x)\right) M(x) \begin{pmatrix} y_m(x) \\ y_m'(x) \\ \vdots \\ y_m^{(\xi)}(x) \end{pmatrix} w(x)dx = A_n \delta_{n,m}, \tag{49}$$
$$n, m \in \mathbb{Z}_+,$$

where

$$M(x) := \begin{pmatrix} d_0(x) \\ d_1(x) \\ \vdots \\ d_\xi(x) \end{pmatrix} (d_0(x), d_1(x), \dots, d_\xi(x)), \qquad x \in (\mathbf{a}, \mathbf{b}). \tag{50}$$

Moreover, if p_n satisfy a differential equation, then y_n satisfy a differential equation as well. Question: when Condition 1 is satisfied? An answer is given by the following proposition.

Proposition 1 ([28, Proposition 2.1]) Let D be a linear differential operator of order $r \in \mathbb{N}$, with complex polynomial coefficients:

$$D = \sum_{k=0}^{r} d_k(z) \frac{d^k}{dz^k}, \quad d_k(z) \in \mathbb{P}.$$

Let $\{u_n(z)\}_{n=0}^{\infty}$, $\deg u_n = n$, be an arbitrary set of complex polynomials. The following statements are equivalent:

(A) The following equation:

$$Dy(z) = u_n(z), \tag{51}$$

for each $n \in \mathbb{Z}_+$, has a complex polynomial solution $y(z) = y_n(z)$ of degree n;

(B) Dz^n is a complex polynomial of degree n, $\forall n \in \mathbb{Z}_+$;

(C) The following conditions hold:

$$\deg d_k \leq k, \qquad 0 \leq k \leq r; \tag{52}$$

$$\sum_{j=0}^{r} [n]_j d_{j,j} \neq 0, \qquad n \in \mathbb{Z}_+, \tag{53}$$

where $d_{j,l}$ means the coefficient by z^l of the polynomial d_j.

If one of the statements (A), (B), (C) holds true, then for each $n \in \mathbb{Z}_+$, the solution of (51) is unique.

Observe that condition (53) holds true, if the following simple condition holds:

$$d_{0,0} > 0, \quad d_{j,j} \geq 0, \qquad j \in \mathbb{Z}_{1,r}. \tag{54}$$

Thus, *there exists a big variety of linear differential operators with polynomial coefficients that have property* (A). This leads to various Sobolev orthogonal polynomials.

In ref. [37] there were constructed families of Sobolev orthogonal polynomials on the real line, depending on an arbitrary finite number of complex parameters. Namely, we considered the following hypergeometric polynomials:

$$\begin{gathered} \mathbf{L}_n(x) = \mathbf{L}_n(x; \alpha, \kappa_1, \ldots, \kappa_\delta) = \\ = {}_{\delta+1}F_{\delta+1}(-n, 1, \ldots, 1; \alpha+1, \kappa_1+1, \ldots, \kappa_\delta+1; x), \end{gathered} \tag{55}$$

$$\begin{gathered} \mathbf{P}_n(x) = \mathbf{P}_n(x; \alpha, \beta, \kappa_1, \ldots, \kappa_\delta) = \\ = {}_{\delta+2}F_{\delta+1}(-n, n+\alpha+\beta+1, 1, \ldots, 1; \alpha+1, \kappa_1+1, \ldots, \kappa_\delta+1; x), \\ \alpha, \beta, \kappa_1, \ldots, \kappa_\delta > -1, \ n \in \mathbb{Z}_+. \end{gathered} \tag{56}$$

Here ${}_pF_q$ is a usual notation for the generalized hypergeometric function, and δ is a positive integer. These families obey differential equations. As for recurrence relations, they were only constructed for the case $\delta = 1$.

In ref. [29] a family of hypergeometric Sobolev orthogonal polynomials on the unit circle was considered:

$$y_n(x) = \frac{(-1)^\rho}{n!} x^n {}_2F_0\left(-n, \rho; -; -\frac{1}{x}\right),$$

depending on a parameter $\rho \in \mathbb{N}$. Observe that the reversed polynomials to y_n appeared in numerators of some biorthogonal rational functions, see [38].

Let $\{g_n(t)\}_{n=0}^\infty$ be a system of OPRL or OPUC, having a generating function of the following form:

$$G(t, w) = f(w) e^{tu(w)} = \sum_{n=0}^\infty g_n(t) \frac{w^n}{n!}, \qquad t \in \mathbb{C}, \quad |w| < R_0, \quad (R_0 > 0), \tag{57}$$

where f, u are analytic functions in the circle $\{|w| < R_0\}$, $u(0) = 0$. Such generating functions for OPRL were studied by Meixner, see, for instance, ref. [39, p. 273]. In the case of OPUC, we do not know any such a system, besides $\{z^n\}_{n=0}^\infty$. Consider the following function:

$$\begin{gathered} F(t, w) = \frac{1}{p(u(w))} G(t, w) = \frac{1}{p(u(w))} f(w) e^{tu(w)}, \\ t \in \mathbb{R}, \quad |w| < R_1 < R_0, \quad (R_1 > 0), \end{gathered} \tag{58}$$

where $p \in \mathbb{P}$: $p(0) \neq 0$. In the case $u(z) = z$, one should take $R_1 \leq |z_0|$, where z_0 is a root of p with the smallest modulus. This ensures that $F(t, w)$ is an analytic function of two variables in any polydisk $C_{T_1, R_1} = \{(t, w) \in \mathbb{C}^2 : |t| < T_1, \ w| < R_1|\}$, $T_1 > 0$. In the general case, since $p(u(0)) = p(0) \neq 0$, there also exists a suitable R_1, which guarantees that F is analytic in C_{T_1, R_1}. Expand the function $F(t, w)$ in Taylor's series by w with a fixed t:

$$F(t,w) = \sum_{n=0}^{\infty} \varphi_n(t) \frac{w^n}{n!}, \qquad (t,w) \in C_{T_1,R_1}, \tag{59}$$

where $\varphi_n(t)$ are some complex-valued functions. Then the function $\varphi_n(t)$ is a complex polynomial of degree n, $\forall n \in \mathbb{Z}_+$, see [28, Lemma 3.5]. Suppose that $\deg p \geq 1$, and

$$p(z) = \sum_{k=0}^{d} c_k z^k, \qquad c_k \in \mathbb{C},\ c_d \neq 0;\ c_0 \neq 0;\ d \in \mathbb{N}. \tag{60}$$

Theorem 1.4 ([28, Theorem 3.7]) Let $d \in \mathbb{N}$, and $p(z)$ be as in (60). Let $\{g_n(t)\}_{n=0}^{\infty}$ be a system of OPRL or OPUC, having a generating function $G(t,w)$ from (57) and $F(t,w)$ be given by (58). Fix some positive T_1, R_1, such that $F(t,w)$ is analytic in the polydisk C_{T_1,R_1}. Polynomials

$$\varphi_n(z) = \sum_{j=0}^{n} \binom{n}{j} b_j g_{n-j}(t), \qquad n \in \mathbb{Z}_+, \tag{61}$$

where $b_j = \left(\frac{1}{p(u(w))}\right)^{(j)}(0)$, have the following properties:

i. Polynomials φ_n are Sobolev orthogonal polynomials:

$$\int \left(\varphi_n(t), \varphi_n'(t), \ldots, \varphi_n^{(d)}(t)\right) \widetilde{M} \overline{\begin{pmatrix} \varphi_m(t) \\ \varphi_m'(t) \\ \vdots \\ \varphi_m^{(d)}(t) \end{pmatrix}} d\mu_g = \tau_n \delta_{n,m},$$
$$\tau_n > 0, \quad n,m \in \mathbb{Z}_+,$$

where

$$\widetilde{M} = (c_0, c_1, \ldots, c_d)^T (\overline{c_0}, \overline{c_1}, \ldots, \overline{c_d}).$$

Here $d\mu_g$ is the measure of orthogonality of g_n.

i. Polynomials φ_n have the generating function $F(t,w)$, and relation (59) holds.

ii. Polynomials φ_n have the following integral representation:

$$\varphi_n(t) = \frac{n!}{2\pi i} \oint_{|w|=R_2} \frac{1}{p(u(w))} f(w) e^{tu(w)} w^{-n-1} dw, \qquad n \in \mathbb{Z}_+, \tag{62}$$

where R_2 is an arbitrary number, satisfying $0 < R_2 < R_1$.

There are two cases of g_n, which lead to additional properties of φ_n, namely, to differential equations and recurrence relations. The next two corollaries are devoted to them.

Corollary 1 ([28]) In conditions of Theorem 1.4 suppose that $g_n(t) = t^n$, $n \in \mathbb{Z}_+$; $f(w) = 1$, $u(w) = w$. Polynomials $\{\varphi_n(t)\}_{n=0}^{\infty}$ satisfy the following recurrence relation:

$$(n+1)\sum_{k=0}^{d}\varphi_{n+1-k}(t)\frac{c_k}{(n+1-k)!}= \\ =t\left(\sum_{k=0}^{d}\varphi_{n-k}(t)\frac{c_k}{(n-k)!}\right),\qquad n\in\mathbb{Z}_+, \tag{63}$$

where $\varphi_r{:=}0$, $r!{:=}1$, for $r\in\mathbb{Z}:\ r<0$.

Polynomials $\{\varphi_n(t)\}_{n=0}^{\infty}$ obey the following differential equation:

$$t\sum_{k=0}^{d}c_k\varphi_n^{(k+1)}(t)=n\left(\sum_{k=0}^{d}c_k\varphi_n^{(k)}(t)\right),\qquad n\in\mathbb{Z}_+.$$

Corollary 2 ([28]) In conditions of Theorem 1.4 suppose that $g_n(t)=H_n(t)$, $n\in\mathbb{Z}_+$, are Hermite polynomials; $f(w)=e^{-w^2}$, $u(w)=2w$. Polynomials $\{\varphi_n(t)\}_{n=0}^{\infty}$ satisfy the following recurrence relation:

$$(n+1)\sum_{k=0}^{d}\varphi_{n+1-k}(t)\frac{c_k2^k}{(n+1-k)!}+2\sum_{k=0}^{d}\varphi_{n-1-k}(t)\frac{c_k2^k}{(n+1-k)!}= \\ =2t\left(\sum_{k=0}^{d}\varphi_{n-k}(t)\frac{c_k2^k}{(n-k)!}\right),\qquad n\in\mathbb{N}, \tag{64}$$

where $\varphi_r{:=}0$, $r!{:=}1$, for $r\in\mathbb{Z}:\ r<0$; and

$$c_0\varphi_1(t)+2c_1\varphi_0(t)=2c_0t\varphi_0(t). \tag{65}$$

Polynomials $\{\varphi_n(t)\}_{n=0}^{\infty}$ obey the following differential equation:

$$\sum_{k=0}^{d}c_k\varphi_n^{(k)}(t)-2t\sum_{k=0}^{d}c_k\varphi_n^{(k+1)}(t)=-2n\left(\sum_{k=0}^{d}c_k\varphi_n^{(k)}(t)\right),\qquad n\in\mathbb{Z}_+. \tag{66}$$

Observe that polynomials φ_n from the last two corollaries fit into the scheme of Problem 1.

5. Conclusion

The theory of orthogonal polynomials is closely related to semi-infinite matrices, as well as to their finite truncations. This interplay has shown its productivity in classical results. Nowadays there appeared new kinds of orthogonality, such as Sobolev orthogonality. It is not yet clear what kind of matrices can be attributed to them. One of candidates is a pencil of matrices, since it appeared in examples. In Section 3 there appeared a pencil of semi-infinite symmetric matrices, while in Section 4 it was a pencil of some banded matrices. In Section 2 we also met a pencil, but it was more close to classical eigenvalue problems of single operators.

The above-mentioned examples of Sobolev orthogonal polynomials also showed that pencils of differential equations appeared here in a natural way. Moreover, there is a large number of differential operators, which have polynomial solutions with

Sobolev orthogonality. This fact promises that Sobolev orthogonal polynomials can find their applications in mathematical physics.

We think that Problem 1 is an appropriate framework for a search and a construction of new Sobolev orthogonal polynomials having nice properties. Notice that one can produce such systems using classical OPRL or OPUC. The differential equation, if it existed, is inherited by new systems of polynomials. The more complicated question is the existence of a recurrence relation.

Besides new families of Sobolev orthogonal polynomials, it is of a big interest finding classes of systems of Sobolev orthogonal polynomials, having recurrence relations. One such a class (orthogonal polynomials on radial rays) was described in Section 2. Thus, it looks reasonable to start not only from Sobolev orthogonality, but from the other side, i.e., from recurrent relations. One such an example of derivation was given by orthogonal polynomials on radial rays from Section 2.

Another possible way was given in Section 3, where we described Jacobi-type pencils. The associated polynomials of a Jacobi type pencil have special orthogonality relations. The associated operator yet has not a suitable functional calculus. As we have seen, under some conditions this operator is a perturbation of a symmetric operator. However, it is not clear how to calculate effectively a polynomial of this operator.

In general, it is a classical situation that the operator theory stands behind special classes of semi-infinite matrices and related objects. The operator theory of single operators is well promoted and it is well recognized by any mathematician. It seems that the theory of operator pencils is less known to the mathematical community. This fact can explain the situation that pencils of semi-infinite matrices and related polynomials appeared on a mathematical scene just recently. We hope that, as in the classical case, these new orthogonal polynomial systems will shed some new light on the theory of operator pencils.

Acknowledgements

The author is grateful to Professors Zolotarev and Yantsevich for their permanent support.

Author details

Sergey Zagorodnyuk
V.N. Karazin Kharkiv National University, Kharkiv, Ukraine

*Address all correspondence to: sergey.m.zagorodnyuk@gmail.com

References

[1] Markus AS. Introduction to the Spectral Theory of Polynomial Operator Pencils. With an appendix by M. V. Keldysh. In: Translations of Mathematical Monographs. Vol. 71. Providence, RI: American Mathematical Society; 1988. pp. iv+250

[2] Rodman L. An Introduction to Operator Polynomials. In: Operator Theory: Advances and Applications. Vol. 38. Basel: Birkhäuser Verlag; 1989. pp. xii +389

[3] Parlett BN. The symmetric eigenvalue problem. In: Corrected reprint of the 1980 original. Classics in Applied Mathematics. Vol. 20. Philadelphia, PA: Society for Industrial and Applied Mathematics (SIAM); 1998 xxiv+398 pp

[4] Gantmacher FR. The Theory of Matrices. Vol. 2. New York: Chelsea Publishing Co; 1959. pp. ix+276

[5] Szegö G. Orthogonal Polynomials. Fourth ed. Providence, R.I: American Mathematical Society, Colloquium Publications, Vol. XXIII; 1975. pp. xiii+432

[6] Akhiezer NI. The Classical Moment Problem and Some Related Questions in Analysis. New York: Hafner Publishing Co.; pp. 1965 x+253

[7] Geronimus JL. Polynomials, Orthogonal on a Circumference and on an Interval. Estimates, Asymptotic Formulas, Orthogonal Series (in Russian), Sovremennye Problemy Matematiki, Gosudarstv. Moscow: Izdat. Fiz.-Mat. Lit; 1958. pp. 240

[8] Simon B. Orthogonal Polynomials on the Unit Circle. Part 1. Classical Theory. Vol. 54, Part 1. Providence, RI: American Mathematical Society Colloquium Publications; 2005. pp. xxvi+466

[9] Bultheel A, González-Vera P, Hendriksen E, Njåstad O. Orthogonal Rational Functions. Cambridge Monographs on Applied and Computational Mathematics. Vol. 5. Cambridge: Cambridge University Press; 1999. pp. xiv+407

[10] Zhedanov A. Biorthogonal rational functions and the generalized eigenvalue problem. Journal of Approximation Theory. 1999;**101**(2):303-329

[11] Krein M. Infinite J-matrices and a matrix-moment problem. (Russian). Doklady Akad. Nauk SSSR (N.S.). 1949; **69**:125-128

[12] Durán AJ, Van Assche W. Orthogonal matrix polynomials and higher-order recurrence relations. Linear Algebra and its Applications. 1995;**219**: 261-280

[13] Zagorodnyuk SM. On generalized Jacobi matrices and orthogonal polynomials. New York Journal of Mathematics. 2003;**9**:117-136

[14] Zagorodnyuk SM. Orthogonal polynomials associated with some Jacobi-type pencils (Russian). Ukraïn. Mathematical. Journal. 2017;**68**(9): 1353-1365 translation in Ukrainian Math. J

[15] Marcellán F. Xu, Yuan.: On Sobolev orthogonal polynomials. Expositiones Mathematicae. 2015;**33**(3):308-352

[16] Durán AJ. A generalization of Favard's theorem for polynomials satisfying a recurrence relation. Journal of Approximation Theory. 1993;**74**(1): 83-109

[17] Durán AJ. On orthogonal polynomials with respect to a positive definite matrix of measures. Canadian

Journal of Mathematics. 1995;**47**(1): 88-112

[18] Milovanović, GV. Orthogonal polynomials on the radial rays in the complex plane and applications. Proceedings of the Fourth International Conference on Functional Analysis and Approximation Theory. Vol. I(Suppl. 2) (Potenza, 2000). Rend. Circ. Mat. Palermo. 2002, no. 68, part I, 65–94

[19] Durán AJ, de la Iglesia MD. Differential equations for discrete Laguerre-Sobolev orthogonal polynomials. Journal of Approximation Theory. 2015;**195**:70-88

[20] Durán AJ, de la Iglesia MD. Differential equations for discrete Jacobi-Sobolev orthogonal polynomials. Journal of Spectral Theory. 2018;**8**(1): 191-234

[21] Zagorodnyuk SM. Orthogonal polynomials on rays: properties of zeros, related moment problems and symmetries. Zh. Mat. Fiz. Anal. Geom. 2008;**4**(3):395-419

[22] Choque Rivero AE, Zagorodnyuk SM. Orthogonal polynomials on rays: Christoffel's formula. Bol. Soc. Mat. Mexicana. 2009; **15**(2):149-164

[23] Ben AJ, Vladimirov AA, Shkalikov AA. Spectral and oscillatory properties of a linear pencil of fourth-order differential operators. Mathematical Notes. 2013;**94**(1):49-59

[24] Zagorodnyuk SM. The inverse spectral problem for Jacobi-type pencils. SIGMA Symmetry Integrability Geom. Methods Appl. 2017;**13**. Paper No. 085, 16 pp

[25] Zagorodnyuk SM. On series of orthogonal polynomials and systems of classical type polynomials. Ukr. Math. J. 2021;**73**(6):799-810 translation from Ukr. Mat. Zh

[26] Zagorodnyuk SM. Difference equations related to Jacobi-type pencils. J. Difference Equ. Appl. 2018;**24**(10): 1664-1684

[27] Littlejohn LL, Mañas-Mañas JF, Moreno-Balcázar JJ, Wellman R. Differential operator for discrete Gegenbauer-Sobolev orthogonal polynomials: Eigenvalues and asymptotics. Journal of Approximation Theory. 2018;**230**:32-49

[28] Zagorodnyuk SM. On some Sobolev spaces with matrix weights and classical type Sobolev orthogonal polynomials. J. Difference Equ. Appl. 2021;**27**(2): 261-283

[29] Zagorodnyuk SM. On a family of hypergeometric Sobolev orthogonal polynomials on the unit circle. Constr. Math. Anal. 2020;**3**(2):75-84

[30] Koekoek R, Lesky PA, Swarttouw RF. Hypergeometric Orthogonal Polynomials and their q-Analogues. With a foreword by Tom H. Koornwinder. In: Springer Monographs in Mathematics. Berlin: Springer-Verlag; 2010. pp. xx+578

[31] Horozov E. Automorphisms of algebras and Bochner's property for vector orthogonal polynomials. SIGMA Symmetry Integrability Geom. Methods Appl. 2016;**12**. Paper No. 050, 14 pp

[32] Horozov E. d-orthogonal analogs of classical orthogonal polynomials. SIGMA Symmetry Integrability Geom. Methods Appl. 2018;**14**. Paper No. 063, 27 pp

[33] Spiridonov V, Zhedanov A. Classical biorthogonal rational functions on elliptic grids. Comptes Rendus

Mathématiques des l'Académie des Sciences. 2000;**22**(2):70-76

[34] Duistermaat JJ, Grünbaum FA. Differential equations in the spectral parameter. Communications in Mathematical Physics. 1986;**103**(2): 177-240

[35] Everitt WN, Kwon KH, Littlejohn LL, Wellman R. Orthogonal polynomial solutions of linear ordinary differential equations. Proceedings of the Fifth International Symposium on Orthogonal Polynomials, Special Functions and their Applications (Patras, 1999). Journal of Computational and Applied Mathematics. 2001;**133**(1-2): 85-109

[36] Horozov E. Vector orthogonal polynomials with Bochner's property. Constructive Approximation. 2018; **48**(2):201-234

[37] Zagorodnyuk SM. On some classical type Sobolev orthogonal polynomials. Journal of Approximation Theory. 2020; **250**(105337) 14 pp

[38] Hendriksen E, van Rossum H. Orthogonal Laurent polynomials. Nederl. Akad. Wetensch. Indag. Math. 1986;**48**(1):17-36

[39] Erdélyi A, Magnus W, Oberhettinger F, Tricomi FG. Higher Transcendental Functions. Vol. III. Based, in part, on notes left by Harry Bateman. New York-Toronto-London: McGraw-Hill Book Company, Inc.; 1955. pp. xvii+292

Chapter 3

Matrix as an Alternative Solution for Evaluating Sentence Reordering Tasks

Amma Kazuo

Abstract

Although sentence reordering is a popular practice in educational contexts its scoring method has virtually remained 'all-or-nothing'. The author proposed a more psychologically valid means of partial scoring called MRS (Maximal Relative Sequence) where a point is counted for each ascending run in the answer sequence allowing gaps and the final score reflects the length of the longest sequence of ascending elements. This scoring method, together with an additional consideration of recovery distances, was woven into an executable programme, and then transplanted to Excel without having to understand a programming language. However, the use of Excel was severely limited by the number of columns available. This chapter reviews the past practices of evaluating partial scoring of reordering tasks and proposes an alternative solution LM (Linearity Matrix), also executable on Excel, with far smaller consumption of columns and with the idea of calculating the recovery distances as well as MRS scores. Although LM and MRS are different scoring procedures, they both reflect psychological complexity of the task involved. Furthermore, LM is versatile as to the adjustability of adjacency weights as an extended model of Kendall's tau. Some reflections on practical application are referred to as well as future directions of the study.

Keywords: reordering, partial scoring, recovery distance, Excel, Kendall's tau

1. Introduction

Sentence reordering is one of the popular tasks in reading comprehension [1, 2]. Regrettably, in the field of language testing, the scoring method, in practice, has been overwhelmingly 'all or nothing', i.e., one can get a full score only if his/her answer matches the correct sequence perfectly. 'All or nothing' is simple enough, but excludes the idea of partial correctness, which Alderson, Percsich, and Szabo see as unfair [3]. There is no consideration of the difference in the test-taker's degrees of performance.

Consider first a reordering task as an example of a reading comprehension question. **Text 1** is taken from Japan's National Centre Examination (2013) with option [b] being the correct answer [4]. The test-takers were told to choose the correct order of the illustrations of a movie story they were presented.

Which of the following shows the order of the scenes as they appear in the movie?

[a] (A) → (C) → (B) → (D)

[b] * (A) → (B) → (C) → (D)

[c] (B) → (D) → (A) → (C)

[d] (B) → (A) → (D) → (C)

Text 1.
*Sample reordering task for reading comprehension. *: correct answer; element codes are rearranged for convenience.*

According to the test manual, only [b] gets the point; the other options make no point. The correct sequence [b] contains three continuous ascending runs: A-B, B-C, and C-D, and three discrete ascending runs: A-C, A-D, and B-D. Of these six sequences [a] satisfies 5 matches, [c] 3 matches, and [d] 4 matches. By another criterion, [a] is reached by dislocating one element (either C or B) from the correct sequence, [c] by two elements, and [d] by two elements. By either case, the three distractors are gradable in terms of proximity to the correct sequence. The 'All or nothing' scoring method accepts only the perfect answer and ignores the differences in the test-taker's partially formed construct.

What should be sought is a rational evaluation method for a partial achievement of item reordering tasks. This paper first reviews some past literature concerning this issue, followed by an overview of a stretch of alternative measurement methods developed by the author. The core issue is a new measurement scheme Linearity Matrix (LM), which can compensate for the shortcomings of the present practices, ensuring quicker and light-weight processing for non-specialists to handle.

2. Literature overview

Alderson, et al. examine four alternative methods to seek fairness and high discriminability: (1) 'Exact matching', (2) 'Previous', (3) 'Next', and (4) 'Edges' [3]. They were all in a test stage and no clear conclusion was reached. Above all, these methods were empirically designed on the basis of ad hoc assumptions. Of them (4) 'Edges' is a linear extension of (1) 'Exact matching', and (2) 'Previous' and (3) 'Next' are variations of 'Adjacent matching'. 'Exact matching' requires each element to be located exactly in the same position as in the correct answer. In option [a] of **Text 1**, A and D will get points; the other options [c] and [d] will get no points because there is no element in the correct position. In 'Adjacent matching' each of the three adjacent pairs will get 1 point. None of the three distractors [a], [c], and [d] will get a point in our example. But if we had an option

[e] (C) → (D) → (A) → (B)

the initial pair C-D and the final pair A-B would each get 1 point.

Kendall's coefficient tau is originally one of the measurement methods of rank-order correlation [5]. Kendall's tau is defined as:

$$\text{tau} = ([\text{number of concordant pairs}] - [\text{number of discordant pairs}]) / [\text{binomial of choosing 2 from } n]$$

In other words,

$$\tau = \frac{4\sum P}{n(n-1)} - 1 \tag{1}$$

where

P is the total number of items ranked after a given item by both rankings, and
n is the number of items

Kendall's tau is a popular tool in evaluating correspondence between machine-translated passages and human translations [6–8]. Papineni, Roukos, Ward and Zhu, for example, measured tau for the degree of correspondence between the n-gram of the reference text and that of the target text produced as a result of machine translation [9]. This study as well as other papers of the same interest deals with open-ended elements for comparison where additions and reductions of words and phrases naturally occur, hence irrelevant to the present scope of item reordering in which the elements are closed.

Bollegala, Okazaki and Ishizuka's 'Average continuity' [10] is a variation of 'Adjacent matching':

$$AC = \exp\left(\frac{1}{k-1} \bullet \sum_{i=2}^{k} \log\left(P_i + \alpha\right)\right) \tag{2}$$

where

k is the maximum number of continuous elements to be considered for calculation,
α is any small number (e.g., 0.001 in Bollegala, et al's example),
and 'precision of n continuous sentences':

$$P_n = m/(N - n + 1) \tag{3}$$

where

n is the length of continuous elements,
m is the number of continuous elements in the correct order, and
N is the number of elements in the correct sequence.

For example, when evaluating an answer CDABE for a correct sequence ABCDE, N = 5 (length of ABCDE), k = N = 5, m = 2 (count of CD and AB), and n = 2 (length of CD or AB).

Their method is sensitive to continuously running elements such as CD or AB, or ABCD in ABCDE with few disorderly elements. In our previous example of **Text 1**, none of the distractors contains a sequence of elements long enough to get a P_n, resulting invariably in AC =exp(log α).

Below is a simulation of AC against the exclusive permutations of five elements (A, B, C, D, E) where the sample size is 120, given the correct sequence ABCDE. It is only when the length of the continuous sequence is 4 (i.e., ABCD and BCDE) that the AC value appears reasonable (= 0.407 when α = 0.001); otherwise, the values are generally low[1]. This tendency is enhanced when α takes a smaller value (**Table 1**)[2].

Therefore, AC is not an appropriate measurement tool when shorter continuous sequences (i.e., two or three consecutive elements such as AB and ABC) are not infrequent. Furthermore, AC cannot count the cases where ascending elements are not adjacent (e.g., ADBEC, where AC = 0.050)[3].

Alpha	Mean AC	SD of AC
0.001	0.087	0.057
0.0001	0.043	0.045
0.000001	0.012	0.020
0.00000001	0.004	0.015

Table 1.
Mean and SD of AC scores for all permutations of five elements. (A perfect sequence is excluded as an outlier).

Lapata prepared all permutations of orders of eight sentences and calculated the tau value for each text, and compared it with the human rating of comprehensibility [11]. She obtained a significant correlation coefficient of $r = 0.45$ ($N = 64$) (p. 478). She also claimed that she confirmed that Kendall's tau was able to predict the text cohesion by measuring the reading time of the target texts. Although this was one of the few studies validating the effect of randomised sentence order, the final coefficient value alone is not convincing enough to ensure the effect of particular text disturbance on comprehension. Other measurement methods for evaluating disrupted orders could have been equally significant while comprising substantial linguistic differences beneath the surface. Therefore, the next step of this study would be to analyse how the target sequences of sentences are created and organised.

In conclusion, both 'Exact matching' and 'Adjacent matching' are incomplete and counter to intuition. The problem with 'Exact matching' is that it is sensitive to the absolute location of elements, and relative sequence is out of consideration. This scoring method may be appropriate for a question in which absolute location is significant, e.g., topic sentence in a paragraph, and initial cause of sequential events. However, if we consider a sequence of combined cause-effect instances, relative locations should also be rewarded with partial points. In contrast, 'Adjacent matching' weighs much on local adjacency. Even though [e] gets 2 points, the two pairs are twisted in relative order.

3. Maximal relative sequence

A new measurement method called 'Maximal Relative Sequence' (MRS) was proposed by the author [12–15]. It was intended to capture the longest possible sequence of ascending run within the answer while allowing gaps of adjacent elements. The MRS score is the number of transitions, i.e., the number of elements in the MRS–1. In an answer CDABE, for example, for which the correct sequence is ABCDE, the longest possible ascending sequence or MRS is either CDE or ABE, and the score is 2. Note that there may be multiple MRS for a single score. It is a special case of Levenshtein distance [16] in the sense that there is no addition or reduction of elements.

MRS is logically and psychometrically endorsed with reference to MED (Minimal Edit Distance) in the following simple relationship.

$$\text{MRS} + \text{MED} = \text{full score} = [\text{number of elements}] - 1 \tag{4}$$

By MED we mean the minimal number of displacement of elements in the answer sequence required to recover the correct sequence, i.e., the number of displacement from the correct sequence to arrive at a certain answer sequence. Thus the MED for BACDE is 1, and that for EDCBA is 4. Take CDABE, for example. It is reachable by dislocating either (C and D) or (A and B) from the correct sequence ABCDE, with a displacement count of two for each case. Since measuring MED is to count the number of elements subject for displacement (e.g., C and D), counting the number of intact elements (i.e., A, B, and E), namely, the elements for MRS is a complement to MED, hence the Eq. (4). The more displacement from the correct sequence is involved, the remoter the answer sequence is from the correct sequence, both logically and psychometrically. Thus, measurement by MRS is practically equivalent to measurement by MED, bearing an advantage over 'Exact matching' and 'Adjacent matching' with respect to the cognitive load needed for recovery.

4. Maximal relative sequence with recovery distance

Talking about recovery, MRS is still a rough indicator of partial achievement, however. The major flaw is that it does not consider the precise recovery distance of the elements involved. Two answers B**A**CDE and BCDE**A** would have the same MRS score of 3 (or 1 displacement of **A**), but the degree of distortion from the correct answer is obviously different. The author's MRS + Dist model was an attempt of incorporating the recovery distance, i.e., the total number of elements that the elements subject for recovery have to jump over [17, 18]. The final score will be calculated as follows:

$$\text{Adjusted score} = \text{MRS} \times \left(1 - [\text{Penalty rate}]\right) \tag{5}$$

where

Penalty rate $= [\text{Recovery distance}]/[\text{Maximal recovery distance}]$

where

Maximal recovery distance $= n \times (n-1)/2$

where

n is the number of elements in the sequence.

Table 2 shows some sample recovery effects.

The author coded a script of computer programme using Xojo [19] enabling machine calculation since calculating recovery distance seemed beyond ocular calculation. One reason for this complication was the need to handle crossing constraints. A crossing constraint prevents redundant recovery moves. In the case of CDABE, the correct procedure is:

$$\begin{aligned} &\text{Step 1:} \quad \text{C}\mathbf{D}\text{ABE} \rightarrow \text{CAB}\mathbf{D}\text{E (Distance} = 2) \\ &\text{Step 2:} \quad \mathbf{C}\text{ABDE} \rightarrow \text{AB}\mathbf{C}\text{DE (Distance} = 2) \\ &\qquad\qquad\qquad\qquad \text{Total distance} = 4 \end{aligned}$$

However, if we started with C, the process would incur an unnecessary extra movement:

Answer	MRS	Elements for recovery	Distance	Penalty rate	MRS + Dist
ABC**E**D	3	E	1	0.1	3 × (1–0.1) = 2.7
AB**E**CD	3	E	2	0.2	3 × (1–0.2) = 2.4
EABCD	3	E	4	0.4	3 × (1–0.4) = 1.8
CDABE	2	C, D	4	0.4	2 × (1–0.4) = 1.2
DCBAE	1	B, C, D	6	0.6	1 × (1–0.6) = 0.4
E**DCBA**	0	A, B, C, D	10	1	0 × (1–1) = 0

Table 2.
Sample recovery effects. (Bold letters indicate elements for recovery or 'disruptors').

$$\begin{aligned} &\text{Step 1}: \quad \textbf{C}\text{DABE} \rightarrow \text{DAB}\textbf{C}\text{E (Distance} = 3) \\ &\text{Step 2}: \quad \textbf{D}\text{ABCE} \rightarrow \text{ABC}\textbf{D}\text{E (Distance} = 3) \\ &\qquad\qquad\qquad\qquad \text{Total distance} = 6 \end{aligned}$$

In the first step, C jumped over D, which jumped over C in the second step. This redundancy has occurred because C jumped over an older element D when it moved right, the relative order having been reversed, which had to be rectified in the second step by making D jump over C. We need a set of constraints as follows:

$$\text{No element can cross over an older element when moving right.} \tag{6}$$

$$\text{No element can cross over a younger element when moving left.} \tag{7}$$

Figure 1 is a flow chart illustrating the procedure of calculating MRS (and Distance as a supplement). The core mechanism of creating MRS is to concatenate ascending elements in the answer sequence into a possible pair and connect its tail with the head of another ascending pair. In the case of CDABE, the first seeds are CD, DE, AB, BE and AE. Some of them grow into larger sequences CDE and ABE. The growth stops here since no more concatenation is possible. These are the MRS and the count of concatenation steps is the MRS score (= 2). A full set of codes is included in [18].

5. MRS by Excel

Resorting to a computer programme meant that the protocol turned opaque in a black box. In order to secure transparency, the author devised an Excel spreadsheet where he transplanted the computer programme to combinations of Excel functions [20].

Readers can trace the concatenation steps in Sections 2 and 3 in **Figure 2**.

As for the recovery distance, the author introduced the use of Kendall's tau instead of a special computer programme [21]. Given a definition of tau as (1), P is the total number of 'behind' elements. When the correct sequence is ABCDE and the target sequence is CDABE, for example, the elements behind A in the ascending order in the target sequence are B and E (2 elements); behind B comes E (1 element); behind C come D and E (2 elements); behind D comes E (1 element). P, in this case, makes 6.

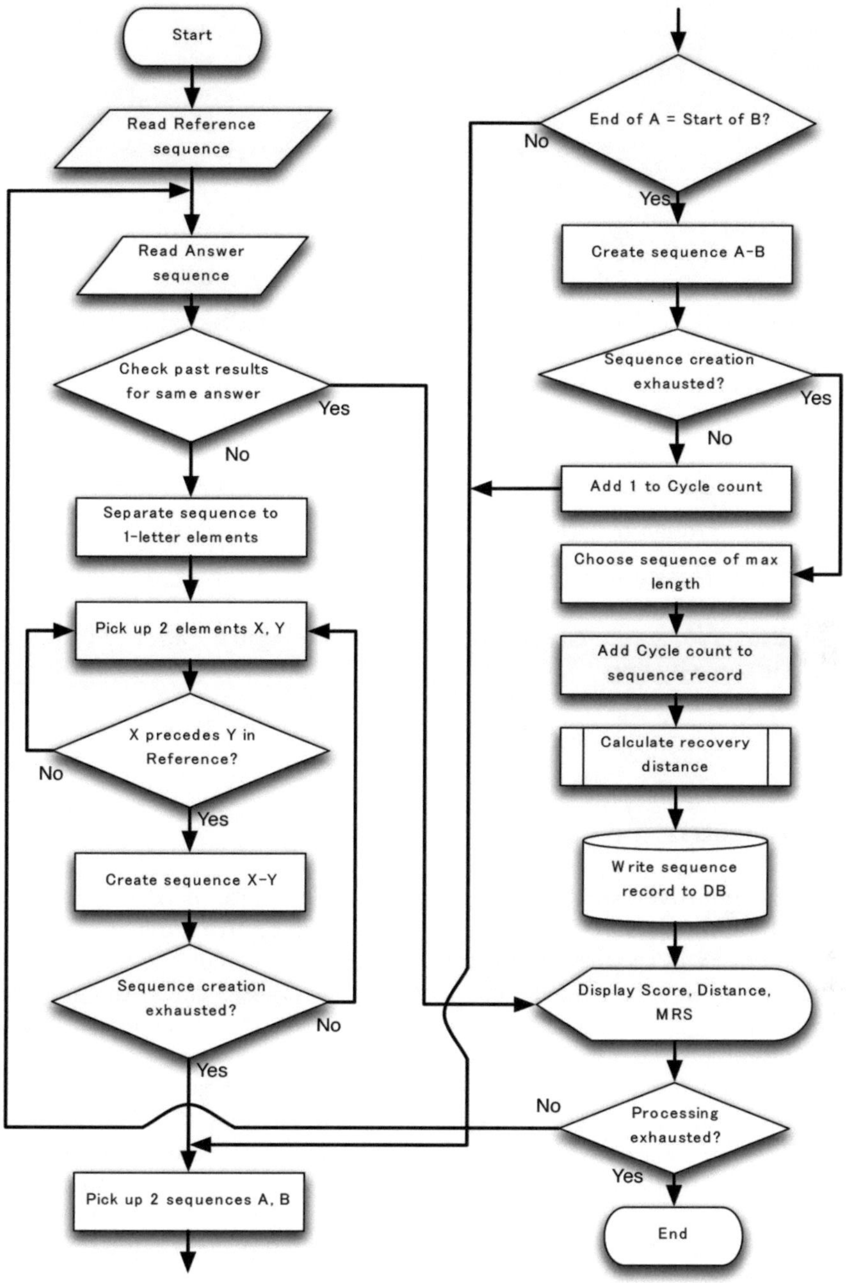

Figure 1.
Flow chart of MRS algorithm.

Since P indicates the number of elements that each element has to jump over to make a complete reverse sequence (EDCBA), the maximum of P is

$$P_{\max} = n \times (n-1)/2 \tag{8}$$

where

n is the number of elements.

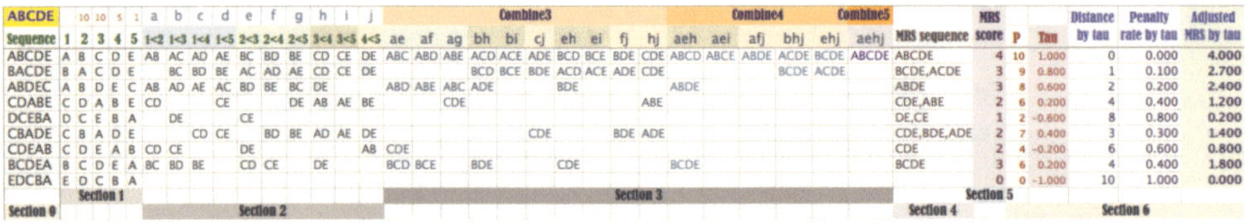

ABCDE		10	10	5	1	a	b	c	d	e	f	g	h	i	j
Sequence	1	2	3	4	5	1<2	1<3	1<4	1<5	2<3	2<4	2<5	3<4	3<5	4<5
ABCDE	A	B	C	D	E	AB	AC	AD	AE	BC	BD	BE	CD	CE	DE
BACDE	B	A	C	D	E		BC	BD	BE	AC	AD	AE	CD	CE	DE
ABDEC	A	B	D	E	C	AB	AD	AE	AC	BD	BE	BC	DE		
CDABE	C	D	A	B	E	CD			CE			DE	AB	AE	BE
DCEBA	D	C	E	B	A		DE			CE					
CBADE	C	B	A	D	E			CD	CE		BD	BE	AD	AE	DE
CDEAB	C	D	E	A	B	CD	CE			DE					AB
BCDEA	B	C	D	E	A	BC	BD	BE		CD	CE		DE		
EDCBA	E	D	C	B	A										
		Section 1													
Section 0										Section 2					

	Combine3										Combine4					Combine5
Sequence	ae	af	ag	bh	bi	cj	eh	ei	fj	hj	aeh	aei	afj	bhj	ehj	aehj
ABCDE	ABC	ABD	ABE	ACD	ACE	ADE	BCD	BCE	BDE	CDE	ABCD	ABCE	ABDE	ACDE	BCDE	ABCDE
BACDE				BCD	BCE	BDE	ACD	ACE	ADE	CDE				BCDE	ACDE	
ABDEC	ABD	ABE	ABC	ADE			BDE				ABDE					
CDABE			CDE							ABE						
DCEBA																
CBADE						CDE			BDE	ADE						
CDEAB	CDE															
BCDEA	BCD	BCE		BDE			CDE				BCDE					
EDCBA																
									Section 3							

Sequence	MRS sequence	MRS score	P	Tau	Distance by tau	Penalty rate by tau	Adjusted MRS by tau
ABCDE	ABCDE	4	10	1.000	0	0.000	4.000
BACDE	BCDE,ACDE	3	9	0.800	1	0.100	2.700
ABDEC	ABDE	3	8	0.600	2	0.200	2.400
CDABE	CDE,ABE	2	6	0.200	4	0.400	1.200
DCEBA	DE,CE	1	2	-0.600	8	0.800	0.200
CBADE	CDE,BDE,ADE	2	7	0.400	3	0.300	1.400
CDEAB	CDE	2	4	-0.200	6	0.600	0.800
BCDEA	BCDE	3	6	0.200	4	0.400	1.800
EDCBA		0	0	-1.000	10	1.000	0.000
		Section 5					
	Section 4				Section 6		

Figure 2.
Sample excel sheet of MRS and MRS + Dist.

The answer CDABE is distant from the correct answer by dislocating C and D by 4 occasions of jumping over (**Table 2**), and it is still distant from the complete reverse by additional 6 occasions of jumping over. It means the sum of recovery distance and P is always constant, $P_{\max}$.

$$\begin{array}{lcccc} \text{ABCDE} \longleftarrow & \longrightarrow \text{CDABE} \longleftarrow & & \longrightarrow \text{EDCBA} \\ \text{Items to} \quad 4 & + & 6 & = 10 \\ \text{jump over} & & & = P_{\max} \end{array}$$

Therefore recovery distance can be calculated by tau. **Figure 2** already includes the column of Distance by tau.

$$\text{Recovery distance} = (1 - \tau) \times n \times (n - 1)/4 \tag{9}$$

6. Linearity matrix

Despite the improved accessibility, a major problem with the Excel calculation is its consumption of columns. If we follow the layout in **Figure 2** the number of columns required will soon reach the limit of 16,384 as we increase the size of the sequence. It means n = 12 is the largest possible size. **Table 3** shows the number of entire columns required for n = 4 to n = 13, including columns for calculating recovery distance by Kendall's tau.

Yet another idea for representing the mechanism of MRS is to make use of a matrix. **Table 4** shows the framework of matrix (Linearity Matrix = LM) in which the relationships of elements are indicated. The value '1' indicates that the row element is correctly followed by the column element. The value will be '0' when the relative order is disrupted. In **Table 5** for CDABE, C-A, C-B, D-A and D-B are in the wrong order. The sum of the values is Linearity Matrix score, representing the wellformedness of the answer sequence.

The greatest advantage of LM is its efficiency in column consumption. When transplanted to Excel the core part of LM requires $n \times (n-1)/2$ columns, resulting in a

n	4	5	6	7	8	9	10	11	12	13
C	54	102	198	390	774	1,542	3,078	6,150	12294	24582

Table 3.
Size of entire columns (C) required by MRS + Dist.

A	1	1	1	1	
	B	1	1	1	
		C	1	1	
			D	1	
				E	LM = 10

Table 4.
Linearity matrix for correct answer ABCDE.

C	1	0	0	1	
	D	0	0	1	
		A	1	1	
			B	1	
				E	LM = 6

Table 5.
Linearity matrix for partially correct answer CDABE.

n	4	5	6	7	8	9	10	11	12	13
C	14	19	25	32	40	49	59	70	82	95

Table 6.
Size of entire columns (C) required by LM.

small size of the entire columns (**Table 6**). Whereas MRS + Dist by Excel would need 3,080,634 columns (in theory) when $n = 20$, LM requires only 214.

An additional advantage of LM is that it can calculate the recovery distance by counting the number of zero values. Since a pair with '1' mark indicates an ascending run, the sum of values (= LM) is equivalent to P of Kendall's tau. It means that the count of zero values represents the recovery distance. In **Table 5**, either C and D have to jump over A and B or A and B have to jump over C and D. In either case, the disruption is solved by removing the elements with zero values.

Furthermore, since zero marks are caused by the elements in incorrect positions, these disruptors are subject to displacement for the entire sequence to be corrected. Without these disruptors, the remaining elements should all be arranged in ascending orders in any combination. Thus we can identify the elements of MRS by identifying a minimal number of steps to remove disruptors. In **Table 7**, we can get B and E as MRS

D	0	1	0	0	
	B	1	1	0	
		E	0	0	
			C	0	
				A	LM = 3, Distance = 7

Table 7.
Linearity matrix for partially correct answer DBECA.

by removing D, C and A; or B and C by removing D, E and A; or D and E by removing B, C and A. These alternative MRSs are obtainable through different procedures, but the count of removal steps stays the same. **Tables 8–10** show the process of the first case.

The optimal strategy for removing the disruptors on Excel has not been found at the moment, but we know in theory that LM provides information of both MRS and recovery distance. By definition, LM counts all pairs of ascending order whereas MRS picks up elements that can form the longest sequence. Therefore in the above sample DE, BE, and BC are all counted for the LM score (=3) but only one of them constitutes an MRS. Similarly, CBAED has also one MRS (CE, CD, BE, BD, AE, or AD), but LM = 6. The different LM scores of the two sequences bearing the same MRS score indicate that the internal structure is different.

As another potential LM can vary the weights for ascending pairs. We could value larger weights for closer pairs and smaller weights for remoter pairs (**Table 11** for the correct sequence and **Table 12** for a sample answer).

Figure 3 shows the correlation of scores between gradient LM and of MRS + Dist. The data were taken from a test of reading comprehension in English as a foreign language for Japanese university students (N = 149). They were asked to reorder eight descriptions of events after watching a video of an expository story. The correlation r = 0.950 and the coefficient of determination R^2 = 0.902 suggest that LM is a highly reliable alternative to MRS + Dist.

D	0	1	0		
	B	1	1		
		E	0		
			C		
				A	LM = 3, Distance = 3

Table 8.
Linearity matrix after removing disruptor A (step 1).

D					
	B	1	1		
		E	0		
			C		
				A	LM = 2, Distance = 1

Table 9.
Linearity matrix after removing disruptor D (step 2).

D					
	B	1			
		E			
			C		
				A	LM = 1, Distance = 0

Table 10.
Linearity matrix after removing disruptor C (step 3).

A	4	3	2	1	
	B	4	3	2	
		C	4	3	
			D	4	
				E	Sum = 30

Table 11.
Linearity matrix for correct answer (n = 5) with gradient weights.

C	4	0	0	1	
	D	0	0	2	
		A	4	3	
			B	4	
				E	Sum = 18

Table 12.
Linearity matrix for partially correct answer (n = 5) with gradient weights.

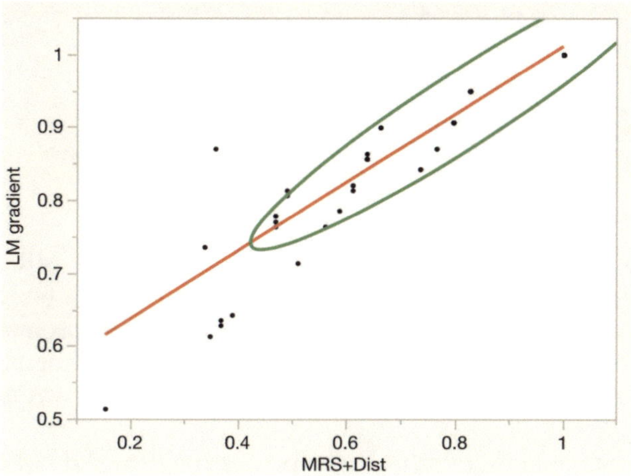

Figure 3.
Plot of scores by gradient LM and MRS + Dist. (The scores are standardised between 0 and 1. The oval represents a density ellipse of probability 0.90.)

For part of the test-takers (N = 31), internal consistency was compared among eight measurement methods where the reordering task was part of a larger reading comprehension test including short-answer questions. **Table 13** indicates that LM methods are stable and can better capture the test-taker performance.

Binary	Exact	Adjacent	MRS	tau	MRS + Dist	LM flat	LM gradient
0.273	0.534	0.479	0.624	0.559	0.560	0.675	0.670

Table 13.
Alpha coefficients of eight measurement methods.

Machine	MRS	MRS + Dist	LM flat	LM gradient
MacBook Pro[4]	3.90	3.80	0.50	0.60
Mac Pro[5]	38.70	107.10	64.90	62.50
HP250 (Windows)[6]	19.45	15.56	14.45	14.45

Table 14.
Processing time of four measurement methods (s).

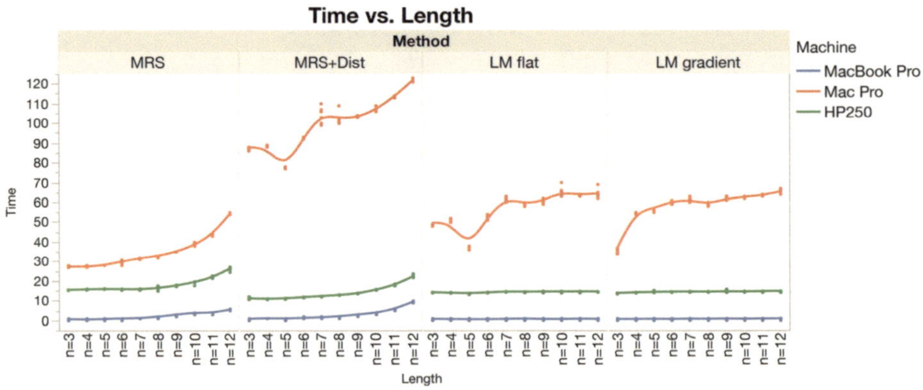

Figure 4.
Processing time(s) by sequence length.

LM is even quicker in processing a vast amount of data than MRS. **Table 14** shows a processing time taken by four methods when $n = 10$. The data (size = 1000) was randomly sampled from all permutations of 10 elements. The programmes for two versions of LM were written by Xojo, the same programming language for MRS. Distance in MRS + Dist used Kendall's tau formula. Each value was an average of ten trials.

It should also be noted that with LM the processing time did not deteriorate as the number of elements increased. **Figure 4** summarises the average processing time of the four measurement methods for sequence length of $n = 3$ to $n = 12$.

7. Applications and limitations

We have seen so far theoretical considerations of evaluating item reordering except for the analyses of correspondence of methods based on actual test results. When constructing an evaluation scheme in reality, however, various non-theoretical factors come in the way. Take an example from Alderson, et al.'s reordering question called 'Compaq task' [3]. **Text 2** is the original; **Text 3** is what they regarded as partially correct; **Text 4** is an alternatively misplaced sequence of my invention. Item codes are rearranged for convenience.

(A) A technician at Compaq Computers told of a frantic call he received on the helpline.
(B) It was from a woman whose new computer simply would not work.
(C) She said she'd taken the computer out of the box, plugged it in, and sat there for 20 minutes waiting for something to happen.
(D) The tech guy asked her what happened when she pressed the power switch.
(E) The woman replied, 'What power switch?'

Text 2.

Correct sequence.

(A) [Same as **Text 2**]
(B) [Same as **Text 2**]
(D) The tech guy asked her what happened when she pressed the power switch.
(E) The woman replied, 'What power switch?'
(**C**) She said she'd taken the computer out of the box, plugged it in, and sat there for 20 minutes waiting for something to happen.

Text 3.

'Partially correct'sequence.

(A) [Same as **Text 2**]
(B) [Same as **Text 2**]
(D) The tech guy asked her what happened when she pressed the power switch.
(C) She said she'd taken the computer out of the box, plugged it in, and sat there for 20 minutes waiting for something to happen.
(E) The woman replied, 'What power switch?'

Text 4.

Incorrect sequence.

Both **Text 3** and **Text 4** differ from the correct sequence by one dislocation of statement (C). However, while **Text 4** is completely unacceptable **Text 3** sounds much more acceptable, if not perfectly. The use of past perfective form in **Text 3** refers back to the point that occurred before (D). The one dubious element is that the tech guy's question in (D) is slightly too specific, which could disrupt the natural flow of discourse. More serious is the fact that **Text 4** is much less acceptable than **Text 3,** even though the recovery distance of **Text 4** is shorter than that of **Text 3**. It means that the recovery distance alone is not necessarily a predictor of penalty.

This type of task may be called an a priori (or jigsaw) task: test-takers must read the fragments at sight and reconstruct the original passage. Because the fragments contain a lot of linguistic clues such as tense, reference words, and definiteness markers, the test-takers can use these clues to connect fragments. Yet another task type (a posteriori task) requires test takers to read or hear the entire passage initially and reconstruct the outline by selecting descriptions in the correct order. Alderson et al.'s 'Queen task' is of this type. In fact, they admit that the a posteriori type might be more appropriate as a measurement tool of reading comprehension (p. 442). **Text 6**, based on intact **Text 5** [22], is another sample of this type where linguistic clues are as much neutralised as possible.

New neighbours

Mr and Mrs Smith married thirty years ago, and they have lived in the same house since then. Mr Smith goes to work at eight o'clock every morning, and he gets home at half past seven every evening, from Monday to Friday.

There are quite a lot of houses in their street, and most of the neighbours are nice. But the old lady in the house opposite Mr and Mrs Smith died, and after a few weeks, a young man and woman came to live in it.

Mrs Smith watched them for a few days from her window and then she said to her husband, 'Bill, the man in that house opposite always kisses his wife when he leaves in the morning and he kisses her again when he comes home in the evening. 'Why don't you do that too?'

'Well,' Mr Smith answered, 'I don't know her very well yet.'

Text 5.

Sample original passage

(1) Mr Smith was a serious man.

(2) An old lady died.

(3) A young couple moved in.

(4) Mrs Smith made repeated observations.

(5) Mrs Smith requested a new action.

Text 6.

Outline items for a *posteriori reordering*.

In this paper, we examined the nature of MRS and LM. Both of them are still in an incubation stage in language testing as well as other psychometric measurements. The differential weighting for ascending pairs in gradient LM is a proposed model without empirical evidence. There might be clusters of items to be fixed together. Alternative exchanges of talks in dialogue (as in the Compaq task) are considered an example of high adhesion whereas some kinds of discourse order may not be as adhesive. Nevertheless, it is meaningful to attempt application of various measurement methods and validate psychometric as well as semantic connectivity. For example, flat LM might be suitable for a task of recollecting historical events because reference to the chronological order of events is relevant to all (or most) pairs. When reconstructing a story, in contrast, MRS or gradient LM might be a better tool, because local connections are considered more important than remote connections, and the wellformedness of the story depends on how much the completed sequence of items looks like a string of stories. Finally, describing MRS by matrix is space and time-saving. LM is like a ripple in the pond; if you observe the wave on the shore you can detect where the stone was cast.

Acknowledgements

The author expresses his deep gratitude to Dokkyo University Information Science Research Institute and to the Multivariate Study Group at SAS Institute Japan for their insightful discussions and suggestions.

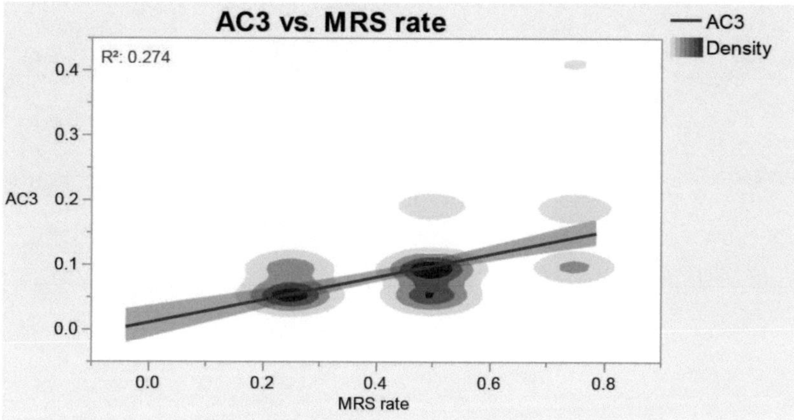

Figure 5.
Comparison of AC and MRS ratings.

Alpha	MRS	MRS + Dist	LM
0.001	0.524	0.368	0.268
0.0001	0.490	0.343	0.253
0.000001	0.416	0.288	0.218
0.00000001	0.350	0.241	0.186

Table 15.
Correlation of AC with MRS, MRS + Dist, and LM. (A perfect sequence is excluded as an outlier).

Conflict of interest

The author declares no conflict of interest.

Notes

1. The correlation of AC and MRS is 0.524, but 89% of AC scores are smaller than 0.1 while 51% of MRS is between 0.5 and 0.6 (**Figure 5**, created by JMP [23]). See Section 2 for MRS.

2. When $N = 5$ (i.e., 5-element sequence), AC is severely affected by the value of α (**Table 15**). See Sections 2 and 3 for MRS, MRS + Dist, and LM.

3. While $AC = 0.050$, MRS + Dist. = 0.350 and LM = 0.667. See Sections 2 and 3 for MRS + Dist and LM.

4. MacBook Pro (8-core Apple M1, 16GB)/OS11.4/Xojo 2021r1

5. Mac Pro (2 × 2.4GHz 6-core Intel Xeon, 25GB, 1.3 MHz DDR3)/OS10.14/Xojo 2017r2.1

6. HP250G7–122 (Intel Core i5-8565U, 1.6GHz, 8GB RAM)/Windows10/Xojo 2021r1

Author details

Amma Kazuo
Dokkyo University, Sōka, Japan

*Address all correspondence to: ammakazuo@mac.com

References

[1] Alderson JC, Clapham C, Wall D. Language Test Construction and Evaluation. Cambridge: Cambridge University Press; 1995. pp. 52-53

[2] Alderson JC. Assessing Reading. Cambridge: Cambridge University Press; 2000. pp. 219-221

[3] Alderson JC, Percsich R, Szabo G. Sequencing as an item type. Language Testing. 2000;**17**(4):423-447

[4] National Centre for University Entrance Examinations. Heisei 25 Nendo Honshiken no Mondai [Examination Questions for Academic Year 2013]. Available from: http://www.dnc.ac.jp/sp/data/shiken_jouhou/h25/jisshikekka/ [Accessed: 08 June, 2019]

[5] Kendall MG. A new measure of rank correlation. Biometrika. 1938;**30**(1–2): 81-93. DOI: 10.1093/biomet/30.1-2.81 [Accessed: 19 January, 2022]

[6] Birch A, Osborne M. Reordering metrics for MT. In: Proceedings of the 49th Annual Meeting of the Association for Computational Linguistics. Stroudsburg, PA, United States: Association for Computational Linguistics; 2011. pp. 1027-1035

[7] Dlougach J, Galinskaya I. Building a reordering system using tree- to-string hierarchical model. Proceedings of COLING 2012; 2013. Available from: https://arxiv.org/abs/1302.3057 [Accessed: 19 January, 2022]

[8] Zechner K. Automatic summarization of open-domain multiparty dialogues in diverse genres. Computational Linguistics. 2002;**28**(4):447-484

[9] Papineni K, Roukos S, Ward T, Zhu W-J. Bleu: A method for automatic evaluation of machine translation. In: Proceedings of the 40th Annual Meeting of the Association for Computational Linguistics (ACL). Philadelphia: ACL; 2002. pp. 311-318

[10] Bollegala D, Okazaki N, Ishizuka M. A bottom-up approach to sentence ordering for multi-document summarization. Information Processing and Management. 2010;**46**:89-109

[11] Lapata M. Automatic evaluation of information ordering: Kendall's tau. Computational Linguistics. 2006;**32**(4): 471-484

[12] Amma K. Partial scoring of sequencing tasks. In: Proceedings of the International Meeting of the Psychometrics Society (IMPS 2007); 9–13 July, 2007. Tokyo, Japan: IMPS; 2007

[13] Amma K. Appraisal of partial scoring in sequencing tasks. In: Proceedings of the JACET 46th Annual Convention (the Japan Association of College English Teachers); 6–8 September 2007. Hiroshima, Japan: The Japan Association of College English Teachers; 2007. pp. 108-109

[14] Amma K. Seijo mondai no bubun saitenho^ to sono programming [partial scoring of sequencing problems and its programming]. Dokkyo Journal of Language Learning and Teaching (Dokkyo University Research Institute of Foreign Language Teaching). 2010;**28**: 1-29

[15] Amma K. Comparison of partial scoring methods in sequencing tasks with reference to internal reliability. In: Proceedings of the JACET 49th Annual Convention (the Japan Association of College English Teachers); 7–9 September 2010. Miyagi, Japan: The

Japan Association of College English Teachers; 2010. pp. 160-161

[16] Levenshtein VI. Binary codes capable of correcting deletions, insertions and reversals. Soviet Physics-Doklady. 1966;**10**(8):707-710

[17] Amma K. Partial scoring of sequencing tasks with distance penalty. In: Abstracts of ALTAANZ Conference 2016 (the Association for Language Testing and Assessment of Australia and New Zealand); 17–19 November 2016. Auckland, New Zealand: The Association for Language Testing and Assessment of Australia and New Zealand; 2016. p. 23

[18] Amma K. Partial scoring of reordering tasks with recovery distance as penalty. Journal of Informatics (Dokkyo University Information Science Research Institute). 2018;7:5-23 J-GLOBAL ID 201802283386410791; Reference number 18A0455950

[19] Xojo. Xojo, Inc. Available from: https://www.xojo.com/ [Accessed: 19 January, 2022]

[20] Amma K. Partial scoring of reordering tasks: Maximal relative sequence by excel. In: Proceedings of 2019 IEEE 2nd International Conference on Information and Computer Technologies (ICICT); 14–17 March 2019. Hawai'i, USA: ICICT. pp. 19-24. DOI: 10.1109/INFOCT.2019.8711372. ISBN: 978-1-7281-3322-5. Available from: https://ieeexplore.ieee.org/document/8711372 [Accessed: 19 January, 2022]

[21] Amma K. Partial scoring of reordering tasks revisited: Linearity matrix by excel. In: Proceedings of 2020 IEEE 3rd International Conference on Information and Computer Technologies (ICICT); 9–12 March 2020. Silicon Valley, USA: ICICT. pp. 1-6. DOI: 10.1109/ICICT50521.2020.00008. ISBN: 978-1-7281-7283-5. Available from: https://ieeexplore.ieee.org/document/9092126, https://conferences.computer.org/icict/2020/pdfs/ICICT2020-sQZ4BHZN9WMCBMwB1asUZ/728300a001/728300a001.pdf [Accessed: 19 January, 2022]

[22] Hill LA. Elementary Steps to Understanding. Oxford: Oxford University Press; 1980. p. 30

[23] JMP (version 13.0). Cary, NC: SAS Institute. Available from: https://www.jmp.com/ [Accessed: 07 April, 2022]

Chapter 4

Weighted Least Squares Perturbation Theory

Aleksandr N. Khimich, Elena A. Nikolaevskaya and Igor A. Baranov

Abstract

The interest in the problem of weighted pseudoinverse matrices and the problem of weighted least squares (WLS) is largely due to their numerous applications. In particular, the problem of WLS is used in the design and optimization of building structures, in tomography, in statistics, etc. The first part of the chapter is devoted to the sensitivity of the solution to the WLS problem with approximate initial data. The second part investigates the properties of a SLAE with approximate initial data and presents an algorithm for finding a weighted normal pseudo solution of a WLS problem with approximate initial data, an algorithm for solving a WLS problem with symmetric positive semidefinite matrices and an approximate right side and also a parallel algorithm for solving a WLS problem. The third part is devoted to the analysis of the reliability of computer solutions of the WLS problem with approximate initial data. Here, estimates of the total error of the WLS problem are presented, and also software-algorithmic approaches to improving the accuracy of computer solutions.

Keywords: weighted least squares problem, error estimates, weighted matrix pseudoinverse, weighted condition number, weighted singular value decomposition

1. Introduction

The interest in the problem of weighted pseudoinverse matrices and the WLS problem is largely due to their numerous applications. In particular, the problem of weighted least squares is used in the design and optimization of building structures, in tomography, in statistics, etc. A number of properties of weighted pseudoinverse matrices underlie the finding of weighted normal pseudosolutions. The field of application of weighted pseudoinverse matrices and weighted normal pseudosolutions is constantly expanding.

The definition of a weighted pseudoinverse matrix with positive definite weights was first introduced by Chipman in article [1]. In 1968, Milne introduced the definition of a skew pseudoinverse matrix in paper [2]. The study of the properties of weighted pseudoinverse matrices and weighted normal pseudosolutions, as well as the construction of methods for solving these and other problems, are devoted to the works of Mitra, Rao, Van Loan, Wang, Galba, Deineka, Sergienko, Ben-Israel, Elden, Wei, Wei, Ward etc. Weighted pseudoinverse matrices and weighted normal pseudosolutions with degenerate weights were studied in [3–5]. The existence and uniqueness of weighted pseudoinverse matrices with indefinite and mixed weights, as well as some of their properties, were described in [6–8]. Application of the weighted pseudoinverse matrix in statistics presented, for example, in [9, 10]. Many results on weighted generalized pseudo-inversions can be found in monographs [11, 12]. Much less work is devoted to the study of weighted pseudoinversion under conditions of approximate initial data. These issues are discussed in [13–17]. Analysis of the properties of weighted pseudoinverses and weighted normal pseudosolutions, as well as the creation of solution methods for these and other problems, are described in [18–20].

When solving applied problems, their mathematical models will have, as a rule, approximate initial data as a result of measurements, observations, assumptions, hypotheses, etc. Later, during discretization ('arithmetization') of the mathematical model, these errors are transformed into the errors of the matrix elements and the right parts of the resolving systems of equations. The input data of systems of linear algebraic equations and WLS problems can be determined directly from physical observations, and therefore they can have errors inherent in all measurements. In this case, the original data we have is an approximation of some exact data. And, finally, the initial data of mathematical models formulated in the form of linear algebra problems can be specified exactly in the form of numbers or mathematical formulas, but, given the finite length of a machine word, it is impossible to work with such an exact model on a computer. The machine model of such a problem in the general case will be approximate either due to errors in converting numbers from the decimal system to binary or due to rounding errors in the implementation of calculations on a computer.

The task is to study the properties of the machine model and to form a model of the problem and an algorithm for obtaining an approximate solution in a machine environment that will approximate the solution of a mathematical problem. The key question of numerical simulation is the reliability of the obtained machine solutions.

The most complete systematic exposition of questions related to the approximate nature of the initial data in problems of linear algebra is given in the monographs [21–24]. Various approaches to the study and solution of ill-posed problems were considered, for example, in [25–28]. Problems of the reliability of a machine solution for problems with approximate initial data, i.e. estimates of the proximity of the machine solution to the mathematical solution, estimates of the hereditary error in the mathematical solution and refinement of the solution were considered in the publications [12, 26, 29–33]. Much less work has been devoted to the study of similar questions for the WLS problem. The sensitivity analysis of a weighted normal pseudosolution under perturbation of the matrix and the right-hand side is the subject of papers [16, 34–36].

The chapter is devoted to the solution of the listed topical problems, namely the development of the perturbation theory for the WLS problem with positive definite weights and the development of numerical methods for the study and solution of mathematical models with approximate initial data.

2. Weighted least squares problem

2.1 Preliminaries

Let the set of all $m \times n$ matrices is denoted by $R^{m\times n}$. Given a matrix $A \in R^{m\times n}$ let A^T is the transpose of A, $\text{rank}(A)$ is the rank of A, $\mathbb{R}(A)$ is the field of values of A and $\mathbb{N}(A)$ is the null space of A. Additionally, let $\|\|$ denote the vector 2-norm and the consistent matrix 2-norm, and let I be an identity matrix.

Given an arbitrary matrix $A \in R^{m\times n}$ and symmetric positive definite matrices M and N of orders m and n, respectively, a unique matrix $X \in R^{m\times n}$, satisfying the conditions:

$$AXA = A, \quad XAX = X, \quad (MAX)^T = MAX, (NXA)^T = NXA, \tag{1}$$

is called the **weighted Moore–Penrose pseudoinverse** of A and is denoted by $X = A^+_{MN}$. Specifically, if $M = I \in R^{m\times m}$ and $N = I \in R^{n\times n}$, then X satisfying conditions (1) is called the **Moore–Penrose pseudoinverse** and is designated as $X = A^+$.

Let $A^\#$ denote the weighted transpose of A, P and Q be idempotent matrices, and $\bar{A} = A + \Delta A$ be a perturbed matrix, i.e.,

$$A^\# = N^{-1}A^TM. \tag{2}$$

$$P = A^+_{MN}A, Q = AA^+_{MN}, \bar{P} = \bar{A}^+_{MN}\bar{A}, \bar{Q} = \bar{A}\bar{A}^+_{MN}. \tag{3}$$

Let $x \in R^m$, $y \in R^n$. The weighted scalar products in R^m and R^n are defined as $(x,y)_M = y^TMx$, $x,y \in R^m$ and $(x,y)_N = y^TNx$, $x,y \in R^n$, respectively. The weighted vector norms are defined as:

$$\begin{aligned} \|x\|_M &= (x,x)_M^{\frac{1}{2}} = \left(x^TMx\right)^{\frac{1}{2}} = \left\|M^{\frac{1}{2}}x\right\|, x \in R^m, \\ \|y\|_N &= (y,y)_N^{\frac{1}{2}} = \left(y^TNy\right)^{\frac{1}{2}} = \left\|N^{\frac{1}{2}}y\right\|, y \in R^n. \end{aligned} \tag{4}$$

Let $x,y \in R^m$ and $(x,y)_M = 0$. Then the vectors x and y are called M-orthogonal, i.e. $M^{\frac{1}{2}}x$- and $M^{\frac{1}{2}}y$-orthogonal. It is easy to show that.

$$\|x+y\|^2_M = \|x\|^2_M + \|y\|^2_M, x,y \in R^m. \tag{5}$$

The weighted matrix norms are defined as:

$$\begin{aligned} \|A\|_{MN} &= \max_{\|x\|_N=1} \|Ax\|_M = \left\|M^{\frac{1}{2}}AN^{-\frac{1}{2}}\right\|, A \in R^{m\times n}, \\ \|B\|_{NM} &= \max_{\|y\|_M=1} \|By\|_N = \left\|N^{\frac{1}{2}}AM^{-\frac{1}{2}}\right\|, B \in R^{n\times m}. \end{aligned} \tag{6}$$

Lemma 1 (see in [37]). Let $A \in R^{m\times n}$, $\text{rank}(A) = k$, M and N are positive definite matrices of orders m and n, respectively. Then, there are matrices $U \in R^{m\times m}$ and $V \in R^{n\times n}$, satisfying $U^TMU = I$ and $V^TN^{-1}V = I$ such that.

$$A = U\begin{pmatrix} D & 0 \\ 0 & 0 \end{pmatrix}V^T, \qquad A^+_{MN} = N^{-1}V\begin{pmatrix} D^{-1} & 0 \\ 0 & 0 \end{pmatrix}U^TM, \tag{7}$$

where $D = \text{diag}(\mu_1, \mu_2, \ldots, \mu_k)$, $\mu_1 \geq \mu_2 \geq \ldots \geq \mu_k > 0$ and μ_i^2 are the nonzero eigenvalues of the matrix $A^\#A$. The nonnegative values μ_i are called the weighted singular values of A, moreover, $\|A\|_{MN} = \mu_1$, $\left\|A^+_{MN}\right\|_{NM} = \frac{1}{\mu_k}$.

The weighted singular value decomposition of A yields an M-orthonormal basis of the vectors of U and an N^{-1}-orthonormal basis of the vectors of V.

2.2 Statement of the problem

In the study of the reliability of the obtained machine results, three linear systems are considered. A system of linear algebraic equations with exact input data

$$Ax = b. \tag{8}$$

We will consider the corresponding weighted least squares problem with positive definite weights M and N:

$$\min_{x \in C} \|x\|_N, C = \{x | \|Ax - b\|_M = \min\}, \tag{9}$$

where $A \in R^{m \times n}$ is a rank-deficient matrix and $b \in R^m$.

Along with (9), we consider the mathematical model with approximately specified initial data.

$$\min_{x \in C} \|\bar{x}\|_N, C = \{\bar{x} | \|(A + \Delta A)\bar{x} - (b + \Delta b)\|_M = \min\}, \tag{10}$$

where.

$$\bar{A} = A + \Delta A, \bar{b} = b + \Delta b, \bar{x} = x + \Delta x. \tag{11}$$

Assume that the errors in the matrix elements and the right-hand side satisfy the relations:

$$\|\Delta A\|_{MN} \le \varepsilon_A \|A\|_{MN}, \quad \|\Delta b\|_M \le \varepsilon_b \|b\|_M. \tag{12}$$

The problem for the approximate solution $\bar{\bar{x}}$ of a system of linear algebraic equations with approximately given initial data

$$\bar{A}\bar{\bar{x}} = \bar{b} + \bar{r}, \tag{13}$$

where $\bar{r} = \bar{A}\bar{\bar{x}} - \bar{b}$ is the residual vector.

The analysis of the reliability of the obtained solution includes an assessment of the hereditary error $\|x - \bar{x}\|_N$, computational error $\|\bar{x} - \bar{\bar{x}}\|_N$ and total error $\|x - \bar{\bar{x}}\|_N$, as well as the refinement of the obtained machine solution to a given accuracy.

2.3 The existence and uniqueness of a weighted normal pseudoinverse

Let linear manifold L be a nonempty subset of space R, closed with respect to the operations of addition and multiplication by a scalar (if x and y are elements of L $\forall \alpha, \beta$, the $\alpha x + \beta y$ is an element of L). Vector x is N-orthogonal to the linear manifold L ($x \perp_N L$) if x is N-orthogonal to each vector from L.

Lemma 2 (see in [38]). There exists a unique decomposition of vector x, namely $x = \hat{x} + \tilde{x}$, where $\hat{x} \in L$, $\tilde{x} \perp_N L$.

Let A is an arbitrary matrix. The kernel of matrix A, denoted by $\mathbb{N}(A)$, is the set of vectors mapped into zero by A: $\mathbb{N}(\mathrm{A}) = \{x : Ax = 0\}$.

The set $\mathbb{R}(A)$ of images of matrix A is the set of vectors that are images of vectors of the space R from the definition domain of A, i.e. $\mathbb{R}(A) = \{b : b = Ax, \forall x\}$.

Let L be a linear manifold in space R, N-orthogonal (M-orthogonal) complement to L, denoted by $L^{\perp_N}$ ($L^{\perp_M}$), defined as the set of vectors in R, each of which is N-orthogonal (M-orthogonal) to L.

Remark 1. If x is a vector from R and $x^T N y = 0$ for any y from R, then $x = 0$.

Theorem 1. Let $A \in R^{m \times n}$, then $\mathbb{N}(A) = \mathbb{R}^{\perp}(A^{\#})$.

Proof. Vector $x \in \mathbb{N}(A)$, if and only if $Ax = 0$. Hence, by virtue of Remark1, we get $x \in \mathbb{N}(A)$, if and only if $y^T MAx = 0$ for any y. Since $y^T MAx = (A^{\#}y)^T Nx$, we get $Ax = 0$ if and only if x is N-orthogonal to all the vectors of the form $A^{\#}y$. Vectors $A^{\#}y$ form $\mathbb{R}(A^{\#})$. The required statement follows from here and from the definition $\mathbb{R}^{\perp_M}(A)$.

Theorem 2 (see in [38]). If A is an $m \times n$ matrix and b is an m-dimensional vector, then the unique decomposition $b = \hat{b} + \tilde{b}$ holds, where $\hat{b} \in \mathbb{R}(A)$ and $\tilde{b} \in \mathbb{N}(A^{\#})$.

Vector $\hat{b}$ is a projection of b on to $\mathbb{R}(A)$, and $\tilde{b}$ is a projection of b on to $\mathbb{N}(A^{\#})$. Vectors $\hat{b}$ and $\tilde{b}$ are M-orthogonal. Hence, $A^{\#}b = A^{\#}\hat{b}$.

By Theorem 1, the following relations hold for the symmetric matrix A: $\mathbb{N}(A) = \mathbb{R}^{\perp}(A), \mathbb{R}(A) = \mathbb{N}^{\perp}(A)$.

Theorem 3. Let $A \in R^{m \times n}$, then $\mathbb{R}(A) = \mathbb{R}(AA^{\#})$, $\mathbb{R}(A^{\#}) = \mathbb{R}(A^{\#}A)$, $\mathbb{N}(A) = \mathbb{N}(A^{\#}A)$ and $\mathbb{N}(A^{\#}) = \mathbb{N}(AA^{\#})$.

Proof. It will be to establish that $\mathbb{N}(A^{\#}) = \mathbb{N}(AA^{\#})$ and $\mathbb{N}(A) = \mathbb{N}(A^{\#}A)$.

For this purpose, we will use Theorem 1. To prove the coincidence of $\mathbb{N}(A^{\#})$ and $\mathbb{N}(AA^{\#})$, note that $AA^{\#}x = 0$ if $A^{\#}x = 0$. On the other hand, if $AA^{\#}x = 0$, then $x^T AA^{\#}x = 0$, i.e. $\|A^{\#}x\|_M = 0$, which entails equality $A^{\#}x = 0$. So, $A^{\#}x = 0$ if and only if $x^T AA^{\#}x = 0$. We can similarly establish that $\mathbb{N}(A) = \mathbb{N}(A^{\#}A)$.

Then let us prove the theorem about the existence and uniqueness of the solution vector that minimizes the norm of the residual $\|Ax - b\|_M$ by the technique proposed in [39] for the least-squares problem.

Theorem 4. Let $A \in R^{m \times n}$, $b \in R^m$, $b \notin \mathbb{R}(A)$. Then there exists a vector $\hat{x}$, that minimizes the norm of the residual $\|Ax - b\|_M$ and vector $\hat{x}$ is a unique vector from $\mathbb{R}(A^{\#})$, that satisfies the equation $Ax = \hat{b}$, where $\hat{b} = AA^{+}_{MN}b$ is the projection of b onto $\mathbb{R}(A)$.

Proof. By virtue of Theorem 2, we get $b = \hat{b} + \tilde{b}$, where $\tilde{b} = \left(I - AA^{+}_{MN}\right)b$ is the projection of b on to $\mathbb{N}(A^{\#})$. Since for every $x, Ax \in \mathbb{R}(A)$ and $\tilde{b} \in \mathbb{R}^{\perp_M}(A)$, then $\hat{b} - Ax \in \mathbb{R}(A)$ and $\tilde{b} \perp \hat{b} - Ax$. Therefore

$$\|b - Ax\|_M^2 = \left\|\hat{b} - Ax + \tilde{b}\right\|_M^2 = \left\|\hat{b} - Ax\right\|_M^2 + \left\|\tilde{b}\right\|_M^2 \geq \left\|\tilde{b}\right\|_M^2. \tag{14}$$

This lower bound is attained since $\hat{b}$ belongs to the set of images A, i.e. $\hat{b}$ is an image of some x_0: $\hat{b} = Ax_0$.

Thereby, for this x_0 the greatest lower bound is attainable:

$$\|b - Ax_0\|_M^2 = \left\|b - \hat{b}\right\|_M^2 = \left\|\tilde{b}\right\|_M^2. \tag{15}$$

It was shown earlier that

$$\|b - Ax\|_M^2 = \left\|\hat{b} - Ax\right\|_M^2 + \left\|\tilde{b}\right\|_M^2 \tag{16}$$

and hence, the lower bound can only be attained for x^*, for which $Ax^* = \hat{b}$. According to Theorem 2, each vector x^*, can be presented as a sum of two orthogonal vectors: $x^* = \hat{x}^* + \tilde{x}^*$, where $\hat{x}^* \in \mathbb{R}(A^\#)$, $\tilde{x}^* \in \mathbb{N}(A)$.

Therefore, $Ax^* = A\hat{x}^*$ and hence, $\|b - Ax^*\|_M^2 = \|b - A\hat{x}^*\|_M^2$. Note that

$$\|x^*\|_{N,}^2 = \|\hat{x}^*\|_N^2 + \|\tilde{x}^*\|_N^2 \geq \|\hat{x}^*\|_N^2, \tag{17}$$

where strict inequality is possible when $x^* \neq \hat{x}^*$ (i.e. if x^* does not coincide with its projection on to $\mathbb{R}(A^\#)$).

It was shown above, that x_0 minimizes $\|Ax - b\|_M$, if and only if $Ax_0 = \hat{b}$, and among the vectors that minimize $\|Ax - b\|_M$, each vector with the minimum norm should belong to the set of images $A^\#$. To establish the uniqueness of a minimum-norm vector, assume that $\hat{x}$ and x^* belong to $\mathbb{R}(A^\#)$ and that $A\hat{x} = Ax^* = \hat{b}$. Then $x^* - \hat{x} \in \mathbb{R}(A^\#)$, however $A(x^* - \hat{x}) = 0$, so $x^* - \hat{x} \in \mathbb{N}(A) = \mathbb{R}^{\perp_N}(A^\#)$.

As vector $x^* - \hat{x}$ is N-orthogonal to itself $\|x^* - \hat{x}\|_N = 0$, i.e. $x^* = \hat{x}$.

Remark 2. There is another assertion that is equivalent to Theorem 4. There exists an n-dimensional vector y such that

$$\|b - AA^\# y\|_M = \inf_x \|b - Ax\|_M. \tag{18}$$

If

$$\|b - Ax_0\|_M = \inf_x \|b - Ax\|_M, \tag{19}$$

then $\|x_0\|_N \geq \|A^\# y\|_N$ with strict inequality for $x_0 \neq \|A^\# y\|_N$.

Vector y satisfies the equation $AA^\# y = \hat{b}$, here $\hat{b}$ is the projection of b onto $\mathbb{R}(A)$.

Theorem 5. Among all the vectors x that minimize the residual $\|Ax - b\|_M$, vector $\hat{x}$, which has the minimum norm $\hat{x} = \min \|x\|_N$, is a unique vector of the form

$$\hat{x} = N^{-1} A^T M y = A^\# y, \tag{20}$$

satisfying the equation

$$A^\# A x = A^\# b, \tag{21}$$

i.e. $\hat{x}$ can be obtained by means of any vector y_0, that satisfies the equation $A^\# A A^\# y = A^\# b$ by the formula $\hat{x} = A^\# y_0$.

Proof. According to the condition of Theorem 3 $\mathbb{R}(A^\#) = \mathbb{R}(A^\# A)$. Since vector $A^\# b$ belongs to the set of images $A^\#$, it should belong to the set of images $A^\# A$ and thus should be an image of some vector x with respect to the transformation $A^\# A$. In other words, Eq. (21) (with respect to x) has at least one solution. If x is a solution of Eq. (21), then $\hat{x}$ is the projection of x on to $\mathbb{R}(A^\#)$, since $Ax = A\hat{x}$ according to Theorem 2. Since $\hat{x} \in \mathbb{R}(A^\#)$, vector $\hat{x}$ is an image of some vector y with respect to the transformation $A^\#$: $\hat{x} = A^\# y$.

Thus, we have established that there exists at least one solution of Eq. (21) in the form of (20). To establish the uniqueness of this solution, we assume that $\hat{x}_1 = A^\# y_1$ and $\hat{x}_2 = A^\# y_2$ satisfy Eq. (21). Then $A^\# A (A^\# y_1 - A^\# y_2) = 0$, therefore, $A^\# (y_1 - y_2) \in \mathbb{N}(A^\# A) = \mathbb{N}(A)$, where from the equality $AA^\# (y_1 - y_2) = 0$ follows.

Therefore $y_1 - y_2 \in \mathbb{N}(AA^{\#}) = \mathbb{N}(A^{\#})$; hence, $\hat{x}_1 = A^{\#}y_1 = A^{\#}y_2 = \hat{x}_2$.

Thus, there exists exactly one solution of Eq. (21) in the form (20). The proof of Theorem 5 will be completed if we can show that by virtue of Theorem 1 the solution found in the form (14) is also a solution of the equation $Ax = \hat{b}$, where $\hat{b}$ is a weighted projection of b on to $\mathbb{R}(A)$, i.e. $A^{\#}b = A^{\#}\hat{b}$.

Theorem 4 establishes that there is a unique, from $\mathbb{R}(A^{\#})$ solution of the equation

$$Ax = \hat{b}. \tag{22}$$

Hence, this unique solution satisfies the equation $A^{\#}Ax = A^{\#}\hat{b}$.

According to the equality $A^{\#}b = A^{\#}\hat{b}$ the unique solution of Eq. (22) belonging to $\mathbb{R}(A^{\#})$, should coincide with $\hat{x}$ which is a unique solution of Eq. (21), which also belongs to $\mathbb{R}(A^{\#})$. Finally, vector $\hat{x}$, mentioned in the proof of Theorem 5 exactly coincides with the vector $\hat{x}$ from Theorem 4. Using the representation of the Moore–Penrose weighted pseudoinverse from [38].

$$A^{+}_{MN} = A^{\#}(A^{\#}AA^{\#})^{+}A^{\#}, \tag{23}$$

we can formulate the following theorem for problem (9) in a shorter form.

Theorem 6. Let $A \in R^{m\times n}$, then $x = A^{+}_{MN}b$—is M-weighted least squares solution with the minimum N-norm of the system $Ax = b$.

Note that in [18] a slightly different mathematical apparatus was used to prove the existence and uniqueness of the M-weighted least squares solution with the minimum N-norm of the system $Ax = b$.

3. Error estimates for the weighted minimum-norm least squares solution

3.1 Estimates of the hereditary error of a weighted normal pseudosolution

Consider some properties of the weighted Moore–Penrose pseudoinverse.

Lemma 3 (see in [16]). Let $A, \Delta A \in R^{m\times n}$, $\mu_i(A)$ and $\mu_i(\bar{A})$ denote the weighted singular values of A and $\bar{A}$ respectively. Then,

$$\mu_i(A) - \|\Delta A\|_{MN} \le \mu_i(\bar{A}) \le \mu_i(A) + \|\Delta A\|_{MN}. \tag{24}$$

Lemma 4 (see [40]). Let $A, \Delta A \in R^{m\times n}$, $\operatorname{rank}(\bar{A}) = \operatorname{rank}(A)$ and $\|\Delta A\|_{MN}\|A^{+}_{MN}\|_{NM} < 1$. Then

$$\left\|\bar{A}^{+}_{MN}\right\|_{NM} \le \frac{\|A^{+}_{MN}\|_{NM}}{1 - \|\Delta A\|_{MN}\|A^{+}_{MN}\|_{NM}}. \tag{25}$$

Lemma 5. Let $G = \bar{A}^{+}_{MN} - A^{+}_{MN}$, $\bar{A} = A + \Delta A$ and $\operatorname{rank}(\bar{A}) = \operatorname{rank}(A)$. Then G can be represented as the sum of three matrices $G = G_1 + G_2 + G_3$, where

$$G_1 = -\bar{A}^{+}_{MN}\Delta A A^{+}_{MN}, \tag{26}$$

$$G_2 = -(I - \bar{P})N^{-1}\Delta A^T A^{+T}_{MN} N A^{+}_{MN} = -(I - \bar{P})\Delta A^{\#}\left(A^{+}_{MN}\right)^{\#} A^{+}_{MN}, \tag{27}$$

$$G_3 = \bar{A}^+_{MN}(I - Q). \tag{28}$$

Proof. Following [26], G can be represented as the sum of the following matrices.

$$\begin{aligned} G = &[\bar{P} + (I - \bar{P})]\left(\bar{A}^+_{MN} - A^+_{MN}\right)[Q + (I - Q)] = \\ &= \bar{P}\bar{A}^+_{MN}Q + \bar{P}\bar{A}^+_{MN}(I - Q) - \bar{P}A^+_{MN}Q - \bar{P}A^+_{MN}(I - Q) + (I - \bar{P})\bar{A}^+_{MN}Q + \\ &+ (I - \bar{P})\bar{A}^+_{MN}(I - Q) - (I - \bar{P})A^+_{MN}Q + (I - \bar{P})A^+_{MN}(I - Q). \end{aligned} \tag{29}$$

Since.

$$\bar{P}\bar{A}^+_{MN} = \bar{A}^+_{MN}, (I - \bar{P})\bar{A}^+_{MN} = 0, A^+_{MN}Q = A^+_{MN}, A^+_{MN}(I - Q) = 0, \tag{30}$$

we obtain

$$\begin{aligned} G &= \bar{A}^+_{MN}Q + \bar{A}^+_{MN}(I - Q) - \bar{P}A^+_{MN} + (I - \bar{P})\bar{A}^+_{MN} = \\ &= \left(\bar{A}^+_{MN}Q - \bar{P}A^+_{MN}\right) - (I - \bar{P})A^+_{MN} + \bar{A}^+_{MN}(I - Q). \end{aligned} \tag{31}$$

Consider each term in this equality separately

$$G_1 = \bar{A}^+_{MN}Q - \bar{P}A^+_{MN} = \bar{A}^+_{MN}AA^+_{MN} - \bar{A}^+_{MN}\bar{A}A^+_{MN} = \bar{A}^+_{MN}(A - \bar{A})A^+_{MN} = \bar{A}^+_{MN}\Delta AA^+_{MN}. \tag{32}$$

To estimate the second term, we use properties (1)

$$\begin{aligned} A^+_{MN} &= (A^+_{MN}A)A^+_{MN} = N^{-1}(NA^+_{MN}A)^TA^+_{MN} = \\ &= N^{-1}A^TA^{+T}_{MN}NA^+_{MN} = N^{-1}\bar{A}^TA^{+T}_{MN}NA^+_{MN} - N^{-1}\Delta A^TA^{+T}_{MN}NA^+_{MN}. \end{aligned} \tag{33}$$

Substituting (33) into the second term of (31) gives

$$G_2 = (I - \bar{P})A^+_{MN} = (I - \bar{P})\left(N^{-1}\bar{A}^TA^{+T}_{MN}NA^+_{MN} - N^{-1}\Delta A^TA^{+T}_{MN}NA^+_{MN}\right). \tag{34}$$

Since,

$$\begin{aligned} (I - \bar{P})N^{-1}\bar{A}^TA^{+T}_{MN}NA^+_{MN} &= N^{-1}\bar{A}^TA^{+T}_{MN}NA^+_{MN} - \bar{A}^+_{MN}\bar{A}N^{-1}\bar{A}^TA^{+T}_{MN}NA^+_{MN} = \\ &= N^{-1}\bar{A}^TA^{+T}_{MN}NA^+_{MN} - N^{-1}\bar{A}^TA^{+T}_{MN}NA^+_{MN} = 0 \end{aligned} \tag{35}$$

we obtain

$$G_2 = (I - \bar{P})A^+_{MN} = -(I - \bar{P})N^{-1}\Delta A^TA^{+T}_{MN}NA^+_{MN} = -(I - \bar{P})\Delta A^{\#}\left(A^+_{MN}\right)^{\#}A^+_{MN}. \tag{36}$$

Finally,

$$G = \bar{A}^+_{MN} - A^+_{MN} = -\bar{A}^+_{MN}\Delta AA^+_{MN} - (I - \bar{P})\Delta A^{\#}\left(A^+_{MN}\right)^{\#}A^+_{MN} + \bar{A}^+_{MN}(I - Q). \tag{37}$$

Lemma 6 (see in [41]). If $\operatorname{rank}(\bar{A}) = \operatorname{rank}(A) = k$, then

$$\|\bar{Q}(I - Q)\|_{MM} = \|Q(I - \bar{Q})\|_{MM}, \tag{38}$$

where Q and $\bar{Q}$ are defined in (3).

Lemma 7. Let $A, \Delta A \in R^{m\times n}$, $\operatorname{rank}(\bar{A}) = \operatorname{rank}(A)$ and $\|\Delta A\|_{MN}\|A^+_{MN}\|_{NM} < 1$.

Then the relative estimate of the hereditary error of the weighted pseudoinverse matrix has the form

$$\frac{\left\|\bar{A}^+_{MN} - A^+_{MN}\right\|_{NM}}{\|A^+_{MN}\|_{NM}} \le C\frac{\varepsilon_A h}{1-\varepsilon_A h}, \tag{39}$$

where $h = h(A) = \|A\|_{MN}\|A^+_{MN}\|_{NM}$ and the estimate of the absolute error

$$\left\|\bar{A}^+_{MN} - A^+_{MN}\right\|_{NM} \le C\frac{(\varepsilon_A h)^2}{1-\varepsilon_A h}, \tag{40}$$

moreover.

if A is not a full rank matrix, then $C = 3$,

if $m > n = k$ or $n > m = k$, then $C = 2$,

if $m = n = k$, then $C = 1$.

Proof. To obtain estimates, we use the results of Lemma 5:

$$\bar{A}^+_{MN} - A^+_{MN} = -\bar{A}^+_{MN}\Delta A A^+_{MN} - (I-\bar{P})\Delta A^{\#}(A^+_{MN})^{\#}A^+_{MN} + \bar{A}^+_{MN}(I-Q). \tag{41}$$

Passing to the weighted norms, we obtain.

$$\left\|\bar{A}^+_{MN} - A^+_{MN}\right\|_{NM} \le \left\|\bar{A}^+_{MN}\Delta A A^+_{MN}\right\|_{NM} + \|\Delta A^{\#}(A^+_{MN})^{\#}A^+_{MN}\|_{NM} + \left\|\bar{A}^+_{MN}\bar{Q}(I-Q)\right\|_{NM}. \tag{42}$$

Using the results of Lemma 6, we can estimate the last summand

$$\begin{aligned}\left\|\bar{A}^+_{MN}\bar{Q}(I-Q)\right\|_{N} &= \left\|\bar{A}^+_{MN}\bar{A}\bar{A}^+_{MN}(I-Q)\right\|_{N} \le \left\|\bar{A}^+_{MN}\right\|_{NM}\|\bar{Q}(I-Q)\|_{MM} \\ &= \left\|\bar{A}^+_{MN}\right\|_{NM}\|Q(I-\bar{Q})\|_{MM}.\end{aligned} \tag{43}$$

According to (38) and (43), we can rewrite (42) in the form

$$\begin{aligned}\left\|\bar{A}^+_{MN} - A^+_{MN}\right\|_{NM} &\le \left\|\bar{A}^+_{MN}\Delta A A^+_{MN}\right\|_{NM} + \|\Delta A(A^+_{MN})A^+_{MN}\|_{NM} + \|A^+_{MN}Q(I-\bar{Q})\|_{NM} \le \\ &\le \left\|\bar{A}^+_{MN}\right\|_{NM}\|\Delta A\|_{MN}\|A^+_{MN}\|_{NM} + \|\Delta A\|_{MN}\|A^+_{MN}\|^2_{NM} + \left\|\bar{A}^+_{MN}\right\|_{NM}\|A^+_{MN}\|_{NM}\|\Delta A\|_{MN}.\end{aligned} \tag{44}$$

Using the results of Lemma 4, we obtain an estimate for the absolute error of the weighted pseudoinverse matrix A.

$$\begin{aligned}\left\|\bar{A}^+_{MN} - A^+_{MN}\right\|_{NM} &\le \frac{\|A^+_{MN}\|_{NM}}{1-\|\Delta A\|_{MN}\|A^+_{MN}\|_{NM}}\Big(\|\Delta A\|_{MN}\|A^+_{MN}\|_{NM} + \|\Delta A\|_{MN}\|A^+_{MN}\|_{NM} + \\ &+\|A^+_{MN}\|_{NM}\|\Delta A\|_{MN}\Big) = \frac{h\varepsilon_A}{1-h\varepsilon_A}(h\varepsilon_A + h\varepsilon_A + h\varepsilon_A) = C\frac{(h\varepsilon_A)^2}{1-h\varepsilon_A}, C = 1,2,3.\end{aligned} \tag{45}$$

To estimate the relative error, we have

$$\frac{\left\|\bar{A}^{+}_{MN} - A^{+}_{MN}\right\|_{NM}}{\left\|A^{+}_{MN}\right\|_{NM}} \leq 3\|\Delta A\|_{MN}\left\|A^{+}_{MN}\right\|_{NM}\frac{1}{1-\|\Delta A\|_{MN}\left\|A^{+}_{MN}\right\|_{NM}} = C\frac{h\varepsilon_A}{1-h\varepsilon_A}, C = 1, 2, 3. \tag{46}$$

Let us estimate the error of the weighted minimum-norm least squares solution. Let us introduce the following notation:

$$\alpha = \frac{\|\Delta b\|_M}{\|A\|_{MN}\|x\|_N}, \beta = \frac{\|r\|_M}{\|x\|_N\|A\|_{MN}}, \gamma = \frac{\|\bar{r}\|_M}{\|A\|_{MN}\|x\|_N}\alpha_l = \frac{\|\Delta b\|_M}{\|A\|_{MN}\|x_l\|_N},$$
$$\beta_l = \frac{\|r\|_M}{\|x_l\|_N\|A\|_{MN}}, \gamma_l = \frac{\|\bar{r}_l\|_M}{\|A\|_{MN}\|x_l\|_N}, \gamma_k = \frac{\|\bar{r}_k\|_M}{\|A\|_{MN}\|x\|_N}. \tag{47}$$

Consider the following three cases.

Case 1. *The rank of the original matrix A remains the same under its perturbation, i.e.,* $\mathrm{rank}(A) = \mathrm{rank}(\bar{A})$.

Theorem 7. Assume that $\|\Delta A\|_{MN}\left\|A^{+}_{MN}\right\|_{NM} < 1$, $\mathrm{rank}(\bar{A}) = \mathrm{rank}(A)$. Then

$$\frac{\|x-\bar{x}\|_N}{\|x\|_N} \leq \frac{h}{1-h\varepsilon_A}(2\varepsilon_A + \alpha + h\varepsilon_A\beta), \tag{48}$$

where $h(A) = \|A\|_{MN}\left\|A^{+}_{MN}\right\|_{NM}$ is the weighted condition number of A, the symbols $\|\|_{MN}$ and $\|\|_{NM}$ denote the weighted matrix norms defined by Eq. (4)–(6), and A^{+}_{MN} is the weighted Moore–Penrose pseudoinverse.

Proof. The error estimate follows from the relation:

$$x - \bar{x} = \left(A^{+}_{MN} - \bar{A}^{+}_{MN}\right)b + \bar{A}^{+}_{MN}\left(b - \bar{b}\right). \tag{49}$$

For the error of the matrix pseudoinverse, we use the representation

$$\bar{A}^{+}_{MN} - A^{+}_{MN} = -\bar{A}^{+}_{MN}\Delta AA^{+}_{MN} - (I-\bar{P})N^{-1}\Delta A^T A^{+T}_{MN}NA^{+}_{MN} + \bar{A}^{+}_{MN}(I-Q). \tag{50}$$

Then,

$$\begin{aligned} x - \bar{x} &= \left[\bar{A}^{+}_{MN}\Delta AA^{+}_{MN} + (I-\bar{P})N^{-1}\Delta A^T A^{+T}_{MN}NA^{+}_{MN} - \bar{A}^{+}_{MN}(I-Q)\right]b + \bar{A}^{+}_{MN}\left(b-\bar{b}\right) = \\ &= \bar{A}^{+}_{MN}\Delta AA^{+}_{MN}b + (I-\bar{P})N^{-1}\Delta A^T A^{+T}_{MN}NA^{+}_{MN}b - \bar{A}^{+}_{MN}(I-Q)b + \bar{A}^{+}_{MN}\left(b-\bar{b}\right) \\ &= \bar{A}^{+}_{MN}\Delta Ax + (I-\bar{P})N^{-1}\Delta A^T A^{+T}_{MN}Nx - \bar{A}^{+}_{MN}(I-Q)b \quad + \bar{A}^{+}_{MN}\left(b-\bar{b}\right) \end{aligned} \tag{51}$$

Thus,

$$\bar{x} - x = \bar{A}^{+}_{MN}\Delta Ax + (I-\bar{P})N^{-1}\Delta A^T A^{+T}_{MN}Nx - \bar{A}^{+}_{MN}(I-Q)b + \bar{A}^{+}_{MN}\left(b-\bar{b}\right). \tag{52}$$

Passing to the weighted norms yields

$$\begin{aligned}\|\bar{x}-x\|_N &= \left\|\bar{A}_{MN}\Delta Ax + (I-\bar{P})N^{-1}\Delta A^T A_{MN}^{+T}Nx - \bar{A}_{MN}^{+}(I-Q)b + \bar{A}_{MN}^{+}(b-\bar{b})\right\|_N \leq \\ &\leq \|\bar{A}_{MN}\Delta Ax\|_N + \|(I-\bar{P})N^{-1}\Delta A^T A_{MN}^{+T}Nx\|_N + \|\bar{A}_{MN}^{+}(I-Q)b + \bar{A}_{MN}^{+}(b-\bar{b})\|_N.\end{aligned} \tag{53}$$

By taking into account the relations

$$(I-Q)b = (I-Q)r = r, r = b - Ax, x = A_{MN}^{+}b \tag{54}$$

and applying Lemma 6, the weighted norm of each term in (27) can be rearranged as follows

a.
$$\begin{aligned}&\left\|\bar{A}_{MN}^{+}\Delta Ax\right\|_N = \left\|N^{1/2}\bar{A}_{MN}^{+}M^{-1/2}M^{1/2}\Delta AN^{-1/2}N^{1/2}x\right\| \leq \\ &\leq \left\|N^{1/2}\bar{A}_{MN}^{+}M^{-1/2}\right\|\|M^{1/2}\Delta AN^{-1/2}\|\|N^{1/2}x\| = \left\|\bar{A}_{MN}^{+}\right\|_{NM}\|\Delta A\|_{MN}\|x\|_N.\end{aligned} \tag{55}$$

b.
$$\begin{aligned}&\|(I-\bar{P})N^{-1}\Delta A^T A_{MN}^{+T}Nx\|_N = \|N^{1/2}(I-\bar{P})N^{-1/2}N^{-1/2}\Delta A^T M^{1/2}M^{-1/2}A_{MN}^{+T}N^{1/2}N^{1/2}x\| \\ &\leq \|N^{1/2}(I-\bar{P})N^{-1/2}\|\|M^{1/2}\Delta AN^{-1/2}\|\|N^{1/2}A_{MN}^{+}M^{-1/2}\|\|N^{1/2}x\| \\ &= \|(I-\bar{P})\|_{NN}\|\Delta A\|_{MN}\|A_{MN}^{+}\|_{NM}\|x\|_N\end{aligned} \tag{56}$$

c. Using Lemma 6, and (28) we can write

$$\begin{aligned}\left\|\bar{A}_{MN}^{+}\bar{Q}(I-Q)b\right\|_N &= \left\|\bar{A}_{MN}^{+}\bar{A}\bar{A}_{MN}^{+}(I-Q)r\right\|_N \leq \left\|\bar{A}_{MN}^{+}\right\|_{NM}\|\bar{Q}(I-Q)\|_{MM}\|r\|_M \\ &= \left\|\bar{A}_{MN}^{+}\right\|_{NM}\|Q(I-\bar{Q})\|_{MM}\|r\|_M\end{aligned} \tag{57}$$

where

$$\begin{aligned}\|Q(I-\bar{Q})\|_{MM} &= \|AA_{MN}^{+}(I-\bar{Q})\|_{MM} = \|M^{1/2}AA_{MN}^{+}(I-\bar{Q})M^{-1/2}\| = \\ &= \left\|M^{-1/2}(MAA_{MN}^{+})^T(I-\bar{Q})M^{-1/2}\right\| = \\ &= \left\|M^{-1/2}(A_{MN}^{+})^T\left(A^T-\bar{A}^T\right)M^{1/2}M^{1/2}(I-\bar{Q})M^{-1/2}\right\| = \\ &= \left\|M^{-1/2}(A_{MN}^{+})^T\Delta A^T M(I-\bar{Q})M^{-1/2}\right\| \leq \left\|M^{-1/2}(A_{MN}^{+})^T\Delta A^T M^{1/2}\right\| = \\ &= \left\|M^{1/2}\Delta AN^{-1/2}N^{1/2}A_{MN}^{+}M^{-1/2}\right\| \leq \|M^{1/2}\Delta AN^{-1/2}\|\|N^{1/2}A_{MN}^{+}M^{-1/2}\| \leq \\ &\leq \|\Delta A\|_{MN}\|A_{MN}^{+}\|_{NM}.\end{aligned} \tag{58}$$

Substituting this result into (31) gives the inequality

$$\left\|\bar{A}_{MN}^{+}\bar{Q}(I-Q)b\right\|_N \leq \left\|\bar{A}_{MN}^{+}\right\|_{NM}\|\Delta A\|_{MN}\|A_{MN}^{+}\|_{NM}\|r\|_M. \tag{59}$$

d. $\left\|\bar{A}^{+}_{MN}(b-\bar{b})\right\|_N = \left\|N^{1/2}\bar{A}^{+}_{MN}M^{-1/2}M^{1/2}(b-\bar{b})\right\| \leq$
$\left\|N^{1/2}\bar{A}^{+}_{MN}M^{-1/2}\right\| \|M^{1/2}(b-\bar{b})\| = \left\|\bar{A}^{+}_{MN}\right\|_{NM} \|(b-\bar{b})\|_M$ (60)

Taking into account $\|I-\bar{P}\| < 1$, and applying Lemma 4, we obtain the following weighted-norm estimate for the relative error

$$\frac{\|x-\bar{x}\|_N}{\|x\|_N} \leq \frac{\left\|\bar{A}^{+}_{MN}\right\|_{NM}\|\Delta A\|_{MN}\|x\|_N}{\|x\|_N} + \frac{\|\Delta A\|_{MN}\|A^{+}_{MN}\|_{NM}\|x\|_N}{\|x\|_N} +$$
$$+\frac{\left\|\bar{A}^{+}_{MN}\right\|_{NM}\|A^{+}_{MN}\|_{NM}\|\Delta A\|_{MN}\|r\|_M}{\|x\|_N} + \frac{\left\|\bar{A}^{+}_{MN}\right\|_{NM}\|\Delta b\|_M}{\|x\|_N} \leq$$
$$\leq \left\|\bar{A}^{+}_{MN}\right\|_{NM}\|\Delta A\|_{MN} + \|\Delta A\|_{MN}\|A^{+}_{MN}\|_{NM} +$$
$$+\frac{\left\|\bar{A}^{+}_{MN}\right\|_{NM}\|A^{+}_{MN}\|_{NM}\|\Delta A\|_{MN}\|r\|_M}{\|x\|_N} + \frac{\left\|\bar{A}^{+}_{MN}\right\|_{NM}\|\Delta b\|_M}{\|x\|_N} \leq \tag{61}$$
$$\leq \frac{\|A^{+}_{MN}\|_{NM}\|A\|_{MN}}{1-\|\Delta A\|_{MN}\|A^{+}_{MN}\|_{NM}}\left(2\frac{\|\Delta A\|_{MN}}{\|A\|_{MN}} + \frac{\|\Delta b\|_M}{\|A\|_{MN}\|x\|_N} +\right.$$
$$\left. +\|A^{+}_{MN}\|_{NM}\|A\|_{MN}\frac{\|\Delta A\|_{MN}}{\|A\|_{MN}}\frac{\|r\|_M}{\|A\|_{MN}\|x\|_N}\right) \leq$$
$$\leq \frac{h(A)}{1-h(A)\varepsilon_A}\left(2\varepsilon_A + \frac{\|\Delta b\|_M}{\|A\|_{MN}\|x\|_N} + h(A)\varepsilon_A\frac{\|r\|_M}{\|x\|_N\|A\|_{MN}}\right).$$

as required.

Specifically, if $M = I \in R^{m\times m}$ and $N = I \in R^{n\times n}$, then the estimates of the hereditary error of normal pseudosolutions of systems of linear algebraic equations follow from next theorem.

Theorem 8 (see in [32]). Let $\|\Delta A\|\,\|A^{+}\| < 1$, $\text{rank}(\bar{A}) = \text{rank}(A) = k$. Then

$$\frac{\|x-\bar{x}\|}{\|x\|} \leq \frac{h}{1-h\varepsilon_A}\left(2\varepsilon_A + \varepsilon_{b_k} + h\varepsilon_A\frac{\|b-b_k\|}{\|b_k\|}\right), \tag{62}$$

where b_k is the projection of the right-hand side of problem (8) onto the principal left singular subspace of the matrix A [42], i.e., $b_k \in \text{Im A}$, $h = h(A) = \|A\|\|A^{+}\|$ is condition number of A, the symbol $\|\|$, unless otherwise stated, denotes the Euclidean vector norm and the corresponding spectral matrix norm, A^{+} is the Moore–Penrose pseudoinverse.

Case 2. *The rank of the perturbed matrix is larger than that of the original matrix A, i.e.* $\text{rank}(\bar{A}) > rank(A) = k$.

Define the idempotent matrices:

$$P = A^{+}_{MN}A, Q = AA^{+}_{MN}, \quad \bar{P}_k = \bar{A}^{+}_{kMN}\bar{A}, \quad \bar{Q}_k = \bar{A}\bar{A}^{+}_{kMN}, \tag{63}$$

where k is the rank of A.

Theorem 9. Assume that $\|\Delta A\|_{MN}\|A^{+}_{MN}\|_{NM} < \frac{1}{2}$, $\text{rank}(\bar{A}) > \text{rank}(A) = k$. Then

$$\frac{\|x-\bar{x}_k\|_N}{\|x\|_N} \le \frac{h}{1-2h\varepsilon_A}\left(2\varepsilon_A + \alpha + h\varepsilon_A\beta\right). \tag{64}$$

where $h(A) = \|A\|_{MN}\|A^{+}_{MN}\|_{NM}$ is the weighted condition number of A, the symbols $\|\|_{MN}$ and $\|\|_{NM}$ denote the weighted matrix norms defined Eq. (4)–(6), and A^{+}_{MN} is the weighted Moore–Penrose pseudoinverse.

Proof. The desired estimate is derived using the method of [32], which is based on the singular value decomposition of matrices. Specifically, $\bar{A}$ is represented as a weighted singular value decomposition:

$$\bar{A} = \bar{U}\bar{D}\bar{V}^T. \tag{65}$$

Along with (38), we consider the decomposition

$$\bar{A}_k = \bar{U}\bar{D}_k\bar{V}^T, \tag{66}$$

where $\bar{D}_k$ is a rectangular matrix whose first k diagonal elements are nonzero and equal to the corresponding elements of $\bar{D}$, while all the other elements are zero.

The weighted minimum-norm least squares solution to problem (10) is approximated by the weighted minimum-norm least squares solution $\bar{x}_k$ to the problem

$$\min_{x\in C}\|\bar{x}\|_N, C = \left\{x \middle| \left\|\bar{A}_k\bar{x}-\bar{b}\right\|_M = \min\right\}. \tag{67}$$

The matrix $\bar{A}_k$ is defined by (48) and has the same rank k as the matrix of the unperturbed problem.

Thus, the error estimation of the least-squares solution for matrices with a modified rank is reduced to the case of the same rank. This fact is used to estimate $\|x-\bar{x}_k\|_N/\|x\|_N$. The error of the weighted pseudoinverse matrix then becomes:

$$\begin{aligned} G_k &= [\bar{P}_k + (I-\bar{P}_k)]\left(\bar{A}^{+}_{kMN} - A^{+}_{MN}\right)[Q+(I-Q)] = \bar{P}_k\bar{A}^{+}_{kMN}Q + \bar{P}_k\bar{A}^{+}_{kMN}(I-Q) - \\ &-\bar{P}_kA^{+}_{MN}Q - \bar{P}_kA^{+}_{MN}(I-Q) - (I-\bar{P}_k)\bar{A}^{+}_{kMN}Q + (I-\bar{P}_k)\bar{A}^{+}_{kMN}(I-Q) - \\ &-(I-\bar{P}_k)A^{+}_{MN}Q + (I-\bar{P}_k)A^{+}_{MN}(I-Q) = \left(\bar{A}^{+}_{kMN}Q - \bar{P}_kA^{+}_{MN}\right) - (I-\bar{P}_k)A^{+}_{MN} + \\ &+\bar{A}^{+}_{kMN}(I-Q) = \bar{A}^{+}_{kMN}AA^{+}_{MN} - \bar{A}^{+}_{kMN}\bar{A}A^{+}_{MN} - (I-\bar{P}_k)A^{+}_{MN} + \bar{A}^{+}_{kMN}(I-Q) = \\ &= \bar{A}^{+}_{kMN}(A-\bar{A})A^{+}_{MN} - (I-\bar{P}_k)A^{+}_{MN} + \bar{A}^{+}_{kMN}(I-Q), \end{aligned} \tag{68}$$

Applying Lemma 5 yields

$$G_k = \bar{A}^{+}_{kMN} - A^{+}_{MN} = -\bar{A}^{+}_{kMN}\Delta AA^{+}_{MN} - (I-\bar{P}_k)N^{-1}\Delta A^TA^{+T}_{MN}NA^{+}_{MN} + \bar{A}^{+}_{kMN}(I-Q_k). \tag{69}$$

For the error of the WLS solution, we obtain

$$\bar{x}_k - x = \bar{A}^{+}_{kMN}\Delta Ax + (I-\bar{P}_k)N^{-1}\Delta A^TA^{+T}_{MN}Nx - \bar{A}^{+}_{kMN}(I-Q_k)b + \bar{A}^{+}_{kMN}(b-\bar{b}). \tag{70}$$

Passing to the weighted norms and applying Lemma 4 gives

$$\begin{aligned}
\frac{\|x-\bar{x}_k\|_N}{\|x\|_N} &\leq \frac{\left\|\bar{A}_{kMN}^{+}\right\|_{NM}\|\Delta A\|_{MN}\|x\|_N}{\|x\|_N}+\frac{\|\Delta A\|_{MN}\|A_{MN}^{+}\|_{NM}\|x\|_N}{\|x\|_N}+\\
&+\frac{\left\|\bar{A}_{kMN}^{+}\right\|_{NM}\|A_{MN}^{+}\|_{NM}\|\Delta A\|_{MN}\|r\|_M}{\|x\|_N}+\frac{\left\|\bar{A}_{kMN}^{+}\right\|_{NM}\|\Delta b\|_M}{\|x\|_N}\leq\\
&\leq \left\|\bar{A}_{kMN}^{+}\right\|_{NM}\|\Delta A\|_{MN}+\|\Delta A\|_{MN}\|A_{MN}^{+}\|_{NM}+\\
&+\frac{\left\|\bar{A}_{kMN}^{+}\right\|_{NM}\|A_{MN}^{+}\|_{NM}\|\Delta A\|_{MN}\|r\|_M}{\|x\|_N}+\frac{\left\|\bar{A}_{kMN}^{+}\right\|_{NM}\|\Delta b\|_M}{\|x\|_N}\leq\\
&\leq \frac{\|A_{MN}^{+}\|_{NM}\|A\|_{MN}}{1-\|\Delta A_k\|_{MN}\|A_{MN}^{+}\|_{NM}}\left(2\frac{\|\Delta A\|_{MN}}{\|A\|_{MN}}+\frac{\|\Delta b\|_M}{\|A\|_{MN}\|x\|_N}+\right.\\
&\left.+\|A_{MN}^{+}\|_{NM}\|A\|_{MN}\frac{\|\Delta A\|_{MN}}{\|A\|_{MN}}\frac{\|r\|_M}{\|A\|_{MN}\|x\|_N}\right)
\end{aligned} \quad (71)$$

Let estimate $\Delta A_k = A - \bar{A}_k$:

$$\begin{aligned}
\|\Delta A_k\|_{MN} &= \left\|\bar{A}_k - A\right\|_{MN} = \left\|\bar{A}_k - \bar{A} + \Delta A\right\|_{MN} \leq \left\|\bar{A}_k - \bar{A}\right\|_{MN} + \|\Delta A\|_{MN} = \\
&= \left\|\bar{U}\begin{pmatrix} 0 & 0 \\ 0 & D_{k+1}\end{pmatrix}\bar{V}^T\right\|_{MN} + \|\Delta A\|_{MN} \leq 2\|\Delta A\|_{MN}.
\end{aligned} \quad (72)$$

Moreover, the theorem condition $\|\Delta A\|_{MN}\|A_{MN}^{+}\|_{NM} < \frac{1}{2}$ leads to $\|\Delta A_k\|_{MN}\|A_{MN}^{+}\|_{NM} < 1$, which is necessary for expression (51) to be well defined. In view of this, (51) yields estimate (33) for the error of the minimum-norm weighted least squares solution.

Specifically, if $M = I \in R^{m\times m}$ and $N = I \in R^{n\times n}$, then the estimates of the hereditary error of normal pseudosolutions of systems of linear algebraic equations for case $\operatorname{rank}(\bar{A}) > \operatorname{rank}(A) = k$ follows from next theorem.

Theorem 10 (see in [32]). Let $\|\Delta A\|\|A^{+}\| < \frac{1}{2}$, $\operatorname{rank}(\bar{A}) > \operatorname{rank}(A) = k$. Then

$$\frac{\|x-\bar{x}_k\|}{\|x\|} \leq \frac{h}{1-2h\varepsilon_A}\left(2\varepsilon_A + \varepsilon_{b_k} + h\varepsilon_A\frac{\|b-b_k\|}{\|b_k\|}\right). \quad (73)$$

Case 3. *The rank of the original matrix is larger than that of the perturbed matrix, i.e.,* $\operatorname{rank}(A) > \operatorname{rank}(\bar{A}) = l$.

By analogy with (33), we define the idempotent matrices:

$$P_l = A_{lMN}^{+}A,\ Q_l = AA_{lMN}^{+},\quad \bar{P} = \bar{A}_{MN}^{+}\bar{A},\quad \bar{Q} = \bar{A}\bar{A}_{MN}^{+}, \quad (74)$$

Theorem 11. Assume that $\mathrm{rank}(A) > \mathrm{rank}(\bar{A}) = l$, $\frac{\|\Delta A\|_{MN}}{\mu_l} < \frac{1}{2}$. Then,

$$\frac{\|x_l - \bar{x}\|_N}{\|x_l\|_N} \le \frac{\mu_1/\mu_l}{1 - 2\|\Delta A\|_{MN}/\mu_l}\left(2\varepsilon_A + \alpha_l + \frac{\mu_1}{\mu_l}\varepsilon_A\beta_l\right), \tag{75}$$

where μ_i are the weighted singular values of A.
Proof. Along with (9), we consider the problem

$$\min_{x\in C} \|x_l\|_N, C = \{x \| \|A_l x - b\|_M = \min\} \tag{76}$$

with the matrix $A_l = UD_lV^T$ of rank l.

Similarly, writing (27) for problems (10) and (54), whose matrix ranks coincide, we obtain

$$G_l = \bar{A}^+_{MN} - A^+_{lMN} = -\bar{A}^+_{MN}\Delta A A^+_{lMN} - (I - \bar{P})N^{-1}\Delta A^T A^{+T}_{lMN} N A^+_{lMN} + \bar{A}^+_{MN}(I - Q_l), \tag{77}$$

$$\bar{x} - x_l = \bar{A}^+_{MN}\Delta A x + (I - \bar{P})N^{-1}\Delta A^T A^{+T}_{lMN} N x - \bar{A}^+_{MN}(I - Q_l)b + \bar{A}^+_{MN}(b - \bar{b}). \tag{78}$$

Applying Lemma 4 and passing to the weighted norms yields the estimate

$$\begin{aligned}\frac{\|x_l - \bar{x}\|_N}{\|x\|_N} &\le \left\|\bar{A}^+_{MN}\right\|_{NM}\|\Delta A\|_{MN} + \|\Delta A\|_{MN}\|A^+_{lMN}\|_{NM} + \\ &+ \frac{\left\|\bar{A}^+_{MN}\right\|_{NM}\|A^+_{lMN}\|_{NM}\|\Delta A\|_{MN}\|r\|_M}{\|x\|_N} + \frac{\left\|\bar{A}^+_{MN}\right\|_{NM}\|\Delta b\|_M}{\|x\|_N} \le \\ &\le \frac{\|A^+_{lMN}\|_{NM}\|A\|_{MN}}{1 - \|\Delta A_l\|_{MN}\|A^+_{lMN}\|_{NM}}\left(2\frac{\|\Delta A\|_{MN}}{\|A\|_{MN}} + \frac{\|\Delta b\|_M}{\|A\|_{MN}\|x\|_N} + \right. \\ &\left. + \|A^+_{lMN}\|_{NM}\|A\|_{MN}\frac{\|\Delta A\|_{MN}}{\|A\|_{MN}}\frac{\|r\|_M}{\|A\|_{MN}\|x\|_N}\right),\end{aligned} \tag{79}$$

which implies (52). This completes the proof of Theorem 11.

For approximately given initial data, the rank of the original matrix should be specified as the numerical rank of the matrix (see in [28]).

Specifically, if $M = I \in R^{m\times m}$ and $N = I \in R^{n\times n}$, then the estimates of the hereditary error of normal pseudosolutions of systems of linear algebraic equations for case $\mathrm{rank}(A) > \mathrm{rank}(\bar{A}) = l$ follows from next theorem.

Theorem 12 (see in [32]). Let $\mathrm{rank}(A) > \mathrm{rank}(\bar{A}) = l$, $\frac{\|\Delta A\|}{\mu_l} < \frac{1}{2}$. Then

$$\frac{\|x_l - \bar{x}\|}{\|x_l\|} \le \frac{\mu_1/\mu_l}{1 - 2\|\Delta A\|/\mu_l}\left(2\varepsilon_A + \varepsilon_{b_l} + \varepsilon_A\frac{\mu_1}{\mu_l}\frac{\|b - b_l\|}{\|b_l\|}\right), \tag{80}$$

where x_l is the projection of the normal pseudosolution of problem (8) onto the right principal singular subspace of the matrix A of dimension l, b_l is projection of the right-hand side b onto the principal left singular subspace of dimension l of the matrix A, μ_i is singular values of the matrix A.

3.2 Estimates of the hereditary error of a weighted normal pseudosolution for full rank matrices

For matrices of full rank, it is essential that their rank does not change due to the perturbation of the elements if the condition $\|\Delta A\|_{MN}\|A^+_{MN}\|_{NM} < 1$ is met.

In addition, in what follows we will use the following property of matrices of full rank [28]

$$A^+_{MN} = \left(A^T MA\right)^{-1} A^T M \text{ for } m \geq n \text{ and } A^+_{MN} = N^{-1}A^T\left(AN^{-1}A^T\right)^{-1} \text{ for } n \geq m. \quad (81)$$

If $m \geq n$, then problem (9) is reduced to a problem of the form

$$\min_{x \in R^n} \|Ax - b\|_M. \quad (82)$$

For such a problem, the following theorem is true.

Theorem 13. Let $\|\Delta A\|_{MI}\|A^+_{MN}\|_{IM} < 1$, $m > n = k$. Then

$$\frac{\|x - \bar{x}\|}{\|x\|} \leq \frac{h}{1 - h\varepsilon_A}\left(\varepsilon_A + \frac{\|\Delta b\|_M}{\|A\|_{MI}\|x\|} + h\varepsilon_A \frac{\|r\|_M}{\|x\|\|A\|_{MI}}\right), \quad (83)$$

where $h = \|A\|_{MI}\|A^+_{MN}\|_{IM}$.

Proof. To prove Theorem 13, as before, we will use relation (49). By (81) $\bar{P} = \bar{A}^+_{MN}\bar{A} = I$, so that from (50) we have the equality

$$\bar{A}^+_{MN} - A^+_{MN} = -\bar{A}^+_{MN}\Delta A A^+_{MN} + \bar{A}^+_{MN}(I - Q), \quad (84)$$

using which we obtain (83).

If $n \geq m$, then problem (9) is reduced to a problem of the form

$$\min_{x \in C} \|x\|_N, C = \{x | Ax = b\} \quad (85)$$

and the following theorem holds for it.

Theorem 14. Let $\|\Delta A\|_{IN}\|A^+_{MN}\|_{NI} < 1$, $n > m = k$. Then

$$\frac{\|x - \bar{x}\|_N}{\|x\|_N} \leq \frac{h}{1 - h\varepsilon_A}\left(2\varepsilon_A + \frac{\|\Delta b\|}{\|A\|_{IN}\|x\|_N}\right), \quad (86)$$

where $h = \|A\|_{IN}\|A^+_{MN}\|_{NI}$.

Proof. Since in this case $Q = AA^+_{MN} = I$, then the expression for $\bar{A}^+_{MN} - A^+_{MN}$ by (81) takes the form

$$\bar{A}^+_{MN} - A^+_{MN} = -\bar{A}^+_{MN}\Delta A A^+_{MN} - (I - \bar{P})N^{-1}\Delta A^T A^{+T}_{MN} N A^+_{MN}. \quad (87)$$

Further calculations are similar to the previous ones. As a result, we come to estimate (86).

Remark 3. The relationship between the condition number of the problem with exact initial data $h(A)$ and the condition number of the matrix of the system with approximately given initial data $h(\bar{A})$ is established by the estimates

$$\sigma_k - \|\Delta A\|_{MN} \le \bar{\sigma}_k \le \sigma_k + \|\Delta A\|_{MN}, \sigma_1 - \|\Delta A\|_{MN} \le \bar{\sigma}_1 \le \sigma_1 + \|\Delta A\|_{MN},$$
$$\frac{\sigma_1 - \|\Delta A\|_{MN}}{\sigma_k + \|\Delta A\|_{MN}} \le \frac{\bar{\sigma}_1}{\bar{\sigma}_k} \le \frac{\sigma_1 + \|\Delta A\|_{MN}}{\sigma_k - \|\Delta A\|_{MN}}, \frac{1-\varepsilon_A}{1+\varepsilon_A h} \le \frac{h(\mathrm{A})}{h(\bar{\mathrm{A}})} \le \frac{1+\varepsilon_A}{1-\varepsilon_A h}, \tag{88}$$

which are easy to obtain for the weighted matrix norm based on the perturbation theory for symmetric matrices.

Lemma 7. Let $A, \Delta A \in R^{m\times n}$, $\operatorname{rank}(\bar{A}) = \operatorname{rank}(A)$ and $\|\Delta A\|_{MN}\|A^+_{MN}\|_{NM} < 1$. Then the estimate of the relative error of the condition number of the matrix A has the form

$$\left|\frac{\bar{h}-h}{h}\right| \le \varepsilon_A \frac{1+h}{1-\varepsilon_A h} \tag{89}$$

where $h = h(A) = \|A\|_{MN}\|A^+_{MN}\|_{NM}$ is weighted condition number of matrix A, $\bar{h} = \bar{h}(A) = \|\bar{A}\|_{MN}\left\|\bar{A}^+_{MN}\right\|_{NM}$ is weighted condition number of the perturbed matrix $\bar{A} = A + \Delta A$.

Proof of Lemma 7 is easy to obtain using the inequality (25).

Theorem 15. Let $\|\Delta A\|_{MN}\left\|\bar{A}^+_{MN}\right\|_{NM} < 1$, $\|\Delta A\|_{MN} \le \varepsilon_{\bar{A}}\|\bar{A}\|_{MN}$, $\operatorname{rank}(\bar{A}) = \operatorname{rank}(A)$. Then,

$$\frac{\|\bar{x}-x\|_N}{\|\bar{x}\|_N} \le \frac{h(\bar{A})}{1-h(\bar{A})\varepsilon_{\bar{A}}}\left(2\varepsilon_{\bar{A}} + \frac{\|\Delta b\|_M}{\|\bar{A}\|_{MN}\|\bar{x}\|_N} + h(\bar{A})\varepsilon_{\bar{A}}\frac{\|\bar{r}\|_M}{\|\bar{x}\|_N\|\bar{A}\|_{MN}}\right), \tag{90}$$

where $h(\bar{A}) = \|\bar{A}\|_{MN}\left\|\bar{A}^+_{MN}\right\|_{NM}$ is weighted matrix condition number $\bar{A}$, the symbols $\|\|_{MN}$ and $\|\|_{NM}$ denote the weighted matrix norms defined by Eq. (4)–(6) and A^+_{MN} is the weighted Moore–Penrose pseudoinverse.

Thus, estimates of the hereditary error, the right-hand side of which is determined by approximate data, can be obtained without inequalities (88). Estimates similar to (90) can be obtained for all the previously considered cases.

Remark 4. Under the conditions of Theorem 15, using the inequality

$$\frac{\|x-\bar{x}\|_N}{\|x\|_N} \le \frac{\|x-\bar{x}\|_N}{\|\bar{x}\|_N}\left(1 + \frac{\|x-\bar{x}\|_N}{\|x\|_N}\right) \tag{91}$$

and inequality (90) we arrive at the estimate in the following theorem.

Theorem 16. Let $\|\Delta A\|_{MN}\left\|\bar{A}^+_{MN}\right\|_{NM} < 1$, $\|\Delta A\|_{MN} \le \varepsilon_{\bar{A}}\|\bar{A}\|_{MN}$, $\operatorname{rank}(\bar{A}) = \operatorname{rank}(A)$. Then

$$\frac{\|\bar{x}-x\|_N}{\|x\|_N} \le \frac{\beta}{1-\beta}, \quad \beta = \frac{h(\bar{A})}{1-h(\bar{A})\varepsilon_{\bar{A}}}\left(2\varepsilon_{\bar{A}} + \frac{\|\Delta b\|_M}{\|\bar{A}\|_{MN}\|\bar{x}\|_N} + h(\bar{A})\varepsilon_{\bar{A}}\frac{\|\bar{r}\|_M}{\|\bar{x}\|_N\|\bar{A}\|_{MN}}\right). \tag{92}$$

Estimates similar to (92) can be obtained for all the previously considered cases.

4. Research and solution of the WLS problem with approximate initial data

4.1 Investigation of the properties of WLS problem with approximate initial data

In the study of the mathematical properties of the weighted least squares problem with approximate initial data associated with computer realization as an approximate model in (10), (11) we will understand exactly the computer model of the problem. We will assume that the error of the initial data ΔA, Δb, in this case, contains in addition to everything, the error that occurs when the matrix coefficients are written to the computer memory or its computing.

Matrix of full rank within the error of initial data, we assume a matrix that cannot change the rank of ΔA change in its elements.

Matrix of full rank within the machine precision, we assume a matrix that cannot change the rank when you change the elements within the machine precision.

Lemma 8. If $\text{rank}(A) = \min(m, n)$, and

$$\|\Delta A\|_{MN}\|A^+_{MN}\|_{NM} < 1, \tag{93}$$

Then $\text{rank}(\bar{A}) = \text{rank}(A)$.

Proof. For proof, let, for example, $\text{rank}(A) = m$. Taking equal$\|\Delta A\|_{MN} = \varepsilon$, in equality (93) can be rewritten as $\frac{\varepsilon}{\mu_m} < 1$, which is equivalent

$$\mu_m - \varepsilon > 0. \tag{94}$$

Let $\bar{\mu}_m$—m-weighted singular value of perturbed matrix $\bar{A} = A + \Delta A$. According to Lemma 3, we can write $\bar{\mu}_m \geq \mu_m - \varepsilon$. Then, taking into account (94), we obtain $\bar{\mu}_m \geq \mu_m - \varepsilon > 0$.

Therefore $\text{rank}(\bar{A}) \geq m$, whence we come to the conclusion that $\text{rank}(\bar{A}) = m$, i.e. $\text{rank}(\bar{A}) = \text{rank}(A)$.

Taking into account the results of Lemma 8, the computer algorithm for studying rank completeness is reduced to checking the two relations

$$\varepsilon_{\bar{A}} h(\bar{A}) < 1, \tag{95}$$

$$1.0 + \frac{1}{h(\bar{A})} \neq 1.0 \tag{96}$$

where $h(\bar{A}) = \|\bar{A}\|_{MN}\left\|\bar{A}^+_{MN}\right\|_{NM}$ is weighted condition number of matrix $\bar{A}$.

The fulfillment of the first condition (95) guarantees that the matrix has a full rank and is within the accuracy of the initial data, and the second (96), which is performed in floating-point arithmetic, means that the matrix has a full rank within the machine precision.

Under these conditions, the solution of the machine problem exists, it is unique and stable. Such a machine problem should be considered as correctly posed within the accuracy of initial data.

Otherwise, the matrix of the perturbed system may be a matrix, not full rank and, therefore, the machine model of the problem (10), (11) should be considered as ill-posed. A key factor in studying the properties of a machine model is the criterion of the correctness of the problem. Thereby, a useful fact is that the condition for studying the machine model of problem (96) includes the value inverse to $h(\bar{A})$. As a result, for large condition numbers of conditionality does not occur an overflow in order. And the disappearance of the order for $1.0/h(\bar{A})$ for large condition numbers is not fatal: the machine result is assumed to be equal to zero, which allows us to make the correct conclusion about the loss of the rank of the matrix of the machine problem.

To analyze the properties of a machine model of problems with matrices of incomplete rank under conditions of approximate initial data, a fundamental role is played definition of the rank of a matrix.

The rank of the matrix in the condition of approximate the initial data (effective rank or δ -rank) is

$$\text{rank}(A, \delta) = \min_{\|A-B\|_{MN} \leq \delta} \text{rank}(B). \tag{97}$$

This means that the δ-rank of the matrix is equal to the minimum rank among all matrices in the neighborhood $\|A - B\|_{MN} \leq \delta$.

From [28] that if $r(\delta)$ is the δ-rank of the matrix, then

$$\mu_1 \geq \ldots \geq \mu_{r(\delta)} > \delta \geq \mu_{r(\delta)+1} \geq \ldots \geq \mu_p, p = \min(m, n). \tag{98}$$

The practical algorithm for finding δ—rank can be defined as follows: find the value of r is equal to the largest value of i, for which the inequality is fulfilled

$$\frac{\delta}{\mu_i} < 1, \mu_i \neq 0, i = 1, 2.. \tag{99}$$

Using the effective rank of a matrix, can always find the number of a stable projection that approximates the solution or projection

To analyze the rank of a matrix of values within the machine precision value δ can be attributed to machine precision, for example, setting it equal macheps$\|B\|$.

4.2 Algorithm for finding a weighted normal pseudosolution of the weighted least squares problem with approximate initial data

The algorithm is based on weighted singular value decomposition of matrices (Lemma 1).

Let $A \in R^{m \times n}$ and $\text{rank}(A) = k$, M- and N-positive-defined matrices of order m and n, respectively.

To solve the ill-posed problems in the formulation (10), (11), the algorithm for obtaining an approximate normal pseudosolution of system (9), depending on the ratio of the ranks of the matrices A and $\bar{A}$ is reduced to the following three cases.

1. If the rank of the matrix has not changed $rank(\bar{A}) = rank(A) = k$, an approximate weighted normal pseudosolution is constructed by the formula

$$\bar{x} = \bar{A}^{+}_{MN}\bar{b}, \tag{100}$$

where $\bar{A}^{+}_{MN}$ is represented as a weighted singular value decomposition (7).

In this case, the weighted normal pseudosolution of system (9) is approximated by the weighted normal pseudosolution of system (10) and, if $\|\Delta A\|_{MN}\|A^+_{MN}\|_{NM} < 1$, then the error of the solution is estimated by formula (48).

If the rank of the matrix is complete and conditions (95), (96) are satisfied, the rank of the matrix does not change, and to estimate the error, one can use formulas (100), (48).

2. Matrix rank increased $\text{rank}(\bar{A}) > \text{rank}(A) = k$. An approximate weighted normal pseudosolution is constructed by the formula

$$\bar{x}_k = \bar{A}^+_{kMN}\bar{b}. \tag{101}$$

Weighted pseudoinverse matrix $\bar{A}^+_{kMN}$ is defined as follows

$$\bar{A}^+_{kMN} = N^{-1}\bar{V}\bar{D}^+_k\bar{U}^T M, \tag{102}$$

where $\bar{D}_k$ is a rectangular matrix, the first k diagonal elements of which are nonzero and coincide with the corresponding elements of the matrix $\bar{D}$ from (7), and all other elements are equal to zero.

In this case, the weighted normal pseudosolution of system (9) is approximated by the projection of the weighted normal pseudosolution of system (10) onto the right principal weighted singular subspace of dimension k of the matrix $\bar{A}$ and, if $\|\Delta A\|_{MN}\|A^+_{MN}\|_{NM} < \frac{1}{2}$, then the error of the solution is estimated by formula (64).

3. If the rank of the matrix has decreased $\text{rank}(A) > \text{rank}(\bar{A}) = l$, an approximation to the projection of a weighted normal pseudosolution of problem (9) is constructed using formula (100). In this case, the projection of the weighted normal pseudosolution of system (9) onto the principal right weighted singular subspace of dimension l of the matrix A is approximated by the weighted normal pseudosolution of system (10) and, if $\frac{\|\Delta A\|_{MN}}{\mu_l} < \frac{1}{2}$, the projection error is estimated by formula (75).

Remark 5. If the rank of the original matrix is unknown, then the δ-rank should be taken as the projection number in (101). In this case, it is guaranteed that a stable approximation is found either to a weighted normal pseudosolution or to a projection, respectively, with error estimates.

If the rank of the original matrix is known, then it is guaranteed to find an approximation to the weighted normal pseudosolution with appropriate estimates.

Remark 6. Because of the zero columns in the matrix D^+, only the largest first n columns of the matrix U can actually contribute to the product (100). Moreover, if some of the weighted singular numbers are equal to zero, then less than n columns of U are needed. If kp is the number of nonzero weighted singular numbers, then U can be reduced to the sizes $m \times kp$, D^+—to the sizes $kp \times kp$, VT—up to size $kp \times n$. Formally, such matrices U and V are not M-orthogonal and N^{-1}-orthogonal, respectively, since they are not square. However, their columns are weighted orthonormal systems of vectors.

5. Analysis of the reliability of computer solutions to the WLS problem with approximate initial data

5.1 Estimates of the total error of a weighted normal pseudosolution for matrices of arbitrary rank

Estimates of the total error take into account both the hereditary error due to the error in the initial data and the computational error due to an approximate method for determining the solution to the problem. In this case, the method of obtaining a solution is not taken into account. The computational error can be a consequence of both an approximate method of obtaining a solution and an error due to inaccuracy in performing arithmetic operations on a computer. The residual vector $\bar{r} = \bar{A}\bar{\bar{x}} - \bar{b}$ takes into account the overall effect of these errors.

Let us obtain estimates for the total error of the weighted normal pseudosolution using the previously introduced notation (47). Let us consider three cases.

Case 1. *The rank of the original matrix A remains the same under its perturbation, i.e.,* $\operatorname{rank}(A) = \operatorname{rank}(\bar{A})$.

Theorem 17. Assume that $\|\Delta A\|_{MN}\|A^{+}_{MN}\|_{NM} < 1$, $\operatorname{rank}(\bar{A}) = \operatorname{rank}(A) = k$ and let $\bar{x} \in \mathbb{R}(\bar{A}^{\#})$. Then

$$\frac{\|\bar{x} - \bar{\bar{x}}\|_N}{\|x\|_N} \leq \frac{h}{1 - h\varepsilon_A}(2\varepsilon_A + \alpha + h\varepsilon_A\beta + \gamma). \tag{103}$$

Proof. For the hereditary error, in this case, estimate (48) holds.

To estimate the computational error $\bar{x} - \bar{\bar{x}}$, we use the relation

$$\bar{A}\ (\bar{x} - \bar{\bar{x}}) = \bar{r} = \bar{b}_k - \bar{A}\bar{\bar{x}}, \tag{104}$$

where $\bar{b}_k$ is projection of the vector $\bar{b}$ on the main left weighted singular subspace of the matrix $\bar{A}$, i.e. $\bar{b}_k \in \mathbb{R}(\bar{A})$.

Considering that $\bar{x} - \bar{\bar{x}} \in \mathbb{R}(\bar{A}^{\#})$ and the fact that $\bar{A}^{+}_{MN}\bar{A}$ is a projector in $\mathbb{R}(\bar{A}^{\#})$, we have

$$\bar{A}^{+}_{MN}\bar{A}\ (\bar{x} - \bar{\bar{x}}) = \bar{x} - \bar{\bar{x}} = \bar{A}^{+}_{MN}\bar{r}. \tag{105}$$

From this, we obtain an estimate of the computational error

$$\frac{\|\bar{x} - \bar{\bar{x}}\|_N}{\|\bar{x}\|_N} \leq \|\bar{A}\|_{MN}\left\|\bar{A}^{+}_{MN}\right\|_{NM}\frac{\|\bar{r}\|_M}{\|\bar{b}_k\|_M}. \tag{106}$$

An estimate of the total error of the normal pseudosolution follows from the relations

$$\frac{\|x - \bar{\bar{x}}\|_N}{\|x\|_N} \leq \frac{\|x - \bar{x}\|_N}{\|x\|_N} + \frac{\|\bar{x} - \bar{\bar{x}}\|_N}{\|x\|_N}, \tag{107}$$

$$\frac{\|\bar{x} - \bar{\bar{x}}\|_N}{\|x\|_N} \leq \|A\|_{MN}\left\|\bar{A}^{+}_{MN}\right\|_{NM}\frac{\|\bar{r}\|_M}{\|b_k\|_M} \tag{108}$$

and estimates (48), (25). The theorem is proved.

Case 2. *The rank of the perturbed matrix is larger than that of the original matrix A, i.e.,* $\mathrm{rank}(\bar{A}) > \mathrm{rank}(A) = k$.

If $\|\Delta A\|_{MN}\|A_{MN}^{+}\|_{NM} < 1$, then from [26], it follows that the rank of the perturbed matrix cannot decrease.

Theorem 18. Assume that $\|\Delta A\|_{MN}\|A_{MN}^{+}\|_{NM} < \frac{1}{2}$, $\mathrm{rank}(\bar{A}) > \mathrm{rank}(A) = k$ and let $\bar{x} \in \mathbb{R}(\bar{A}_k^{\#})$. Then

$$\frac{\|x-\bar{\bar{x}}\|_N}{\|x\|_N} \le \frac{h}{1-h\varepsilon_A}(2\varepsilon_A + \alpha + h\varepsilon_A\beta + \gamma_k). \tag{109}$$

Proof. To estimate the computational error $\|\bar{x}_k - \bar{\bar{x}}\|_N$, we use the fact that $\bar{A}_k\bar{x}_k = \bar{b}_k$. Then for arbitrary vector $\bar{x} \in \mathbb{R}(\bar{A}_k^{\#})$

$$\bar{\mathrm{A}}_k(\bar{x}_k - \bar{\bar{x}}) = \bar{r}_k = \bar{b}_k - \bar{A}_k\bar{\bar{x}}, \bar{A}_k{}^{+}\bar{\mathrm{A}}_k(\bar{x}_k - \bar{\bar{x}}) = \bar{A}_k{}^{+}\bar{r}_k. \tag{110}$$

Considering the fact that $x_k - \bar{\bar{x}} \in \mathbb{R}(\bar{A}_k{}^{\#})$, and operator $\bar{A}_{kMN}^{+}\bar{A}_k$ is the projection operator in $\mathbb{R}(\bar{A}_k{}^{\#})$, we obtain

$$\bar{A}_{kMN}^{+}\bar{A}_k(\bar{x}_k - \bar{\bar{x}}) = \bar{x}_k - \bar{\bar{x}} = \bar{A}_{kMN}^{+}\bar{r}_k, \bar{x}_k - \bar{\bar{x}} = \bar{A}_{kMN}^{+}\bar{r}_k. \tag{111}$$

Hence follows an estimate of the computational error for the projection of the normal pseudosolution

$$\frac{\|\bar{x}_k - \bar{\bar{x}}\|_N}{\|\bar{x}_k\|_N} \le \|\bar{A}_k\|_{MN}\left\|\bar{A}_{kMN}^{+}\right\|_{NM}\frac{\|\bar{r}_k\|_M}{\|\bar{b}_k\|_M}. \tag{112}$$

The estimate of the total error follows from the inequalities

$$\frac{\|x-\bar{\bar{x}}\|_N}{\|x\|_N} \le \frac{\|x-\bar{x}_k\|_N}{\|x\|_N} + \frac{\|\bar{x}_k-\bar{\bar{x}}\|_N}{\|x\|_N}, \frac{\|\bar{x}_k-\bar{\bar{x}}\|_N}{\|x\|_N} \le \frac{\|A\|_{MN}\left\|\bar{A}_{kMN}^{+}\right\|_{NM}\|\bar{r}_k\|_M}{\|b_k\|_M} \tag{113}$$

and estimates (25), (64).

Case 3. *The rank of the original matrix is larger than that of the perturbed matrix, i.e.,* $\mathrm{rank}(A) > \mathrm{rank}(\bar{A}) = l$.

Consider the case when the condition $\|\Delta A\|_{MN}\|A_{MN}^{+}\|_{NM} < 1$ not satisfied and the rank of the perturbed matrix can decrease.

Theorem 19. Assume that $\mathrm{rank}(A) > \mathrm{rank}(\bar{A}) = l$, $\frac{\|\Delta A\|_{MN}}{\mu_l} < \frac{1}{2}$ and let $\bar{x} \in Im(\bar{A}^{\#})$. Then

$$\frac{\|x_l - \bar{\bar{x}}\|_N}{\|x_l\|_N} \le \frac{\mu_1/\mu_l}{1-2\|\Delta A\|_{MN}/\mu_l}\left(2\varepsilon_A + \alpha_l + \frac{\mu_1}{\mu_l}\varepsilon_A\beta_l + \gamma_l\right) \tag{114}$$

Proof. For the proof, along with problem (9), consider the problem

$$\min_{x\in C}\|x_l\|_N, C = \{x \| \|A_l x - b\|_M = \min\} \tag{115}$$

With matrix $A_l = U\Sigma_l V^T$ with rang l.

The estimate of the computational error in this case will be

$$\frac{\|\bar{x}-\bar{\bar{x}}\|_N}{\|\bar{x}\|_N} \le \|\bar{A}\|_{MN}\left\|\bar{A}^+_{MN}\right\|_{NM}\frac{\|\bar{r}\|_M}{\left\|\bar{b}_l\right\|_M}. \tag{116}$$

The estimate of the total error follows from the inequalities

$$\frac{\|x_l-\bar{\bar{x}}\|_N}{\|x_l\|_N} \le \frac{\|x_l-\bar{x}\|_N}{\|x_l\|_N} + \frac{\|\bar{x}-\bar{\bar{x}}\|_N}{\|x_l\|_N}, \ \frac{\|\bar{x}-\bar{\bar{x}}\|_N}{\|x\|_N} \le \frac{\|A\|_{MN}\left\|\bar{A}^+_{lMN}\right\|_{NM}\|\bar{r}_l\|_M}{\|b_l\|_M}, \tag{117}$$

obvious relationships $\|A\|_{MN} = \|A_l\|_{MN}$, $\|A^+_{lMN}\|_{NM} = 1/\mu_l$, estimates of the hereditary error (75) and the inequality $\|\Delta A_l\|_{MN} \le 2\|\Delta A\|_{MN}$.

5.2 Estimates of the total error of the weighted normal pseudosolution for matrices of full rank

In the following Theorems 20 and 21, the weighted pseudoinverse A^+_{MN} is represented in accordance with the properties of the full rank matrix (81).

Theorem 20. Let $\|\Delta A\|_{MI}\|A^+_{MN}\|_{IM} < 1$, $m > n = k$ and $\bar{\bar{x}} \in R(\bar{A}^\#)$. Then

$$\frac{\|x-\bar{\bar{x}}\|}{\|x\|} \le \frac{h}{1-h\varepsilon_A}\left(\varepsilon_A + \frac{\|\Delta b\|_M}{\|A\|_{MI}\|x\|} + h\varepsilon_A\frac{\|r\|_M}{\|A\|_{MI}\|x\|} + \frac{\|\bar{r}_k\|_M}{\|A\|_{MI}\|x\|}\right). \tag{118}$$

Proof. The estimate of the computational error is determined by formula (106), namely

$$\frac{\|\bar{x}-\bar{\bar{x}}\|}{\|\bar{x}\|} \le \|\bar{A}\|_{MN}\left\|\bar{A}^+_{MN}\right\|_{NM}\frac{\|\bar{r}\|_M}{\left\|\bar{b}_k\right\|_M}. \tag{119}$$

The estimate for the total error (118) follows from the inequalities

$$\frac{\|x-\bar{\bar{x}}\|}{\|x\|} \le \frac{\|x-\bar{x}\|}{\|x\|} + \frac{\|\bar{x}-\bar{\bar{x}}\|}{\|x\|}, \ \frac{\|\bar{x}-\bar{\bar{x}}\|}{\|x\|} \le \|A\|_{MI}\left\|\bar{A}^+_{MN}\right\|_{IM}\frac{\|\bar{r}\|_M}{\|b_k\|_M} \tag{120}$$

and estimates for the pseudoinverse matrix (25) and the hereditary error (83).

Theorem 21. Let $\|\Delta A\|_{IN}\|A^+_{MN}\|_{NI} < 1$, $n > m = k$ and $\bar{\bar{x}} \in R(\bar{A}^\#)$. Then

$$\frac{\|x-\bar{\bar{x}}\|_N}{\|x\|_N} \le \frac{h}{1-h\varepsilon_A}\left(2\varepsilon_A + \frac{\|\Delta b\|}{\|A\|_{IN}\|x\|_N} + \frac{\|\bar{r}\|}{\|b_k\|}\right), \tag{121}$$

The proof of Theorem 21 is similar to the proof of the previous theorem, taking into account the estimate for the hereditary error (86).

Remark 7. Here, we did not indicate a method for obtaining an approximate weighted normal pseudosolution $\bar{\bar{x}}$, satisfying the conditions of the theorems. Algorithms for obtaining such approximations are considered, for example, in Section 4.2.

Remark 8. Along with estimates (103), (109), (114), (118), (121), error estimates can be obtained, the right-hand sides of which depend on the input data of systems of linear algebraic equations with approximately given initial data. For example, the following theorem holds.

Theorem 22. Let $\|\Delta A\|_{MN}\left\|\bar{A}^{+}_{MN}\right\|_{NM} < 1$, and $\bar{\bar{x}} \in R(\bar{A}^{\#})$. Then, for the total error of the normal pseudosolution, the following estimate is fulfilled

$$\frac{\|x-\bar{\bar{x}}\|_N}{\|\bar{x}\|_N} \le \frac{h(\bar{A})}{1-h(\bar{A})\varepsilon_{\bar{A}}}\left(2\varepsilon_{\bar{A}} + \frac{\|\Delta b\|_M}{\|\bar{A}\|_{MN}\|\bar{x}\|_N} + h(\bar{A})\varepsilon_{\bar{A}}\frac{\|\bar{r}\|_M}{\|\bar{A}\|_{MN}\|\bar{x}\|_N} + \frac{\|\bar{r}\|_M}{\|\bar{b}_k\|_M}\right). \tag{122}$$

Estimate (122) can be obtained from the inequality

$$\frac{\|x-\bar{\bar{x}}\|_N}{\|\bar{x}\|_N} \le \frac{\|x-\bar{x}\|_N}{\|\bar{x}\|_N} + \frac{\|\bar{x}-\bar{\bar{x}}\|_N}{\|\bar{x}\|_N} \tag{123}$$

and estimates (90), (106).

If the weighted pseudoinverse matrix is known or its weighted singular value decomposition is obtained during the process of solving the problem, then a practical estimate of the computational error can be obtained using (104). When calculating the residual $\bar{r} = \bar{b}_k - \bar{A}\bar{\bar{x}}$, the explicit form of the projection operator onto $R(\bar{A}^{\#})$ is used.

In conclusion, we note that the determining factor for obtaining estimates is the use of a weighted singular value decomposition [37] and the technique of reducing the problem of estimating the error of a pseudosolution to an estimate of the error [32] for problems with matrices of the same rank. Based on the results obtained, an algorithm for finding the effective rank of matrices can be developed, as well as an algorithm for calculating stable projections of a weighted normal pseudosolution.

5.3 Software-algorithmic methods for increasing the accuracy of computer solutions

The numerical methods we have considered for solving systems of linear algebraic equations and WLS problems have one common property. Namely, the actually calculated solution (pseudosolution) is exact in accordance with the inverse analysis of errors [43] for some perturbed problem. These perturbations are very small and are often commensurate with the rounding errors of the input data. If the input data is given with an error (measurements, calculations, etc.), then usually they already contain significantly larger errors than rounding errors. In this case, any attempt to improve the machine solution (pseudosolution) without involving additional information about the exact problem or errors of the input data errors will be untenable.

The situation changes significantly if a mathematical problem with accurate input data is considered. Now the criterion of bad or good conditionality of the computer model of the problem depends on the mathematical properties of the computer model of the problem and the mathematical properties of the processor (length of the computer word), and it becomes possible in principle to achieve any given accuracy of the computer solution. In this case, as follows from estimates (48), (64), (75), (83), (86), it is obviously possible to refine the computer solution by solving a system with increased bit depth, in particular, using the GMP library [44] for implementation of computations with arbitrary bit depth.

To predict the length of the mantissa (machine word) that provides a given accuracy for a solution (joint systems), you can use the following rule of thumb: the number of correct decimal significant digits in a computer solution is $\mu - \alpha$, where μ is the decimal order of the mantissa of a floating-point number ε, α is the decimal order of the condition number. Thus, knowing the conditionality of the matrix of the system and the accuracy of calculations on a computer, it is possible to determine the required bit depth to obtain a reliable solution.

The GMP library is used to work on integers, rational numbers and floating-point numbers. The main feature of the library is the bitness of numbers (precision) is practically unlimited. Therefore, the main field of application is computer algebraic calculations, cryptography, etc. The functions of the GMP library allow not only setting the bit depth at the beginning of the program and performing calculations with this bit depth, but also changing the bit width as needed in the computation process, i.e. execute different fragments of the algorithm with different bit depths.

The library's capabilities were tested in the study of solutions to degenerate and ill-conditioned systems in [45].

6. Conclusions

In the framework of these studies, estimates of the hereditary error of the weighted normal pseudosolution for matrices of arbitrary form and rank are obtained, including when the rank of the perturbed matrix may change. Three cases are considered: the rank of the matrix does not change when the data is disturbed, the rank increases and the rank decreases. In the first case, the weighted normal pseudosolution of the approximate problem is taken as an approximation to the weighted normal pseudosolution, in the other two, the problem is reduced to the case when the ranks of the matrices are the same. Also, the estimates of the error for the weighted pseudoinverse matrix and the weighted condition number of the matrix are obtained, the existence and uniqueness of the weighted normal pseudosolution are investigated and proved. Estimates of the total error of solving the weighted least squares problem with matrices of arbitrary form and rank are established.

The results obtained in the perturbation theory of weighted least squares problem can be a theoretical basis for further research into various aspects of the WLS problem and the development of methods for calculating weighted pseudoinverse matrices and weighted normal pseudosolutions with approximate initial data, in particular, in the design and optimization of building structures, in tomography, in the calibration of viscometers, in statistics. The results of the research can be used in the educational process when reading special courses on this section of the theory of matrices.

Author details

Aleksandr N. Khimich, Elena A. Nikolaevskaya* and Igor A. Baranov
V.M. Glushkov Institute of Cybernetics of NAS of Ukraine, Kyiv, Ukraine

*Address all correspondence to: elena_nea@ukr.net

References

[1] Chipman JS. On least squares with insufficient observation. Journal of the American Statistical Association. 1964; **59**(308):1078-1111

[2] Milne RD. An oblique matrix pseudoinverse. SIAM Journal on Applied Mathematics. 1968;**16**(5):931-944

[3] Ward JF, Boullion TL, Lewis TO. Weighted pseudoinverses with singular weights. SIAM Journal on Applied Mathematics. 1971;**21**(3):480-482

[4] Galba EF, Deineka VS, Sergienko IV. Weighted pseudoinverses and weighted normal pseudosolutions with singular weights. Computational Mathematics and Mathematical Physics. 2009;**49**(8): 1281-1297

[5] Sergienko IV, Galba EF, Deineka VS. Existence and uniqueness of weighted pseudoinverse matrices and weighted normal pseudosolutions with singular weights. Ukrainian Mathematical Journal. 2011;**63**(1):98-124

[6] Varenyuk NA, Galba EF, Sergienko IV, Khimich AN. Weighted pseudoinversion with indefinite weights. Ukrainian Mathematical Journal. 2018;**70**(6):866-889

[7] Galba EF, Varenyuk NA. Representing weighted pseudoinverse matrices with mixed weights in terms of other pseudoinverses. Cybernetics and System Analysis. 2018;**54**(2):185-192

[8] Galba EF, Varenyuk NA. Expansions of weighted pseudoinverses with mixed weights into matrix power series and power products. Cybernetics and System Analysis. 2019;**55**:760-771. DOI: 10.1007/s10559-019-00186-9

[9] Goldman AJ, Zelen M. Weak generalized inverses and minimum variance linear unbiased estimation. Journal of Research of the National Bureau of Standards. 1964;**68B**(4):151-172

[10] Rao CR, Mitra SK. Generalized Inverse of Matrices and its Applications. New York: Wiley; 1971. p. 240

[11] Nashed MZ. Generalized Inverses and Applications. New York: Academic Press; 1976. p. 1068

[12] Ben-Israel A, TNE G. Generalized Inverses: Theory and Applications. New York: Springer-Verlag; 2003. p. 420. DOI: 10.1007/b97366

[13] Khimich AN. Perturbation bounds for the least squares problem. Cybernetics and System Analysis. 1996; **32**(3):434-436

[14] Khimich AN, Nikolaevskaya EA. Reliability analysis of computer solutions of systems of linear algebraic equations with approximate initial data. Cybernetics and System Analysis. 2008; **44**(6):863-874

[15] Nikolaevskaya EA, Khimich AN. Error estimation for a weighted minimum-norm least squares solution with positive definite weights. Computational Mathematics and Mathematical Physics. 2009;**49**(3): 409-417

[16] Wei Y, Wang D. Condition numbers and perturbation of the weighted Moore-Penrose inverse and weighted linear least squares problem. Applied Mathematics and Computation. 2003;**145**(1):45-58

[17] Wei Y. A note on the sensitivity of the solution of the weighted linear least squares problem. Applied Mathematics and Computation. 2003;**145**(2–3):481-485

[18] Molchanov IN, Galba EF. A weighed pseudoinverse for complex matrices.

Ukrainian Mathematical Journal. 1983; **35**(1):46-50

[19] Elden L. A weighted pseudoinverse, generalized singular values and constrained least squares problems. BIT. 1982;**22**:487-502

[20] Wei Y. The weighted Moore–Penrose inverse of modified matrices. Applied Mathematics and Computation. 2001;**122**(1):1-13. DOI: 10.1016/S0096-3003(00)00007-2

[21] Voevodin VV. On the regularization method. Zhurnal Vychislitel'noy Matematiki i Matematicheskoy Fiziki. 1969;**9**:673-675

[22] Tikhonov AN. Regularization of ill-posed problems. Doklady Akademii Nauk SSSR. 1963;**153**:42-52. DOI: 10.1016/S0096-3003(00)00007-2

[23] Ivanov VK, Vasin VV, Tanana VP. Theory of Linear Ill-Posed Problems and Applications. Moscow: Nauka; 1978. 206 p

[24] Morozov VA. Regularization Methods for Unstable Problems. Moscow: Moscow State University; 1987 [in Russian]

[25] Albert AE. Regression and the Moore–Penrose Pseudoinverse. New York: Academic Press; 1972. p. 180

[26] Lawson CL, Hanson RJ. Solving Least Squares Problems. Moscow: Nauka; 1986. p. 232

[27] Kirichenko NF. Analytical representation of perturbations of pseudoinverses. Kibernetika i Sistemnyi Analiz. 1997;**2**:98-107

[28] Golub GH, Van Loan CF. Matrix Computations. Baltimore: Johns Hopkins University Press; 1996. p. 694.

[29] Elden L. Perturbation theory for the least squares problem with linear equality constraints. SIAM Journal on Numerical Analysis. 1980;**17**:338-350

[30] Bjork A. Numerical Methods for Least Squares Problems. Linköping: Linköping University; 1996. p. 407

[31] Voevodin VV. Computational Foundations of Linear Algebra. Moscow: Nauka; 1977

[32] Khimich AN. Estimates of perturbations for least squares solutions. Kibernetika i Sistemnyi Analiz. 1996;**3**: 142-145

[33] Khimich AN, Voitsekhovskii SA, Brusnikin VN. The reliability of solutions to linear mathematical models with approximately given initial data. Mathematical Machines and Systems. 2004;**3**:3-17

[34] Wei Y, Wu H. Expression for the perturbation of the weighted Moore–Penrose inverse. Computers & Mathematcs with Applications. 2000;**39**:13-18

[35] Wei M. Supremum and Stability of Weighted Pseudoinverses and Weighted Least Squares Problems: Analysis and Computations. New York: Huntington; 2001. p. 180

[36] Wang D. Some topics on weighted Moore–Penrose inverse, weighted least squares, and weighted regularized Tikhonov problems. Applied Mathematics and Computation. 2004;**157**:243-267

[37] Van Loan CF. Generalizing the singular value decomposition. SIAM Journal on Numerical Analysis. 1976;**13**: 76-83

[38] Wang G, Wei Y, Qiao S. Generalized Inverses: Theory and Computations. Beijing: Science Press; 2004. p. 390

[39] Albert A. Regression, Pseudoinversion, and Recurrent Estimation. Moscow: Nauka; 1977. p. 305

[40] Khimich AN, Nikolaevskaya EA. Error estimation for weighted least squares solutions. Computational Mathematics. 2006;**3**:36-45

[41] Khimich AN, Nikolaevskaya EA. Perturbation analysis of weighted least squares solutions. Theory of Optimal Solutions. 2007;**6**:12-18

[42] Voevodin VV, Kuznecov YA. Matrices and Calculations. Moscow: Nauka; 1984. p. 318

[43] Wilkinson JH, Reinsch C. Handbook Algorithms in ALGOL. Linear Algebra. Mechanical: Moscow; 1976. p. 389

[44] GNU Multiple Precision Arithmetic Library. Availabe from: www.gmplib.org

[45] Nikolaevskaya EA, Khimich AN, Chistyakova TV. Programming with Multiple Precision. Berlin/Heidelberg, Germany: Springer; 2012

Chapter 5

A Study on Approximation of a Conjugate Function Using Cesàro-Matrix Product Operator

Mohammed Hadish

Abstract

In this chapter, we present a study of the inaccuracy estimation of a function $\tilde{\zeta}$ conjugate of a function ζ (2π-periodic) in weighted Lipschitz space $W(L^p, p \geq 1, \xi(\omega))$, by Cesàro-Matrix ($C^\delta T$) product means of its CFS[1]. This chapter is divided into seven section. The first section contains introduction of our chapter, the second section, we introduce some basic definitions and notations. In the third section lemmas and the fourth section contains our main theorems and proofs. In the fifth section, we introduce corollaries, the sixth section contains particular cases of our results and the last section contains exercise of our chapter.

Keywords: weighted Lipschitz class, error approximation, Cesàro (C^δ) means, Matrix (T) means $C^\delta T$ product means, conjugate Fourier series, generalized Minkowski's inequality

1. Introduction

The studies of estimations of conjugate of functions in different Lipschitz classes and Hölder classes using single summability operators, have been made by the researchers like [1–4] etc. in past few decades. The studies of estimation of error of cojugate of functions in different Lipschitz classes and Hölder classes using different product operator, have been made by the researchers like [5–12] etc. in recent past.

In this problem, we andeavour consider more sophisticated class of function in contemplation of reach at the best estimation of function $\tilde{\zeta}$ conjugate of a function ζ (2π − periodic) by trigonometric polynomial of degree more than λ. It can be paid attention the results procure thus far in the route of present work could not lay out the best approximation of the function also, in this work, we have used Cesàro-Matrix $(C^\delta T)$ of product operators which is developed here in order to work using more generalized operator. It is important to mention here that $C^\delta T$ is the more generalized product operator than the product operators Cesàro-Harmonic $(C^\delta H)$, Cesàro-N-örlund $(C^\delta N_p)$, Cesàro-Riesz $(C^\delta \overline{N}_p)$, Cesàro-generalized Nörlund $(C^\delta N_{pq})$ and

[1] CFS denotes Conjugate Fourier series and we use this abbreviation throughout the paper.

Cesàro-Euler $(C^{\delta}H)$ and furthermore C^1H, C^1N_p, C^1N_{pq}, C^1E^q and C^1E^1 product operators are the special cases of $C^{\delta}T$ for $\delta = 1$.

Therefore, we establish two theorems so obtain best inaccuracy estimation of a function $\tilde{\zeta}$, conjugate to a 2π-periodic function ζ in weighted $W(L^p, \xi(\omega))$ space of its CFS. Here we shall consider the two cases (i) $p > 1$ and (ii) $p = 1$ in order to get the Hölder's inequality satisfied. Our theorems generalizes six previously known results. Thus, the results of [5, 8–12] becomes the special cases of our theorem. Some inportant corollaries are also obtained from our theorems.

Note 1 *The CFS is not necessarily a FS.*[2]

Example 1 *The series*

$$\sum_{\lambda=2}^{\infty}\left(\frac{\sin(\lambda x)}{\log\lambda}\right)$$

conjugate to the FS

$$\sum_{\lambda=2}^{\infty}\left(\frac{\cos(\lambda x)}{\log\lambda}\right)$$

is not a FS (Zygmund [13], p. 186).

From above example, we conclude that, a separate study of conjugate series in the present direction of work is quite essential.

2. Definitions and notations

2.1 Lipschitz class

Let $C_{2\pi}$ is a Banach space of all periodic functions with period 2π and continuous on the interval $0 \le x \le 2\pi$ under the supremum norm.

The best λ-order error approximation of a function $\tilde{\zeta} \in C_{2\pi}$ is defined by

$$E_{\lambda}(\tilde{\zeta}) = \inf_{t_\lambda} \|\tilde{\zeta} - t_{\lambda}\|,$$

where t_λ is a trigonometric polynomial of degree λ (Bernstein [14]).

Let us define the L^p space of all 2π-periodic and integrable functions as

$$L^p[0, 2\pi] := \left\{\tilde{\zeta} : [0, 2\pi] \to \mathbb{R} : \int_0^{2\pi} |\tilde{\zeta}(x)|^p\, dx < \infty\right\}, p \ge 1.$$

Now, $\|\cdot\|_p$ is defined as

$$\|\tilde{\zeta}\|_p = \begin{cases} \left\{\frac{1}{2\pi}\int_0^{2\pi} |\tilde{\zeta}(x)|^p\, dx\right\}^{\frac{1}{p}} & \text{for } 1 \le p < \infty \\ \operatorname{ess}\sup\limits_{x \in (0, 2\pi)} |\tilde{\zeta}(x)| & \text{for } p = \infty. \end{cases}$$

---

[2] FS denotes Fourier series and we use this abbreviation throughout the paper.

We define the following Lipschitz classes of function

$$\zeta \in Lip\alpha \text{ if } Lip\alpha := \{\zeta : [0, 2\pi] \to \mathbb{R} : |\zeta(x+\omega) - \zeta(x)| = O(\omega^{\alpha})\}$$

for $0 < \alpha \leq 1$;

$$\zeta \in Lip(\alpha, p) \text{ if } Lip(\alpha, p) := \left\{\zeta \in L^p[0, 2\pi] : \|\zeta(x+\omega) - \zeta(x)\|_p = O(\omega^{\alpha})\right\}$$

for $p \geq 1, 0 < \alpha \leq 1$;

$$\zeta \in Lip(\alpha, \xi(\omega)) \text{ if } Lip(\alpha, \xi(\omega)) := \left\{\zeta \in L^p[0, 2\pi] : \|\zeta(x+\omega) - \zeta(x)\|_p = O(\xi(\omega))\right\}$$

for $p \geq 1, 0 < \alpha \leq 1 \,\&\, \beta \geq 0$;

$$\zeta \in W(L^p \xi(\omega)) \text{if } W(L^p \xi(\omega)) := \{\zeta \in L^p[02\pi] : \left\|(\zeta(x+\omega) - \zeta(x)) \sin^{\beta}\left(\frac{\omega}{2}\right)\right\|_p = O(\xi(\omega))\}$$

where $\xi(\omega) > 0$ and increasing with $\omega > 0$ and L^p space of all 2π-periodic and integrable functions. Under above assumptions for $\alpha \in (0, 1], p \geq 1, \omega > 0$, we observed that

$$W(L^p, \xi(\omega)) \xrightarrow{\beta=0} Lip(\xi(\omega), p) \xrightarrow{\xi(\omega)=\omega^{\alpha}} Lip(\alpha, p) \xrightarrow{p\to\infty} Lip\alpha.$$

Remark 1 *If* $\frac{\xi(\omega)}{\omega}$ *is non-increasing; then* $\frac{\xi\left(\frac{\pi}{\lambda+1}\right)}{\frac{\pi}{\lambda+1}} \leq \frac{\xi\left(\frac{1}{\lambda+1}\right)}{\frac{1}{\lambda+1}}$ *i.e.,* $\xi\left(\frac{\pi}{\lambda+1}\right) \leq \pi\xi\left(\frac{1}{\lambda+1}\right)$.

2.2 Some important single summability

Let

$$\sum_{\lambda=0}^{\infty} v_{\lambda} \tag{1}$$

be an infinite series such that $s_k = \sum_{m=0}^{k} v_m$. Let

$$\sigma_r^{\eta} = \sum_{k=0}^{r} \frac{\binom{r-k+\eta-1}{r-k}}{\binom{r+\eta}{r}} s_k, \text{ for } \eta > -1. \tag{2}$$

If $\lim_{\lambda\to\infty}\sigma_{\lambda}^{\eta} = s$ then we say that the series (1) is (C, η) summable to s or summable by Cesàro mean of order η.If we take $\eta = 0$ in (2), (C, η) summability reduces to an ordinary sum and if we take $\eta = 1$, then (C, η) summability reduces to $(C, 1)$ summability or Cesàro summability of order 1.

Let

$$t_{\lambda}^{E^q} = \frac{1}{(1+q)^{\lambda}} \sum_{k=0}^{\lambda} \binom{\lambda}{k} \frac{1}{q^{k-\lambda}} s_k, q > 0.$$

If $\lim_{\lambda\to\infty} t_\lambda^{E^q} = s$ then we say that the series (1) is (E, q) summable to s or summable by Euler mean (E, q) (Hardy [15]). If $q = 0$, (E, q) method reduces to an ordinary sum and if $q = 1$, (E, q) means reduces to $(E, 1)$ means.

An infinite series (1) with the sequence $\{s_\lambda\}$ of its partial sums is said to be summable by harmonic method (Riesz [16] or simply summable $\left(N, \frac{1}{\lambda+1}\right)$ to sum s, where s is a finite number, if the sequence to sequence transformation

$$t_\lambda = \frac{1}{\log \lambda} \sum_{v=0}^{\lambda} \frac{s_v}{\lambda - v + 1} \text{ as } \lambda \to \infty.$$

Let $\{p_\lambda\}$ be a sequence of constants, real or complex and let

$$P_\lambda = \sum_{k=0}^{\lambda} p_k, (P_\lambda \neq 0).$$

Let

$$t_\lambda^{N_p} = \frac{1}{P_\lambda} \sum_{k=0}^{\lambda} p_{\lambda-k} s_k = \frac{1}{P_\lambda} \sum_{k=0}^{\lambda} p_k s_{\lambda-k}. \tag{3}$$

If

$$\lim_{\lambda\to\infty} t_\lambda^{N_p} = s$$

then we say that the series (1) is (N, p_λ) summable to s or summable by Nörlund (N, p_λ) means.

Let $\{p_\lambda\}$ and $\{q_\lambda\}$, be two sequences of constants, real or complex such that

$$P_\lambda = p_0 + p_1 + \ldots + p_\lambda; P_{-1} = p_{-1} = 0, \tag{4}$$

$$Q_\lambda = q_0 \mid q_1 \mid \ldots \mid q_\lambda; Q_{-1} = q_{-1} = 0. \tag{5}$$

$$R_\lambda = \sum_{k=0}^{\lambda} p_k q_{\lambda-k} \neq 0 \text{ for all } \lambda. \tag{6}$$

Convolution of the two sequences $\{p_\lambda\}$ and $\{q_\lambda\}$, is defined as

$$R_\lambda = (p * q)_\lambda = \sum_{k=0}^{\lambda} p_k q_{\lambda-k}.$$

We write

$$t_\lambda^{N_{pq}} = \frac{1}{R_\lambda} \sum_{k=0}^{\lambda} p_{\lambda-k} q_k s_k;$$

then the generalized Nörlund means $(N_{p,q})$ of the sequence $\{s_\lambda\}$ is denoted by the sequence t_λ^{pq}. If $t_\lambda^{pq} \to s$, as $\lambda \to \infty$ then, the series (1) is said to be summable to s by $N_{p,q}$ method and is denoted by $s_\lambda \to s(N_{p,q})$ ([17]).

Let $\{p_\lambda\}$ be a sequence of real constants such that $p_0 > 0, p_\lambda \geq 0$ and $P_\lambda = \sum_{v=0}^{\lambda} p_v \neq 0$, such that $P_\lambda \to \infty$ as $\lambda \to \infty$.

If

$$t_\lambda = \frac{1}{P_\lambda} \sum p_v s_v \to s, \text{as } \lambda \to \infty,$$

then we say that $\{s_\lambda\}$ is summable by $(\overline{N}, p_\lambda)$ means and we write

$$s_\lambda = s(\overline{N}, p_\lambda),$$

where $\{s_\lambda\}$ is the sequence of λth partial sum of the series (1).

Let $T = (l_{\lambda,k})$ be an infinite triangular matrix satisfying the conditions of regularity [18] i.e.,

$$\begin{cases} \sum_{k=0}^{\lambda} l_{\lambda,k} = 1 \; as \; \lambda \to \infty \\ l_{\lambda,k} = 0 \; for \; k > \lambda \\ \sum_{k=0}^{\lambda} |l_{\lambda,k}| \leq M, a \; finite \; constant \end{cases} \tag{7}$$

The sequence-to-sequence transformation

$$t_\lambda^T(\tilde{\zeta}; x) := \sum_{k=0}^{\lambda} l_{\lambda,k} s_k = \sum_{k=0}^{\lambda} l_{\lambda,\lambda-k} s_{\lambda-k}$$

defines the sequence $t_\lambda^T(\zeta; x)$ of triangular matrix means of the sequence $\{s_\lambda\}$ generated by the sequence of coefficients $(l_{\lambda,k})$.

If $t_\lambda^T(\tilde{\zeta}; x) \to s$ as $\lambda \to \infty$ then the infinite series $\sum_{\lambda=0}^{\infty} v_\lambda$ or the sequence $\{s_\lambda\}$ is summable to s by triangular matrix (T-method) [13].

2.3 $C^\delta T$ product means

we define $C^\delta T$ means as

$$t_\lambda^{C^\delta T}(\tilde{\zeta}; x) := \sum_{r=0}^{\lambda} \frac{\binom{\lambda - r + \delta - 1}{\delta - 1}}{\binom{\delta + \lambda}{\delta}} \sum_{k=0}^{r} l_{r,k} s_k(\tilde{\zeta}; x) \tag{8}$$

If $t_\lambda^{C^\delta T}(\tilde{\zeta}; x) \to s$ as $\lambda \to \infty$, then $\sum_{\lambda=0}^{\infty} v_\lambda$ is summable to s by $C^\delta T$ method.

Note 2 *Since C^δ and T both are regular then $C^\delta T$ method is also regular.*

Remark 2 *The special cases of $C^\delta T$ means: $C^\delta T$ transform reduces to*

i. $C^\delta H$ transform if $l_{\lambda,k} = \frac{1}{(\lambda-k+1)\log(\lambda+1)}$;

ii. $C^\delta N_p$ transform if $l_{\lambda,k} = \frac{p_{\lambda-k}}{P_\lambda}$ where $P_\lambda = \sum_{k=0}^{\lambda} p_k \neq 0$;

iii. $C^{\delta}\overline{N}_p$ transform if $l_{\lambda,k} = \frac{p_k}{P_\lambda}$;

iv. $C^{\delta}E^{q}$ transform when $a_{\lambda,k} = \frac{1}{(1+q)^{\lambda}}\binom{\lambda}{k}q^{\lambda-k}$;

v. $C^{\delta}E^{1}$ when $l_{\lambda,k} = \frac{1}{2^{\lambda}}\binom{\lambda}{k}$;

vi. $C^{\delta}N_{pq}$ transform if $l_{\lambda,k} = \frac{p_{\lambda-k}q_k}{R_\lambda}$ where $R_\lambda = \sum_{k=0}^{\lambda} p_k q_{\lambda-k}$.

In above special case (ii), (iii), and (vi) p_λ and q_λ are two non-negative monotonic non-increasing sequences of real constants.

Remark 3 *C^1H, C^1N_p, C^1N_{pq}, C^1E^q and C^1E^1 transforms are also the special cases of $C^{\delta}T$ for $\delta = 1$.*

Example 2 *we consider*

$$1 - 1574\sum_{\lambda=1}^{\infty}(-1573)^{\lambda-1} \tag{9}$$

The λth partial sum of the series (9) is given by

$$s_\lambda = (-1573)^{\lambda}, \forall\lambda \in \mathbb{N}_0$$

we take $l_{\lambda,k} = \frac{1}{(787)^{\lambda}}\binom{\lambda}{k}(786)^{\lambda-k}$, then

$$\begin{aligned} t_\lambda^T &= l_{\lambda,0}s_0 + l_{\lambda,1}s_1 + \dots + l_{\lambda,\lambda}s_\lambda \\ &= \frac{1}{(787)^{\lambda}}\left[\binom{\lambda}{0}(786)^{\lambda} - \binom{\lambda}{1}(786)^{\lambda-1}.1573 + \dots + \binom{\lambda}{\lambda}(-1573)^{\lambda}\right] \\ &= \frac{1}{(787)^{\lambda}}(-787)^{\lambda} \\ &= \begin{cases} 1, & \lambda \textit{ is even} \\ -1, & \lambda \textit{ is odd} \end{cases} \end{aligned} \tag{10}$$

in above example, we see the series is summable neither by Cesàro means nor Matrix means, but summable by Cesàro-Matrix.

Thus, $C^{\delta}T$ means is more powerfull and effective than single C^{δ} and T means.

Example 3 *we consider another infinite series*

$$1 - 6 + 30 - 150 + 750 - 3750 + 18750 - \dots \tag{11}$$

The λth partial sum of the series (11) is given by

$$s_\lambda = (-5)^{\lambda}, \forall\lambda \in \mathbb{N}_0$$

we take $l_{\lambda,k} = \frac{1}{3^\lambda}\binom{\lambda}{k}2^{\lambda-k}$, then

$$\begin{aligned} t_\lambda^T &= l_{\lambda,0}s_0 + l_{\lambda,1}s_1 + \dots + l_{\lambda,\lambda}s_\lambda \\ &= \frac{1}{3^\lambda}\left[\binom{\lambda}{0}2^\lambda - \binom{\lambda}{1}2^{\lambda-1}.5 + \dots + \binom{\lambda}{\lambda}(-5)^\lambda\right] \\ &= \frac{1}{3^\lambda}(-3)^\lambda \\ &= \begin{cases} 1, & \lambda \in \{2m : m \in Z\} \\ -1, & \lambda \in \{2m+1 : m \in Z\} \end{cases} \end{aligned} \tag{12}$$

in above example, we see the series is summable neither by Cesàro means of order one nor Matrix means, but summable by Cesàro-Matrix.

2.4 Notations

$$\tilde{K}_\lambda^{C^\delta T} = \frac{1}{2\pi}\sum_{r=0}^{\lambda}\frac{\binom{r+\delta-1}{\delta-1}}{\binom{\delta+\lambda}{\delta}}\sum_{k=0}^{r} l_{r,r-k}\frac{\cos\left(r-k+\frac{1}{2}\right)\omega}{\sin\frac{\omega}{2}} \tag{13}$$

$$\varrho = \text{integral part of}\left(\frac{1}{\omega}\right)$$

$$\psi(x,\omega) = \zeta(x+\omega) - \zeta(x-\omega)$$

We use the following in our work.

$$\frac{1}{\sin\left(\frac{\omega}{2}\right)} \le \frac{\pi}{\omega}, 0 < \omega \le \pi \tag{14}$$

$$\sin\omega \le \omega, \omega \ge 0 \tag{15}$$

$$|\cos\lambda\omega| \le 1, \forall\omega \in \mathbb{R} \tag{16}$$

Zygmund ([13]).

Note 3 *Following conditions are used in the proof of the main results*

$$\begin{cases} l_{\lambda,\lambda-k} - l_{\lambda+1,\lambda+1-k} \ge 0 \text{ for } 0 \le k \le \lambda \\ L_{\lambda,k} = \sum_{r=k}^{\lambda} l_{\lambda,\lambda-r} \text{ and } L_{\lambda,0} = 1, \forall\lambda \in \mathbb{N}_0 \end{cases}. \tag{17}$$

Remark 4 *Considering the matrix $T = (l_{\lambda,k})$ as*

$$l_{\lambda,k} = \begin{cases} \frac{2018 \times (2019)^k}{(2019)^{\lambda+1} - 1}, & 0 \le k \le \lambda \\ 0, & k > \lambda \end{cases},$$

we can observe that (7, 17) satisfied.

Remark 5 *Function $\overline{\zeta}$ denotes a conjugate to a 2π-period and Lebesgue integrable function and this notation is used throughout the chapter.*

3. Lemmas

For the proof of our theorems, following lemmas are required:

Lemma 3.1 *If conditions (7, 17) hold for $\{l_{\lambda,k}\}$, then*

$$|\tilde{K}_\lambda^{C^\delta T}(\omega)| = O\left(\frac{1}{\omega}\right) \forall \delta \geq 1, 0 < \omega \leq \frac{\pi}{\lambda+1}.$$

Proof. For $0 < \omega \leq \frac{\pi}{\lambda+1}$, using (14, 15, 16)

$$|\tilde{K}_\lambda^{C^\delta T}(\omega)| = \left| \frac{1}{2\pi} \sum_{r=0}^{\lambda} \frac{\binom{r+\delta-1}{\delta-1}}{\binom{\delta+\lambda}{\delta}} \sum_{k=0}^{r} l_{r,r-k} \frac{\cos\left(r-k+\frac{1}{2}\right)\omega}{\sin\frac{\omega}{2}} \right|$$

$$\leq \frac{1}{2\pi} \sum_{r=0}^{\lambda} \frac{\binom{r+\delta-1}{\delta-1}}{\binom{\delta+\lambda}{\delta}} \sum_{k=0}^{r} l_{r,r-k} \frac{|\cos\left(r-k+\frac{1}{2}\right)\omega|}{|\sin\frac{\omega}{2}|}$$

$$\leq \frac{1}{2\omega} \sum_{r=0}^{\lambda} \frac{\binom{r+\delta-1}{\delta-1}}{\binom{\delta+\lambda}{\delta}} \sum_{k=0}^{r} l_{r,r-k}$$

$$= \frac{1}{2\omega} \sum_{r=0}^{\omega} \frac{(r+\delta-1)!}{(\delta-1)!r!} \times \frac{\delta!\lambda!}{(\delta+\lambda)!} L_{r,0}$$

$$= \frac{\lambda!\delta}{2\omega\,(\delta+\lambda)!} \sum_{r=0}^{\lambda} \frac{(r+\delta-1)!}{r!} \text{ since } \mathrm{L}_{\mathrm{r},0} = 1$$

$$= \frac{\lambda!\delta}{2\omega\,\delta!(\delta+1)\cdots(\delta+\lambda)} \left[\frac{(\delta-1)!}{0!} + \frac{\delta!}{1!} \cdots + \frac{(\lambda+\delta-1)!}{\lambda!}\right]$$

$$\leq \frac{\lambda!\delta}{2\omega\,\delta!(\delta+1)\cdots(\delta+\lambda)} \times (\lambda+1) \frac{\delta!(\delta+1)\cdots(\delta+\lambda-1)}{\lambda!}$$

$$= \frac{\delta}{2\omega} \times \frac{\lambda+1}{\delta+\lambda}$$

$$\leq \frac{\delta}{2\omega}$$

$$= O\left(\frac{1}{\omega}\right) \text{ for all } \delta \geq 1.$$

Lemma 3.2 *If conditions (7, 17) holds for* $\{l_{\lambda,k}\}$, *then*

$$\left|\tilde{K}_{\lambda}^{C^{\delta}T}(\omega)\right| = O\left(\frac{1}{(\lambda+1)\omega^2}\right)\forall\delta\geq 1,\ \frac{\pi}{\lambda+1}\leq\omega\leq\pi.$$

Proof. For $\frac{\pi}{\lambda+1}\leq\omega\leq\pi$, using (14), $l_{r,r-k}\geq l_{r+1,r+1-k}\geq l_{r+1,r-k}$ and $L_{\varrho+1,0}=1$.

$$\begin{aligned}\left|\tilde{K}_{\lambda}^{C^{\delta}T}(\omega)\right| &= \left|(2\pi)^{-1}\sum_{r=0}^{\lambda}\frac{\binom{r+\delta-1}{\delta-1}}{\binom{\delta+\lambda}{\delta}}\sum_{k=0}^{r} l_{r,r-k}\frac{\cos\left(r-k+\frac{1}{2}\right)\omega}{\sin\frac{\omega}{2}}\right| \\ &= O\left(\frac{1}{\omega}\right)\left|\operatorname{Re}\sum_{r=0}^{\lambda}\frac{\binom{r+\delta-1}{\delta-1}}{\binom{\delta+\lambda}{\delta}}\sum_{k=0}^{r} l_{r,r-k}e^{i\left(r-k+\frac{1}{2}\right)\omega}\right|\end{aligned} \tag{18}$$

Now we consider

$$\begin{aligned}\left|\sum_{r=0}^{\lambda}\frac{\binom{r+\delta-1}{\delta-1}}{\binom{\delta+\lambda}{\delta}}\sum_{k=0}^{r} l_{r,r-k}e^{i\left(r-k+\frac{1}{2}\right)\omega}\right| &\leq \left|\sum_{r=0}^{\varrho}\frac{\binom{r+\delta-1}{\delta-1}}{\binom{\delta+\lambda}{\delta}}\sum_{k=0}^{r} l_{r,r-k}e^{i(r-k)\omega}\right| \\ &+ \left|\sum_{r=\varrho+1}^{\lambda}\frac{\binom{r+\delta-1}{\delta-1}}{\binom{\delta+\lambda}{\delta}}\sum_{k=0}^{\varrho} l_{r,r-k}e^{i(r-k)\omega}\right| \\ &+ \left|\sum_{r=\varrho+1}^{\lambda}\frac{\binom{r+\delta-1}{\delta-1}}{\binom{\delta+\lambda}{\delta}}\sum_{k=\varrho+1}^{r} l_{r,r-k}e^{i(r-k)\omega}\right| \\ &= \Lambda_1+\Lambda_2+\Lambda_3,\ (\text{say}).\end{aligned} \tag{19}$$

Now,

$$\Lambda_1 \leq \sum_{r=0}^{\varrho} \frac{\binom{r+\delta-1}{\delta-1}}{\binom{\delta+\lambda}{\delta}} \sum_{k=0}^{r} l_{r,r-k} \left| e^{i(r-k)\omega} \right|$$

$$\leq \sum_{r=0}^{\varrho} \frac{\binom{r+\delta-1}{\delta-1}}{\binom{\delta+\lambda}{\delta}} L_{r,0}$$

$$= \sum_{r=0}^{\varrho} \frac{(r+\delta-1)!}{(\delta-1)!r!} \times \frac{\delta!\lambda!}{(\delta+\lambda)!} \text{ since, } \mathrm{L}_{\mathrm{r},0} = 1$$

$$= \frac{\lambda!}{(\delta+1)\ldots(\delta+\lambda)\,\delta!} \sum_{r=0}^{\varrho} \frac{(r+\delta-1)!\delta}{r!}$$

$$= \frac{\lambda!}{(\delta+1)\ldots(\delta+\varrho)\ldots(\delta+\lambda)} \left[1+\delta+\frac{\delta(\delta+1)}{2!}+\ldots+\frac{(\varrho+\delta-1)!}{\varrho!(\delta-1)!}\right]$$

$$\leq \frac{\lambda!}{(\delta+1)\ldots(\delta+\varrho)\ldots(\delta+\lambda)} \left[(\varrho+1)\frac{\delta(\delta+1)\ldots(\delta+\varrho-1)}{\varrho!}\right]$$

$$= \frac{\varrho!(\varrho+1)\ldots\lambda}{(\delta+\varrho)\ldots(\delta+\lambda-1)} \times \frac{\delta}{\delta+\lambda} \times \frac{\varrho+1}{\varrho!}$$

$$\leq \frac{\delta}{\delta+\lambda} \times (\varrho+1)$$

$$\leq \frac{\delta}{\delta+\lambda} \times \left(\frac{1}{\omega}+1\right)$$

$$= O\left(\frac{1}{\omega(\lambda+1)}(1+\omega)\right) \text{ for all } \delta \geq 1.$$

Changing the order of summation and applying Abel's transformation in Λ_2, we have

$$\Lambda_2 = \left| \sum_{k=0}^{\varrho} \sum_{r=\varrho+1}^{\lambda} \frac{\begin{pmatrix} r+\delta-1 \\ \delta-1 \end{pmatrix}}{\begin{pmatrix} \delta+\lambda \\ \delta \end{pmatrix}} l_{r,r-k} e^{i(r-k)\omega} \right|$$

$$= \frac{1}{\begin{pmatrix} \delta+\lambda \\ \delta \end{pmatrix}} \left| \sum_{k=0}^{\varrho} \left[\left\{ \sum_{r=\varrho+1}^{\lambda-1} \left(\begin{pmatrix} r+\delta-1 \\ \delta-1 \end{pmatrix} l_{r,r-k} - \begin{pmatrix} r+\delta \\ \delta-1 \end{pmatrix} l_{r+1,r+1-k} \right) \sum_{\nu=0}^{r} e^{i(\nu-k)\omega} \right\} \right. \right.$$

$$\left. \left. + \begin{pmatrix} \lambda+\delta-1 \\ \delta-1 \end{pmatrix} l_{\lambda,\lambda-k} \sum_{\nu=0}^{\lambda} e^{i(\nu-k)\omega} - \begin{pmatrix} \delta+\varrho \\ \delta-1 \end{pmatrix} l_{\varrho+1,\varrho+1-k} \right] \right|$$

$$= O(\omega^{-1}) \frac{1}{\begin{pmatrix} \eta+\lambda \\ \eta \end{pmatrix}} \sum_{k=0}^{\varrho} \left[\left| \sum_{r=\varrho+1}^{\lambda-1} \left(\begin{pmatrix} r+\delta-1 \\ \delta-1 \end{pmatrix} l_{r,r-k} - \begin{pmatrix} r+\delta \\ \delta-1 \end{pmatrix} l_{r+1,r+1-k} \right) \right| \right.$$

$$\left. + \left| \begin{pmatrix} \lambda+\delta-1 \\ \delta-1 \end{pmatrix} l_{\lambda,\lambda-k} \right| + \left| \begin{pmatrix} \delta+\varrho \\ \delta-1 \end{pmatrix} l_{\varrho+1,\varrho+1-k} \right| \right]$$

$$= O(\omega^{-1}) \frac{1}{\begin{pmatrix} \delta+\lambda \\ \delta \end{pmatrix}} \sum_{k=0}^{\varrho} \left[\begin{pmatrix} \delta+\varrho \\ \delta-1 \end{pmatrix} l_{\varrho+1,\varrho+1-k} + \begin{pmatrix} \delta+\lambda-1 \\ \delta-1 \end{pmatrix} l_{\lambda,\lambda-k} + \begin{pmatrix} \delta+\lambda-1 \\ \delta-1 \end{pmatrix} l_{\lambda,\lambda-k} \right.$$

$$\left. + \begin{pmatrix} \delta+\varrho \\ \delta-1 \end{pmatrix} l_{\varrho+1,\varrho+1-k} \right]$$

$$= O(\omega^{-1}) \frac{1}{\begin{pmatrix} \delta+\lambda \\ \delta \end{pmatrix}} \sum_{k=0}^{\varrho} \left[\begin{pmatrix} \delta+\varrho \\ \delta-1 \end{pmatrix} l_{\varrho+1,\varrho+1-k} + \begin{pmatrix} \delta+\lambda-1 \\ \delta-1 \end{pmatrix} l_{\lambda,\lambda-k} \right]$$

$$= O(\omega^{-1}) \frac{\lambda!}{(\delta+1)\dots(\delta+\lambda)} \sum_{k=0}^{\varrho} \left[\frac{\delta(\delta+1)\dots(\delta+\varrho)\dots(\delta+\lambda)}{(\delta+\varrho+1)\dots(\delta+\lambda)(\varrho+1)!} l_{\varrho+1,\varrho+1-k} \right.$$

$$\left. + \frac{\delta(\delta+1)\dots(\delta+\lambda-1)(\delta+\lambda)}{(\delta+\lambda)\,\lambda!} l_{\lambda,\lambda-k} \right]$$

$$= O(\omega^{-1}) \frac{\lambda!}{(\delta+1)\dots(\delta+\lambda)} \times \frac{\delta(\delta+1)\dots(\delta+\lambda)}{(\lambda+1)!} \sum_{k=0}^{\varrho} \left(l_{\varrho,\varrho-k} + l_{\lambda,\lambda-k} \right)$$

$$= O\left(\frac{1}{\omega(\lambda+1)} \left(L_{\varrho,0} + L_{\lambda,0} \right) \right)$$

$$= O\left(\frac{1}{\omega(\lambda+1)} \right).$$

Applying Abel's transformation in Λ_3, we have

$$\Lambda_3 = \left| \sum_{r=\varrho+1}^{\lambda} \frac{\binom{r+\delta-1}{\delta-1}}{\binom{\delta+\lambda}{\delta}} \left[\sum_{k=\varrho+1}^{r-1} (l_{r,r-k} - l_{r,r-k+1}) \sum_{\nu=0}^{k} e^{i(r-\nu)\omega} + l_{r,0} \sum_{\nu=0}^{r} e^{\iota(r-\nu)\omega} \right.\right.$$

$$\left.\left. -l_{r,r-\tau-1} \sum_{\nu=0}^{\varrho} e^{i(r-\nu)\omega} \right]\right|$$

$$= O(\omega^{-1}) \sum_{r=\varrho+1}^{\lambda} \frac{\binom{r+\delta-1}{\delta-1}}{\binom{\delta+\lambda}{\delta}} \left[\left| \sum_{k=\varrho+1}^{r-1} (l_{r,r-k} - l_{r,r-k+1}) \right| + l_{r,0} + l_{r,r-\varrho-1} \right]$$

$$= O(\omega^{-1}) \sum_{r=\varrho+1}^{\lambda} \frac{\binom{r+\delta-1}{\delta-1}}{\binom{\delta+\lambda}{\delta}} [\, -l_{r,r-\varrho} + l_{r,1}| + l_{r,0} + l_{r,r-\varrho-1}|]$$

$$= O(\omega^{-1}) \frac{1}{\binom{\delta+\lambda}{\delta}} \sum_{r=\varrho+1}^{\lambda} \binom{r+\delta-1}{\delta-1} l_{r,r-\varrho}$$

$$= \begin{array}{l} O(\omega^{-1}) \dfrac{\lambda!}{(\delta+1)\dots(\delta+\lambda)} \\ \times \dfrac{\delta(\delta+1)\dots(\delta+\lambda)}{\lambda!(\delta+\lambda)} \left[\dfrac{\delta\dots(\delta+\varrho)}{(\varrho+1)!} l_{\varrho+1,1} + \dots + \dfrac{\delta\dots(\delta+\lambda-1)}{\lambda!} l_{\lambda,\lambda-\varrho} \right] \end{array}$$

$$= O(\omega^{-1}) \frac{\lambda!}{(\delta+1)\dots(\delta+\lambda)} \times \frac{\delta(\delta+1)\dots(\delta+\lambda)}{\lambda!(\delta+\lambda)} [l_{\varrho+1,1} + l_{\varrho+2,2} + \dots l_{\lambda,\lambda-\varrho}]$$

$$= O(\omega^{-1}) \frac{\delta}{\delta+\lambda} [l_{\varrho+1,1} + l_{\varrho+1,2} + \dots + l_{\varrho+1,\lambda-\varrho}]$$

$$= O\left(\frac{1}{\omega(\lambda+1)}\right) L_{\varrho+1,1}$$

$$= O\left(\frac{1}{\omega(\lambda+1)}\right).$$

Combining Λ_1, Λ_2 and Λ_3 we have,

$$\Lambda_1+\Lambda_2+\Lambda_3 = O\left[\frac{1}{\omega(\lambda+1)}\times(1+\omega)\right]+O\left[\frac{1}{\omega(\lambda+1)}\right]+O\left[\frac{1}{\omega(\lambda+1)}\right]$$

$$= O\left[\frac{1}{(\lambda+1)}\left(1+\frac{3}{\omega}\right)\right] \tag{20}$$

$$= O\left(\frac{1}{\lambda+1}\times\frac{3+\pi}{\omega}\right)$$

(Let $1+\frac{3}{\omega}\le\frac{k}{\omega}$ for ω fixed $k^{min}=3+\pi$)
Now, from (19, 20) we get

$$\left|\tilde{K}_\lambda^{C^\delta T}(\omega)\right| = O\left(\frac{1}{(\lambda+1)\omega^2}\right)$$

4. Main theorems

Theorem 4.1 *The error approximation of* $\tilde{\zeta}$ *in* $W(L^p,\xi(\omega))$, $(p>1)$, *by* $C^\delta T$ *means of its CFS is given by*

$$\|t_\lambda^{C^\delta T}(\tilde{\zeta};x)-\tilde{\zeta}(x)\|_p = O\left[(\lambda+1)^\beta\xi\left(\frac{1}{\lambda+1}\right)\right],$$

where $0\le\beta<\frac{1}{p}$ and condition (17) holds and positive increasing function $\xi(\omega)$ satisfies the following conditions:

$$\frac{\xi(\omega)}{\omega^{\beta+1-\sigma}} \text{ is non-decreasing;} \tag{21}$$

$$\left\{\int_0^{\frac{\pi}{\lambda+1}}\left(\frac{\lambda^{-\sigma}|\psi(x,\omega)|\sin^\beta\left(\frac{\omega}{2}\right)}{\xi(\omega)}\right)^p d\omega\right\}^{\frac{1}{p}} = O\left((\lambda+1)^{\sigma-\frac{1}{p}}\right), \text{ for } \beta<\sigma<\frac{1}{p}; \tag{22}$$

$$\frac{\xi(\omega)}{\omega} \text{ is non-decreasing;} \tag{23}$$

$$\text{and } \left\{\int_{\frac{\pi}{\lambda+1}}^{\pi}\left(\frac{\omega^{-\eta}|\psi(x,\omega)|\sin^\beta\left(\frac{\omega}{2}\right)}{\xi(\omega)}\right)^p d\omega\right\}^{\frac{1}{p}} = O\left((\lambda+1)^{\eta-\frac{1}{p}}\right), \tag{24}$$

where $\frac{1}{p}<\eta<\beta+\frac{1}{p}$ for η being an arbitrary number and $p+q=pq$.
Conditions (22, 24) hold uniformly in x.

Conditions (22, 24) can be verified by using the fact that $\psi(x,\omega) \in W(L_p, \xi(\omega))$ and $\frac{\psi(x,\omega)}{\xi(\omega)}$ is bounded function.

Proof. The λth partial sums of the CFS is denoted by $s_\lambda(\tilde{\zeta};x)$, and is given by

$$s_\lambda(\tilde{\zeta};x) - \tilde{\zeta}(x) = \frac{1}{2\pi}\int_0^\pi \psi(x,\omega)\frac{\cos\left(\lambda+\frac{1}{2}\right)\omega}{\sin\frac{\omega}{2}}\,d\omega,$$

one can consult [13] for detailed work on FS and CFS.

Denoting $C^\delta T$ means of $\{s_\lambda(\tilde{\zeta}:x)\}$ by $t_\lambda^{C^\delta T}(\tilde{\zeta}:x)$, we get

$$t_\lambda^{C^\delta T}(\tilde{\zeta};x) - \tilde{\zeta}(x) = \sum_{r=0}^{\lambda}\frac{\begin{pmatrix}\lambda-r+\delta-1\\ \delta-1\end{pmatrix}}{\begin{pmatrix}\delta+\lambda\\ \delta\end{pmatrix}}\sum_{k=0}^{r} l_{r,k}\left[s_k(\tilde{\zeta};x)-\tilde{\zeta}(x)\right]$$

$$= \frac{1}{2\pi}\int_0^\pi \psi(x,\omega)\sum_{r=0}^{\lambda}\frac{\begin{pmatrix}r+\delta-1\\ \delta-1\end{pmatrix}}{\begin{pmatrix}\delta+\lambda\\ \delta\end{pmatrix}}\sum_{k=0}^{r} l_{r,r-k}\frac{\cos\left(r-k+\frac{1}{2}\right)\omega}{\sin\left(\frac{\omega}{2}\right)}\,d\omega \tag{25}$$

$$= \int_0^\pi \psi(x,\omega)\tilde{K}_\lambda^{C^\delta T}(\omega)\,d\omega \text{ (By the notation (13))}$$

$$= \int_0^{\frac{\pi}{\lambda+1}} \psi(x,\omega)\tilde{K}_\lambda^{C_\delta T}(\xi)\,d\omega + \int_{\frac{\pi}{\lambda+1}}^{\pi} \psi(x,\omega)\tilde{K}_\lambda^{C_\delta T}(\omega)\,d\omega \tag{26}$$

$$= I_1 + I_2, \text{ say}$$

Applying (14), Lemma 3.1, Hölder's inequality and second mean value theorem for integral, we have

$$I_1 = O(1)\left\{\int_0^{\frac{\pi}{\lambda+1}}\left(\frac{\omega^{-\sigma}|\psi(x,\omega)|\sin^\beta\left(\frac{\omega}{2}\right)}{\xi(\omega)}\right)^p d\omega\right\}^{\frac{1}{p}} \times \left\{\int_0^{\frac{\pi}{\lambda+1}}\left(\frac{\xi(\omega)}{\omega^{-\sigma+1}\sin^\beta\left(\frac{\omega}{2}\right)}\right)^q d\omega\right\}^{\frac{1}{q}}$$

$$= O\left[(\lambda+1)^{\sigma-\frac{1}{p}} \times \left\{\int_0^{\frac{\pi}{\lambda+1}}\left(\frac{\xi(\omega)}{\omega^{\beta+1-\sigma}}\right)^q d\omega\right\}^{\frac{1}{q}}\right]$$

$$= O\left[(\lambda+1)^{\sigma-\frac{1}{p}}(\lambda+1)^{\beta+\frac{1}{p}-\sigma}\xi\left(\frac{\pi}{\lambda+1}\right)\right]$$

$$= O\left[(\lambda+1)^\beta\xi\left(\frac{1}{\lambda+1}\right)\right] \tag{27}$$

in view of condition (22) and $p^{-1}+q^{-1}=1$ and Remark 1.

Again using Lemma 3.2, Hölder's inequality and (14), we have

$$
\begin{aligned}
I_2 &= O\left(\frac{1}{\lambda+1}\right)\int_{\frac{\pi}{\lambda+1}}^{\pi} \frac{|\psi(x,\omega)|}{\omega^2}\,d\omega \\
&= O\left\{\frac{1}{\lambda+1}\int_{\frac{\pi}{\lambda+1}}^{\pi}\left(\frac{\omega-\eta|\psi(x,\omega)|\sin^{\beta}\left(\frac{\omega}{2}\right)}{\xi(\omega)}\right)^{p} d\omega\right\}^{\frac{1}{p}} \times \left\{\int_{\frac{\pi}{\lambda+1}}^{\pi}\left(\frac{\omega^{-1}\xi(\omega)}{\omega^{-\eta+1+\beta}}\right)^{q} d\omega\right\}^{\frac{1}{q}} \\
&= O\left[(\lambda+1)^{-1+\eta-\frac{1}{p}}\xi\left(\frac{\pi}{\lambda+1}\right)\left(\frac{\lambda+1}{\pi}\right)\left(\int_{\frac{\pi}{\lambda+1}}^{\pi}\omega^{-(\beta+1-\eta)q}\,d\omega\right)^{\frac{1}{q}}\right] \\
&= O\left[(\lambda+1)^{\eta-\frac{1}{p}}\xi\left(\frac{\pi}{\lambda+1}\right)(\lambda+1)^{\beta+1-\eta-\frac{1}{q}}\right] \\
&= O\left[(\lambda+1)^{\beta}\xi\left(\frac{1}{\lambda+1}\right)\right]
\end{aligned}
\tag{28}
$$

in view of (23, 24) the second mean value theorem for integrals, $0<\eta<\beta+\frac{1}{p}, p+q=pq$ and Remark 1.

Collecting (26)-(28), we get

$$\left|t_{\lambda}^{C^{\delta}T}(\tilde{\zeta},x)-\tilde{\zeta}(x)\right| = O\left[(\lambda+1)^{\beta}\xi\left(\frac{1}{\lambda+1}\right)\right].$$

Now, using L_p-norm of a function, we get

$$\left\|t_{\lambda}^{C^{\delta}T}(\tilde{\zeta},x)-\tilde{\zeta}(x)\right\|_{p} = O\left[(\lambda+1)^{\beta}\xi\left(\frac{1}{\lambda+1}\right)\right]$$

Now, we establish the following theorem for the case $p=1$:

Theorem 4.2 *The inaccuracy estimation of $\tilde{\zeta}\in W(L^1,\xi(\omega))$, by $C^{\delta}T$ product operatior of its CFS is given by*

$$\|t_{\lambda}^{C^{\delta T}}(\tilde{\zeta};x)-\tilde{\zeta}(x)\|_1 = O\left[(\lambda+1)^{\beta}\xi\left(\frac{1}{\lambda+1}\right)\right],$$

where $0\le\beta<1$, provided (17) holds and increasing function $\xi(\omega)>0$ satisfies conditions (21) to (24) of Theorem 4.1 for $p=1$, $\beta<\sigma<1$ and $1<\eta<\beta+1$.

Proof. Following the proof of Theorem 4.1, for $p=1$, i.e., $q=\infty$, we have

$$
\begin{aligned}
I_1 &= O\left\{\int_0^{\frac{\pi}{\lambda+1}}\left(\frac{\omega^{-\sigma}|\psi(x,\omega)|\sin^{\beta}\left(\frac{\omega}{2}\right)}{\xi(\omega)}\right) d\omega \times \operatorname{ess}\sup_{0<\omega\le\frac{\pi}{\lambda+1}}\left|\frac{\xi(\omega)}{\omega^{-\sigma+1}\sin^{\beta}\left(\frac{\omega}{2}\right)}\right|\right\} \\
&= O\left((\lambda+1)^{\sigma-1}\right)\operatorname{ess}\sup_{0<\omega\le\frac{\pi}{\lambda+1}}\left|\frac{\xi(\omega)}{\omega^{\beta-\sigma+1}}\right|
\end{aligned}
$$

$$= O\left((\lambda+1)^{\sigma-1}\right)\left\{\frac{\xi\left(\frac{\pi}{\lambda+1}\right)}{\left(\frac{\pi}{\lambda+1}\right)^{\beta-\sigma+1}}\right\} \tag{29}$$

$$= O\left((\lambda+1)^{\beta}\xi\left(\frac{1}{\lambda+1}\right)\right)$$

in view of conditions (21, 22) for $p = 1$,

$$I_2 = O\left(\frac{1}{\lambda+1}\right)\int_{\frac{\pi}{\lambda+1}}^{\pi} \frac{|\psi(x,\omega)|}{\omega^2}\,d\omega$$

$$= O\left\{\frac{1}{\lambda+1}\int_{\frac{\pi}{\lambda+1}}^{\pi} \frac{\omega^{-\eta}|\psi(x,\omega)|\sin^{\beta}\left(\frac{\omega}{2}\right)}{\xi(\omega)}\,d\omega\right\} \times \text{ess} \sup_{\frac{\pi}{\lambda+1}\le\omega\le\pi}\left|\frac{\xi(\omega)}{\omega^{-\eta+\beta+2}}\right| \tag{30}$$

$$= O\left[(\lambda+1)^{\eta-2}\xi\left(\frac{\pi}{\lambda+1}\right)\left(\frac{(\lambda+1)^{2+\beta-\eta}}{\pi^{2+\beta-\eta}}\right)\right]$$

$$= O\left[(\lambda+1)^{\beta}\xi\left(\frac{\pi}{\lambda+1}\right)\right]$$

in view of (21, 22). Collecting (28) and (29), we get

$$|t_{\lambda}^{C^{\delta}T}(\tilde{\zeta};x) - \tilde{\zeta}(x)| = O\left[(\lambda+1)^{\beta}\xi\left(\frac{1}{\lambda+1}\right)\right] \tag{31}$$

finally from (31),

$$\|t_{\lambda}^{C^{\delta}T}(\tilde{\zeta};x) - \tilde{\zeta}(x)\|_1 = O\left[(\lambda+1)^{\beta}\xi\left(\frac{1}{\lambda+1}\right)\right]$$

in view of Remark 1.

5. Corollaries

Corollary 5.1 *The inaccuracy estimation of* $\tilde{\zeta} \in Lip(\xi(\omega), p)$ *class by* $C^{\delta}T$ *means of its CFS is given by*

$$\left\|t_{\lambda}^{C^{\delta}T}(\tilde{\zeta},x) - \tilde{\zeta}(x)\right\|_p = O\left[\xi\left(\frac{1}{\lambda+1}\right)\right]$$

where, $C^{\delta}T$ is as defined in (8).

Proof. Considering $\beta = 0$ in Theorem 4.1, we can obtain the proof.

Corollary 5.2 *The inaccuracy estimation of* $\tilde{\zeta} \in Lip(\alpha, p)$ *space by* $C^{\delta}T$ *product means of its CFS is given by*

$$\left\|t_{\lambda}^{C^{\delta}T}(\tilde{\zeta},x) - \tilde{\zeta}(x)\right\|_p = O[(\lambda+1)^{-\alpha}]$$

where, $C^\delta T$ is as defined in (8).

Proof. If we consider $\beta = 0$ & $\xi(\omega) = \omega^\alpha$ in Theorem 4.1, we can obtain the proof.

Corollary 5.3 *The error estimate of $\tilde{\zeta}$ in Lipα $(0<\alpha<1)$ class by $C^\delta T$ product means of its CFS is given by*

$$\left\| t_\lambda^{C^\delta T}(\tilde{\zeta}, x) - \tilde{\zeta}(x) \right\|_p = O[(\lambda+1)^{-\alpha}]$$

where, $C^\delta T$ is as defined in (8).

Proof. If we take $\beta = 0$ & $\xi(\omega) = \omega^\alpha$ & $p \to \infty$ in Theorem 4.1, we can obtain the proof.

For $\alpha = 1$, we can write an independent proof to obtain

$$\left\| t_\lambda^{C^\delta T}(\tilde{\zeta}, x) - \tilde{\zeta}(x) \right\|_\infty = O\left[\frac{\log(\lambda+1)}{\lambda+1}\right]$$

Corollary 5.4 *The error estimate of $\tilde{\zeta} \in W(L^p, \xi(\omega))$ class by $C^\delta H$ means*

$$t_\lambda^{C^\delta H} = \sum_{r=0}^{\lambda} \frac{\binom{\lambda - r + \delta - 1}{\delta - 1}}{\binom{\delta + \lambda}{\delta}} (\log(r+1))^{-1} \sum_{k=0}^{r} \frac{1}{(r-k+1)} s_k,$$

of the CFS is given by

$$\| t_\lambda^{C^\delta H}(\tilde{\zeta}; x) - \tilde{\zeta}(x) \|_p = O\left[(\lambda+1)^\beta \xi\left(\frac{1}{\lambda+1}\right)\right]$$

provided $C^\delta T$ defined in (8) and $\xi(\omega)$ satisfies the conditions (21) to (24).

Corollary 5.5 *The error estimate of $\tilde{\zeta} \in W(L^p, \xi(\omega))$ class by $C^\delta N_p$ means*

$$t_\lambda^{C^\delta N_p} = \sum_{r=0}^{\lambda} \frac{\binom{\lambda - r + \delta - 1}{\delta - 1}}{\binom{\delta + \lambda}{\delta}} \frac{1}{P_r} \sum_{k=0}^{r} p_{r-k} s_k,$$

of the CFS is given by

$$\| t_\lambda^{C^\delta N_p}(\tilde{\zeta}; x) - \tilde{\zeta}(x) \|_p = O\left[(\lambda+1)^\beta \xi\left(\frac{1}{\lambda+1}\right)\right]$$

provided $C^\delta T$ defined in (8) and $\xi(\omega)$ satisfies the conditions (21) to (24).

Corollary 5.6 *The error estimate of $\tilde{\zeta} \in W(L^p, \xi(\omega))$ class by $C^\delta N_{pq}$ means*

$$t_\lambda^{C^\delta N_{pq}} = \sum_{r=0}^{\lambda} \frac{\binom{\lambda - r + \delta - 1}{\delta - 1}}{\binom{\delta + \lambda}{\delta}} \frac{1}{R_r} \sum_{k=0}^{r} p_{r-k} q_k s_k,$$

of the CFS is given by

$$\|t_{\lambda}^{C^{\delta}N_{pq}}(\tilde{\zeta};x) - \tilde{\zeta}(x)\|_p = O\left[(\lambda+1)^{\beta}\xi\left(\frac{1}{\lambda+1}\right)\right]$$

provided $C^{\delta}T$ defined in (8) and $\xi(\omega)$ satisfies the conditions (21) to (24).

Corollary 5.7 *The error approximation of* $\tilde{\zeta} \in W(L^p, \xi(\omega))$ *class by* $C^{\delta}\overline{N}_p$ *means*

$$t_{\lambda}^{C^{\delta}\overline{N}_p} = \sum_{r=0}^{\lambda} \frac{\binom{\lambda - r + \delta - 1}{\delta - 1}}{\binom{\delta + \lambda}{\delta}} \frac{1}{P_r} \sum_{k=0}^{r} p_k s_k,$$

of the CFS is given by

$$\|t_{\lambda}^{C^{\delta}\overline{N}_p}(\tilde{\zeta};x) - \tilde{\zeta}(x)\|_p = O\left[(\lambda+1)^{\beta}\xi\left(\frac{1}{\lambda+1}\right)\right]$$

provided $C^{\delta}T$ defined in (8) and $\xi(\omega)$ satisfies the conditions (21) to (24).

Corollary 5.8 *The error estimate of* $\tilde{\zeta} \in W(L^p, \xi(\omega))$ *class by* $C^{\delta}E^q$ *means*

$$t_{\lambda}^{C^{\delta}E^q} = \sum_{r=0}^{\delta} \frac{\binom{\delta - r + \delta - 1}{\delta - 1}}{\binom{\delta + \delta}{\delta}} \frac{1}{(1+q)^r} \sum_{k=0}^{r} \binom{r}{k} q^{r-k} s_k,$$

of the CFS is given by

$$\|t_{\lambda}^{C^{\delta}E^q}(\tilde{\zeta};x) - \tilde{\zeta}(x)\|_p = O\left[(\lambda+1)^{\beta}\xi\left(\frac{1}{\lambda+1}\right)\right]$$

provided $C^{\delta}T$ defined in (8) and $\xi(\omega)$ satisfies the conditions (21) to (24).

Corollary 5.9 *The error estimate of* $\zeta \in W(L^p, \xi(\omega))$ *class by* $C^{\delta}E^1$ *means*

$$t_{\lambda}^{C^{\delta}E^1} = \sum_{r=0}^{\lambda} \frac{\binom{\lambda - r + \delta - 1}{\delta - 1}}{\binom{\delta + \lambda}{\delta}} \frac{1}{2^r} \sum_{k=0}^{r} \binom{r}{k} s_k,$$

of the FS is given by

$$\|t_{\lambda}^{C^{\delta}E^1}(\tilde{\zeta};x) - \tilde{h}(x)\|_p = O\left[(\lambda+1)^{\beta}\xi\left(\frac{1}{\lambda+1}\right)\right]$$

provided $C^{\delta}T$ defined in (8) and $\xi(\omega)$ satisfies the conditions (21) to (24).

Remark 6 *The corollaries for 5.1 to 5.9 can also be obtained for the special cases* C^1H, C^1N_p, C^1N_{pq}, $C^1\tilde{N}_p$, C^1E^q *and* C^1E^1 *all things considered Remark 3.*

6. Particular cases

The following special cases of our theorems for $\delta = 1$ are.

6.1. If we take Remark 1(iv) and $\beta = 0, \xi(\omega) = \omega^\alpha, 0 < \alpha \leq 1$ in our theorem, then the Theorem 2 of [8] become a special case of our theorem.

6.2. If $\beta = 0, \xi(\omega) = \omega^\alpha, 0 < \alpha \leq 1 \,\&\, p \to \infty$ in our theorem, then the Theorem 3.3 of [9] become a special case of our result.

6.3. If we consider Remark 2(iv) then the main Theorem 2.2. of [5] become a special case of our result.

6.4. The Theorem 2 of [10] become a special case of our result.

6.5. If we consider Remark 2 (iv) then the Theorem 3.1 of [11] become a special case of our result.

6.6. If we consider Remark 2(ii) then the main Theorem 1 of [12] become a special case of our result.

7. Exercise

Q. 7.1. Prove that the infinite series $1 - 4038\sum_{j=1}^{\infty} (-4038)^{j-1}$ is neither summable by matrix means(T) nor Cesáro means of order one (C^1) but it summable by $C^\delta T$ means for $\delta = 1$.

Q. 7.2. Prove that a function f is 2π-periodic and Lebesgue integrable then the error approximation of f in Lipα class by $C^\delta T$ product means of its Fourier series is given by

$$E_n(f) = \begin{cases} O[(n+1)^{-\alpha}], & 0 \leq \alpha < 1 \\ O\Big[(n+1)^{-1}\{\log(n+1)\}\Big], & \alpha = 1, \end{cases}$$

where $C^\delta T$ is as defined in (8) and provided (17) holds.
{Hint: see [19]}.

Q. 7.3. Consider the matrix $T = (a_{n,k})$ as

$$a_{n,k} = \begin{cases} \dfrac{2 \times 3^k}{3^{n+1} - 1}, & 0 \leq k \leq j \\ 0, & k > n \end{cases},$$

check all conditions of T method as defined in (7) and also satisfies condition (17). [Hint: see [19]].

Q. 7.4. If the conditions of (7) and (17) holds for $\{a_{\lambda,k}\}$, then prove that

$$\left| (2\pi)^{-1} \sum_{r=0}^{\lambda} \frac{\binom{r+\delta-1}{\delta-1}}{\binom{\delta+\lambda}{\delta}} \sum_{k=0}^{r} l_{r,r-k} \frac{\cos\left(r-k+\frac{1}{2}\right)\omega}{\sin\frac{\omega}{2}} \right| = \begin{cases} O(\lambda+1), \forall \delta \geq 1, 0 < \omega \leq \dfrac{\pi}{\lambda+1} \\ O\left(\dfrac{1}{(\lambda+1)\omega^2}\right), \forall \delta \geq 1, \dfrac{\pi}{\lambda+1} \leq \omega \leq \pi. \end{cases}$$

Acknowledgements

My heart goes out to acknowledge my indebtedness to my reverend parents for their blessing, sacrifice, affection and giving me enthusiasm at every stage of my study. I am also grateful to all my family and friends specially Mrs. Anshu Rani, Dr. Pradeep Kumar and Niraj Pal for their timely help and giving me constant encouragements.

I also would like to thank the reviewers for their thoughtful and efforts towards improving my chapter.

AMS classification

40C10, 40G05, 40G10, 42A10, 42A24, 40C05, 41A10, 41A25, 42B05, 42A50

Author details

Mohammed Hadish
Government Inter College, Fatehpur, Uttar Pradesh, India

*Address all correspondence to: hadish@cusb.ac.in

References

[1] Rhaodes BE. On the degree of approximation of functions belonging to Lipschitz class by Hausdorff means of its Fourier series. Tamkang Journal of Mathematics. 2003;**34**(3):245-247 MR2001920 (2004g: 41023). Zentralblatt für Mathematik und ihre Grenzgebiete 1039.42001

[2] Nigam HK, Sharma K. A study on degree of approximation by Karamata summability method. J. Inequal. Appl. 2011;**85**(1):1-28

[3] Kushwaha JK. On the approximation of generalized Lipschitz function by Euler means of conjugate series of Fourier series. Scientific World Journal. 2013;**2013**:508026

[4] Mittal ML, Singh U, Mishra VN. approximation of functions (signals) belonging to $W(L^p, \xi(t))$-class by means of conjugate Fourier series using Norlund operators. Varahmihir J. Math. Sci. India. 2006;**6**(1):383-392

[5] Nigam HK. A study on approximation of conjugate of functions belonging to Lipschitz class and generalized Lipschitz class by product summability means of conjugate series of Fourier series. Thai Journal of Mathematics. 2012;**10**(2): 275-287

[6] Nigam HK, Hadish M. Best approximation of functions in generalized Hölder class. Journal of Inequalities and Applications. 2018;**1**:1-15

[7] Nigam HK, Hadish M. A study on approximation of a conjugate function using Cesàro-Matrix product operator. Nonlinear Functional Analysis and Applications. 2020;**25**(4):697-713

[8] Sonkar S, Singh U. Degree of approximation of the conjugate of signals (functions) belonging to Lip (α, r)-class by $(C, 1)(E, q)$ means of conjugate trigonometric Fourier series. Journal of inequalities and applications. 2012;**2012**: 278

[9] Deger U. On approximation to the functions in the $W(L_p, \xi(t))$ class by a new matrix mean. Novi Sad J. Math. 2016;**46**(1):1-14

[10] Singh U, Srivastava SK. Trigonometric approximation of functions belonging to certain Lipschitz classes by C^1T operator. Asian-European J. Math. 2014;7(4):1450064

[11] Mishra VN, Khan HH, Khatri K, Mishra LN. Degree of approximation of conjugate signals (functions) belonging to the generalized weighted $W(L_r, \xi(t), r \geq 1)$ class by $(C, 1)(E, q)$ means of conjugate Trigonometric Fourier series. Bulletin of Mathematical Analysis and Applications. 2013;**5**(4):40-53

[12] Mishra VN, Khan HH, Khatri K, Mishra LN. Approximation of functions belonging to the generalized Lipschitz class by C^1N_p summability method of conjugate series of Fourier series. Mathematicki vesnik. 2014;**66**(2):155-164

[13] Zygmund A. Trigonometric Series. 3rd rev. ed. Cambridge: Cambridge Univ. Press; 2002

[14] Bernshtein SN. On the best approximation of continuous functions by polynomials of given degree, in: Collected Works [in Russian], Izd. Nauk SSSR, Moscow. 1952;**1**:11-1047

[15] Hardy GH. Divergent Series. Oxford University Press; 1949

[16] Riesz M. Sur l'equivalence decertaines method de sommation,

proceeding of London. Mathematical Society. 1924;**22**:412-419

[17] Borwein D. On products of sequences. J. London Math. Soc. 1958;**33**: 352-357

[18] Töeplitz O. Uberallagemeine lineara Mittelbil dunger. P.M.F. 1913;**22**:113-119

[19] Nigam HK, Hadish M. On approximation of function in generalized Zygmund class using $C^{\eta T}$ operator. Journal of Mathematical Inequalities. 2020;**14**(1):273-289

Chapter 6

Quaternion MPCEP, CEPMP, and MPCEPMP Generalized Inverses

Ivan I. Kyrchei

Abstract

A generalized inverse of a matrix is an inverse in some sense for a wider class of matrices than invertible matrices. Generalized inverses exist for an arbitrary matrix and coincide with a regular inverse for invertible matrices. The most famous generalized inverses are the Moore–Penrose inverse and the Drazin inverse. Recently, new generalized inverses were introduced, namely the core inverse and its generalizations. Among them, there are compositions of the Moore–Penrose and core inverses, MPCEP (or MP–Core–EP) and EPCMP (or EP–Core–MP) inverses. In this chapter, the notions of the MPCEP inverse and CEPMP inverse are expanded to quaternion matrices and introduced new generalized inverses, the right and left MPCEPMP inverses. Direct method of their calculations, that is, their determinantal representations are obtained within the framework of theory of quaternion row-column determinants previously developed by the author. In consequence, these determinantal representations are derived in the case of complex matrices.

Keywords: Moore–Penrose inverse, Drazin inverse, generalized inverse, core-EP inverse, quaternion matrix, noncommutative determinant

1. Introduction

The field of complex (or real) numbers is designated by $\mathbb{C}$ ($\mathbb{R}$). The set of all $m \times n$ matrices over the quaternion skew field

$$\mathbb{H} = \left\{h_0 + h_1\mathbf{i} + h_2\mathbf{j} + h_3\mathbf{k} | \mathbf{i}^2 = \mathbf{j}^2 = \mathbf{k}^2 = \mathbf{ijk} = -1, h_0, h_1, h_2, h_3 \in \mathbb{R}\right\},$$

is represented by $\mathbb{H}^{m\times n}$, while $\mathbb{H}_r^{m\times n}$ is reserved for the subset of $\mathbb{H}^{m\times n}$ with matrices of rank r. If $h = h_0 + h_1\mathbf{i} + h_2\mathbf{j} + h_3\mathbf{k} \in \mathbb{H}$, its conjugate is $\overline{h} = h_0 - h_1\mathbf{i} - h_2\mathbf{j} - h_3\mathbf{k}$, and its norm $\|h\| = \sqrt{h\overline{h}} = \sqrt{\overline{h}h} = \sqrt{h_0^2 + h_1^2 + h_2^2 + h_3^2}$. For $\mathbf{A} \in \mathbb{H}^{m\times n}$, its rank and conjugate transpose are given by $\text{rank}(\mathbf{A})$ and $\mathbf{A}^*$, respectively. A matrix $\mathbf{A} \in \mathbb{H}^{n\times n}$ is said to be Hermitian if $\mathbf{A}^* = \mathbf{A}$. Also,

- $\mathcal{C}_r(\mathbf{A}) = \left\{\mathbf{c} \in \mathbb{H}^{m\times 1} : \mathbf{c} = \mathbf{Ad}, \mathbf{d} \in \mathbb{H}^{n\times 1}\right\}$ is the right column space of $\mathbf{A}$;
- $\mathcal{R}_l(\mathbf{A}) = \left\{\mathbf{c} \in \mathbb{H}^{1\times n} : \mathbf{c} = \mathbf{dA}, \mathbf{d} \in \mathbb{H}^{1\times m}\right\}$ is the left row space of $\mathbf{A}$;

- $\mathcal{N}_r(\mathbf{A}) = \{\mathbf{d} \in \mathbb{H}^{n\times 1} : \mathbf{A}\mathbf{d} = 0\}$ is the right null space of $\mathbf{A}$;

- $\mathcal{N}_l(\mathbf{A}) = \{\mathbf{d} \in \mathbb{H}^{1\times m} : \mathbf{d}\mathbf{A} = 0\}$ is the left null space of $\mathbf{A}$.

Let us recall the definitions of some well-known generalized inverses that can be extend to quaternion matrices as follows.

Definition 1.1. *The Moore–Penrose inverse of* $\mathbf{A} \in \mathbb{H}^{n\times m}$ *is the unique matrix* $\mathbf{A}^{\dagger} = \mathbf{X}$ *determined by equations*

$$(1)\ \mathbf{AXA} = \mathbf{A};\ (2)\ \mathbf{XAX} = \mathbf{X};\ (3)\ (\mathbf{AX})^* = \mathbf{AX};\ (4)\ (\mathbf{XA})^* = \mathbf{XA}. \qquad (1)$$

Definition 1.2. *The Drazin inverse of* $\mathbf{A} \in \mathbb{H}^{n\times n}$ *is the unique* $\mathbf{A}^d = \mathbf{X}$ *that satisfying Eq. (2) from* (1) *and the following equations,*

$$(5)\ \mathbf{A}^k = \mathbf{X}\mathbf{A}^{k+1}, \quad (6)\ \mathbf{XA} = \mathbf{AX},$$

where $k = \mathrm{Ind}(\mathbf{A})$ is the index of $\mathbf{A}$, i.e. the smallest positive number such that $\mathrm{rank}\left(\mathbf{A}^{k+1}\right) = \mathrm{rank}\left(\mathbf{A}^{k}\right)$. If $\mathrm{Ind}(\mathbf{A}) \leq 1$, then $\mathbf{A}^d = \mathbf{A}^{\#}$ is the group inverse of $\mathbf{A}$. If $\mathrm{Ind}(\mathbf{A}) = 0$, then $\mathbf{A}^{\#} = \mathbf{A}^{\dagger} = \mathbf{A}^{-1}$.

A matrix $\mathbf{A}$ satisfying the conditions (i), (j), ... is called an $\{i,j, ...\}$-inverse of $\mathbf{A}$, and is denoted by $\mathbf{A}^{(i,j,...)}$. In particular, $\mathbf{A}^{(1)}$ is called the *inner inverse*, $\mathbf{A}^{(2)}$ is called the *outer inverse*, and $\mathbf{A}^{(1,2)}$ is called the *reflexive inverse*, and $\mathbf{A}^{(1,2,3,4)}$ is the Moore–Penrose inverse, etc.

Note that the Moore–Penrose inverse inducts the orthogonal projectors $\mathbf{P}_A = \mathbf{A}\mathbf{A}^{\dagger}$ and $\mathbf{Q}_A = \mathbf{A}^{\dagger}\mathbf{A}$ onto the right column spaces of $\mathbf{A}$ and $\mathbf{A}^*$, respectively.

In [1], the core-EP inverse over the quaternion skew field was presented similarly as in [2].

Definition 1.3. *The core-EP inverse of* $\mathbf{A} \in \mathbb{H}^{n\times n}$ *is the unique matrix* $\mathbf{A}^{\dagger} = \mathbf{X}$ *which satisfies*

$$\mathbf{X} = \mathbf{XAX}, \quad \mathcal{C}_r(\mathbf{X}) = \mathcal{C}_r\left(\mathbf{A}^d\right) = \mathcal{C}_r(\mathbf{X}^*).$$

According to [3], (Theorem 2.3), for $m \geq \mathrm{Ind}(\mathbf{A})$, we have that $\mathbf{A}^{\textcircled{\dagger}} = \mathbf{A}^d\mathbf{A}^m(\mathbf{A}^m)^{\dagger}$. In a special case that $\mathrm{Ind}(\mathbf{A}) \leq 1$, $\mathbf{A}^{\textcircled{\dagger}} = \mathbf{A}^{\textcircled{\#}}$ is the core inverse of $\mathbf{A}$ [4].

Definition 1.4. *The dual core-EP inverse of* $\mathbf{A} \in \mathbb{H}^{n\times n}$ *is the unique matrix* $\mathbf{A}_{\textcircled{\dagger}} = \mathbf{X}$ *for which*

$$\mathbf{X} = \mathbf{XAX}, \quad \mathcal{R}_l(\mathbf{X}) = \mathcal{R}_l\left(\mathbf{A}^d\right) = \mathcal{R}_l(\mathbf{X}^*).$$

Recall that, $\mathbf{A}_{\textcircled{\dagger}} = (\mathbf{A}^m)^{\dagger}\mathbf{A}^m\mathbf{A}^d$ for $m \geq \mathrm{Ind}(\mathbf{A})$.

Since the quaternion core-EP inverse $\mathbf{A}^{\dagger}$ is related to the right space $\mathcal{C}_r(\mathbf{A})$ of $\mathbf{A} \in \mathbb{H}^{n\times n}$ and the quaternion dual core-EP inverse $\mathbf{A}_{\dagger}$ is related to its left space $\mathcal{R}_l(\mathbf{A})$. So, in [1], they are also named the right and left core-EP inverses, respectively.

Various representations of core-EP inverse can be found in [1, 5–7]. In [8], continuity of core-EP inverse was investigated. Bordering and iterative methods to find the core-EP inverse were proved in [9, 10], and its determinantal representation for complex matrices was derived in [2]. New determinantal representations of the complex core-EP inverse and its various generalizations were obtained in [11]. The core-EP inverse was generalized to rectangular matrices [12], Hilbert space operators [13], Banach algebra elements [14], tensors [15], and elements of rings [3]. Combining the core-EP inverse or the dual core-EP inverse with the Moore–Penrose inverse, the MPCEP inverse and CEPMP inverse were introduced in [16] for bounded linear Hilbert space operators.

In the last years, interest in quaternion matrix equations is growing significantly based on the increasing their applications in various fields, among them, robotic manipulation [17], fluid mechanics [18, 19], quantum mechanics [20–22], signal processing [23, 24], color image processing [25–27], and so on.

The main goals of this chapter are investigations of the MPCEP and CEPMP inverses, introductions and representations of new right and left MPCEPMP inverses over the quaternion skew field, and obtaining of their determinantal representations as a direct method of their constructions. The chapter develops and continues the topic raised in a number of other works [28–33], where determinantal representations of various generalized inverses were obtained.

The remainder of our chapter is directed as follows. In Section 2, we introduce of the quaternion MPCEP and CEPMP inverses and give characterizations of new generalized inverses, namely left and right MPCEPMP-inverses. In Section 3, we commence with introducing determinantal representations of the projection matrices inducted by the Moore–Penrose inverse and of core-EP inverse previously obtained within the framework of theory of quaternion row-column determinants and, based of them, determinantal representations of the MPCEP, CEPMP, and left and right MPCEPMP inverses are derived. Finally, the conclusion is drawn in Section 4.

2. Characterizations of the quaternion MPCEP, CEPMP, and MPCEPMP inverses

Analogously as in [16], the MPCEP inverse and CEPMP inverse can be defined for quaternion matrices.

Definition 2.1. *Let* $\mathbf{A}\in\mathbb{H}^{n\times n}$. *The MPCEP (or MP-Core-EP) inverse of* $\mathbf{A}$ *is the unique solution* $\mathbf{A}^{\dagger,\textcircled{\dagger}}=\mathbf{X}$ *to the system*

$$\mathbf{X}=\mathbf{XAX},\qquad \mathbf{XA}=\mathbf{A}^{\dagger}\mathbf{A}\mathbf{A}^{\textcircled{\dagger}}\mathbf{A},\qquad \mathbf{AX}=\mathbf{A}\mathbf{A}^{\textcircled{\dagger}}.$$

The CEPMP (or Core-EP-MP) inverse of $\mathbf{A}$ is the unique solution $\mathbf{A}^{\textcircled{\dagger},\dagger}=\mathbf{X}$ to the system

$$\mathbf{X}=\mathbf{XAX},\qquad \mathbf{AX}=\mathbf{A}\mathbf{A}_{\textcircled{\dagger}}\mathbf{A}\mathbf{A}^{\dagger},\qquad \mathbf{XA}=\mathbf{A}_{\textcircled{\dagger}}\mathbf{A}.$$

We can represent the MPCEP inverse and CEPMP inverse, by [16], as

$$\mathbf{A}^{\dagger,\textcircled{\dagger}}=\mathbf{A}^{\dagger}\mathbf{A}\mathbf{A}^{\textcircled{\dagger}},\tag{2}$$

$$\mathbf{A}^{\textcircled{\dagger},\dagger}=\mathbf{A}_{\textcircled{\dagger}}\mathbf{A}\mathbf{A}^{\dagger}.\tag{3}$$

According to our concepts, we can define the left and right MPCPMP inverses.

Definition 2.2. *Suppose* $\mathbf{A}\in\mathbb{H}^{n\times n}$. *The right MPCEPMP inverse of* $\mathbf{A}$ *is defined as*

$$\mathbf{A}^{\dagger,\textcircled{\dagger},\dagger,r} = \mathbf{A}^{\dagger}\mathbf{A}\mathbf{A}^{\textcircled{\dagger}}\mathbf{A}\mathbf{A}^{\dagger}.$$

The left MPCEPMP inverse of $\mathbf{A}$ is defined as

$$\mathbf{A}^{\dagger,\textcircled{\dagger},\dagger,l} = \mathbf{A}^{\dagger}\mathbf{A}\mathbf{A}_{\textcircled{\dagger}}\mathbf{A}\mathbf{A}^{\dagger}.$$

The following gives the characteristic equations of these generalized inverses.
Theorem 2.3. *Let* $\mathbf{A}, \mathbf{X} \in \mathbb{H}^{n\times n}$. *The following statements are equivalent:*

i. $\mathbf{X}$ *is the right MPCEPMP inverse of* $\mathbf{A}$.

ii.

$$\mathbf{X} = \mathbf{A}^{\dagger,\textcircled{\dagger}}\mathbf{P}_{A}. \tag{4}$$

iii. $\mathbf{X}$ *is the unique solution to the three equations:*

$$1.\mathbf{X} = \mathbf{XAX}, \qquad 2.\mathbf{XA} = \mathbf{A}^{\dagger,\textcircled{\dagger}}\mathbf{A}, \qquad 3.\mathbf{AX} = \mathbf{AA}^{\textcircled{\dagger}}\mathbf{P}_{\mathbf{A}}. \tag{5}$$

Proof. $[(\mathrm{i})] \mapsto [(\mathrm{ii})]$. By Eq. (2) and the denotation of $\mathbf{P}_A$, it is evident that

$$\mathbf{A}^{\dagger,\textcircled{\dagger},\dagger,r} = \mathbf{A}^{\dagger}\mathbf{A}\mathbf{A}^{\textcircled{\dagger}}\mathbf{A}\mathbf{A}^{\dagger} = \mathbf{A}^{\dagger,\textcircled{\dagger}}\mathbf{P}_{A}.$$

$[(\mathrm{i})] \mapsto [(\mathrm{iii})]$. Now, we verify the condition (5). Let $\mathbf{X} = \mathbf{A}^{\dagger,\textcircled{\dagger},\dagger,r} = \mathbf{A}^{\dagger}\mathbf{A}\mathbf{A}^{\textcircled{\dagger}}\mathbf{A}\mathbf{A}^{\dagger}$. Then, from the Definition 1.1 and the representation (2), we have

$$\mathbf{XAX} = \mathbf{A}^{\dagger}\mathbf{A}\mathbf{A}^{\textcircled{\dagger}}\mathbf{A}\left(\mathbf{A}^{\dagger}\mathbf{A}\mathbf{A}^{\dagger}\right)\mathbf{A}\mathbf{A}^{\textcircled{\dagger}}\mathbf{A}\mathbf{A}^{\dagger} = \mathbf{A}^{\dagger}\mathbf{A}\mathbf{A}^{\textcircled{\dagger}}\left(\mathbf{A}\mathbf{A}^{\dagger}\mathbf{A}\right)\mathbf{A}^{\textcircled{\dagger}}\mathbf{A}\mathbf{A}^{\dagger} =$$

$$= \mathbf{A}^{\dagger}\mathbf{A}\left(\mathbf{A}^{\textcircled{\dagger}}\mathbf{A}\mathbf{A}^{\textcircled{\dagger}}\right)\mathbf{A}\mathbf{A}^{\dagger} = \mathbf{A}^{\dagger}\mathbf{A}\mathbf{A}^{\textcircled{\dagger}}\mathbf{A}\mathbf{A}^{\dagger} = \mathbf{X},$$

$$\mathbf{XA} = \mathbf{A}^{\dagger}\mathbf{A}\mathbf{A}^{\textcircled{\dagger}}\left(\mathbf{A}\mathbf{A}^{\dagger}\mathbf{A}\right) = \mathbf{A}^{\dagger,\textcircled{\dagger}}\mathbf{A},$$

$$\mathbf{AX} = \left(\mathbf{A}\mathbf{A}^{\dagger}\mathbf{A}\right)\mathbf{A}^{\textcircled{\dagger}}\mathbf{A}\mathbf{A}^{\dagger} = \mathbf{A}\mathbf{A}^{\textcircled{\dagger}}\mathbf{P}_{A}.$$

To prove that the system (5) has unique solution, suppose that $\mathbf{X}$ and $\mathbf{X}_1$ are two solutions of this system. Then $\mathbf{XA} = \mathbf{A}^{\dagger,\textcircled{\dagger}}\mathbf{A} = \mathbf{X}_1\mathbf{A}$ and $\mathbf{AX} = \mathbf{AA}^{\textcircled{\dagger}}\mathbf{P}_A = \mathbf{AX}_1$, which give $\mathbf{X}(\mathbf{AX}) = (\mathbf{XA})\mathbf{X}_1 = \mathbf{X}_1\mathbf{AX}_1 = \mathbf{X}_1$. Therefore, $\mathbf{X}$ is the unique solution to the system. □

The next theorem can be proved in the same way.
Theorem 2.4. *Let* $\mathbf{A}, \mathbf{X} \in \mathbb{H}^{n\times n}$. *The following statements are equivalent:*

i. $\mathbf{X}$ *is the left MPCEPMP inverse of* $\mathbf{A}$.

ii.

$$\mathbf{X} = \mathbf{Q}_{A}\mathbf{A}^{\textcircled{\dagger},\dagger}. \tag{6}$$

iii. $\mathbf{X}$ *is the unique solution to the system:*

$$1.\mathbf{X} = \mathbf{XAX}, \qquad 2.\mathbf{AX} = \mathbf{AA}^{\textcircled{\dagger},\dagger}, \qquad 3.\mathbf{XA} = \mathbf{A}_{\textcircled{\dagger}}\mathbf{A}.$$

3. Determinantal representations of the quaternion MPCEP and *CEPMP inverses

It is well known that the determinantal representation of the regular inverse is given by the cofactor matrix. The construction of determinantal representations of generalized inverses is not so evident and unambiguous even for matrices with complex or real entries. Taking into account the noncommutativity of quaternions, this task is more complicated due to a problem of defining the determinant of a matrix with noncommutative elements (see survey articles [34–36] for detail). Only now, the solving this problem begins to be decided thanks to the theory of noncommutative column-row determinants introduced in [37, 38].

For arbitrary quaternion matrix $\mathbf{A} \in \mathbb{H}^{n \times n}$, there exists an exact technique to generate n row determinants ($\mathfrak{R}$-determinants) and n column determinants ($\mathfrak{C}$-determinants) by stating a certain order of factors in each term.

Definition 3.1. *Let* $\mathbf{A} = (a_{ij}) \in \mathbb{H}^{n \times n}$.

- *For an arbitrary row index* $i \in I_n$*, the ith* $\mathfrak{R}$*-determinant of* $\mathbf{A}$ *is defined as*

$$\mathrm{rdet}_i \mathbf{A} := \sum_{\sigma \in S_n} (-1)^{n-r} \left(a_{i i_{k_1}} a_{i_{k_1} i_{k_1+1}} \dots a_{i_{k_1+l_1} i} \right) \dots \left(a_{i_{k_r} i_{k_r+1}} \dots a_{i_{k_r+l_r} i_{k_r}} \right),$$

in which S_n denotes the symmetric group on $I_n = \{1, \dots, n\}$, while the permutation σ is defined as a product of mutually disjunct subsets ordered from the left to right by the rules

$$\sigma = \left(i i_{k_1} i_{k_1+1} \dots i_{k_1+l_1} \right) \left(i_{k_2} i_{k_2+1} \dots i_{k_2+l_2} \right) \dots \left(i_{k_r} i_{k_r+1} \dots i_{k_r+l_r} \right),$$

$$i_{k_t} < i_{k_t+s}, \quad i_{k_2} < i_{k_3} < \dots < i_{k_r}, \quad \forall\ t = 2, \dots, r, \quad s = 1, \dots, l_t.$$

- *For an arbitrary column index* $j \in I_n$*, the jth* $\mathfrak{C}$*-determinant of* $\mathbf{A}$ *is defined as the sum*

$$\mathrm{cdet}_j \mathbf{A} = \sum_{\tau \in S_n} (-1)^{n-r} \left(a_{j_{k_r} j_{k_r+l_r}} \cdots a_{j_{k_r+1} j_{k_r}} \right) \cdots \left(a_{j j_{k_1+l_1}} \cdots a_{j_{k_1+1} j_{k_1}} a_{j_{k_1} j} \right),$$

in which a *permutation* τ is ordered from the right to left in the following way:

$$\tau = \left(j_{k_r+l_r} \cdots j_{k_r+1} j_{k_r} \right) \cdots \left(j_{k_2+l_2} \cdots j_{k_2+1} j_{k_2} \right) \left(j_{k_1+l_1} \cdots j_{k_1+1} j_{k_1} j \right), j_{k_t} < j_{k_t+s}, \quad j_{k_2} < j_{k_3} < \cdots < j_{k_r}.$$

It is known that all $\mathfrak{R}$- and $\mathfrak{C}$-determinants are different in general. However, in [37], the following equalities are verified for a Hermitian matrix $\mathbf{A}$ that introduce a determinant of a Hermitian matrix: $\mathrm{rdet}_1 \mathbf{A} = \cdots = \mathrm{rdet}_n \mathbf{A} = \mathrm{cdet}_1 \mathbf{A} = \cdots = \mathrm{cdet}_n \mathbf{A} := \det \mathbf{A} \in \mathbb{R}$.

$\mathfrak{D}$-Representations of various generalized inverses were developed by means of the theory of $\mathfrak{R}$- and $\mathfrak{C}$-determinants (see e.g. [28–31]).

The following notations are used for determinantal representations of generalized inverses.

Let $\alpha := \{\alpha_1, \dots, \alpha_k\} \subseteq \{1, \dots, m\}$ and $\beta := \{\beta_1, \dots, \beta_k\} \subseteq \{1, \dots, n\}$ be subsets with $1 \leq k \leq \min\{m, n\}$. Suppose that $\mathbf{A}^{\alpha}_{\beta}$ is a submatrix of $\mathbf{A} \in \mathbb{H}^{m \times n}$ whose rows and columns are indexed by α and β, respectively. Then, $\mathbf{A}^{\alpha}_{\alpha}$ is a principal submatrix of $\mathbf{A}$

whose rows and columns are indexed by α. If $\mathbf{A}$ is Hermitian, then $|\mathbf{A}|_\alpha^\alpha$ stands for a principal minor of $\det\mathbf{A}$. The collection of strictly increasing sequences of $1 \le k \le n$ integers chosen from $\{1, \dots, n\}$ is denoted by $L_{k,n} := \{\alpha : \alpha = (\alpha_1, \dots, \alpha_k), 1 \le \alpha_1 < \dots < \alpha_k \le n\}$. For fixed $i \in \alpha$ and $j \in \beta$, put $I_{r,m}\{i\} := \{\alpha : \alpha \in L_{r,m}, i \in \alpha\}, J_{r,n}\{j\} := \{\beta : \beta \in L_{r,n}, j \in \beta\}$.

Let $\mathbf{a}_{.j}$ and $\mathbf{a}_{.j}^*$ be the jth columns, $\mathbf{a}_{i.}$ and $\mathbf{a}_{i.}^*$ be the ith rows of $\mathbf{A}$ and $\mathbf{A}^*$, respectively. Suppose that $\mathbf{A}_{i.}(\mathbf{b})$ and $\mathbf{A}_{.j}(\mathbf{c})$ stand for the matrices obtained from $\mathbf{A}$ by replacing its ith row with the row vector $\mathbf{b} \in \mathbb{H}^{1\times n}$ and its jth column with the column vector $\mathbf{c} \in \mathbb{H}^m$, respectively.

Based on determinantal representations of the Moore–Penrose inverse obtained in [28], we have determinantal representations of the projections.

Lemma 3.2. [28] *If* $\mathbf{A} \in \mathbb{H}_r^{m\times n}$, *then the determinantal representations of the projection matrices* $\mathbf{A}^\dagger\mathbf{A} =: \mathbf{Q}_A = \left(q_{ij}^A\right)_{n\times n}$ *and* $\mathbf{A}\mathbf{A}^\dagger =: \mathbf{P}_A = \left(p_{ij}^A\right)_{m\times m}$ *can be expressed as follows*

$$q_{ij}^A = \frac{\sum_{\beta \in J_{r,n}\{i\}} \operatorname{cdet}_i\left((\mathbf{A}^*\mathbf{A})_{.i}(\dot{\mathbf{a}}_{.j})\right)_\beta^\beta}{\sum_{\beta \in J_{r,n}} |\mathbf{A}^*\mathbf{A}|_\beta^\beta} = \frac{\sum_{\alpha \in I_{r,n}\{j\}} \operatorname{rdet}_j\left((\mathbf{A}^*\mathbf{A})_{j.}(\dot{\mathbf{a}}_{i.})\right)_\alpha^\alpha}{\sum_{\alpha \in I_{r,n}} |\mathbf{A}^*\mathbf{A}|_\alpha^\alpha}, \tag{7}$$

$$p_{ij}^A = \frac{\sum_{\alpha \in I_{r,m}\{j\}} \operatorname{rdet}_j\left((\mathbf{A}\mathbf{A}^*)_{j.}(\ddot{\mathbf{a}}_{i.})\right)_\alpha^\alpha}{\sum_{\alpha \in I_{r,m}} |\mathbf{A}\mathbf{A}^*|_\alpha^\alpha} = \frac{\sum_{\beta \in J_{r,m}\{i\}} \operatorname{cdet}_i\left((\mathbf{A}\mathbf{A}^*)_{.i}(\ddot{\mathbf{a}}_{.j})\right)_\beta^\beta}{\sum_{\beta \in J_{r,m}} |\mathbf{A}\mathbf{A}^*|_\beta^\beta}, \tag{8}$$

where $\dot{\mathbf{a}}_{i.}$ and $\dot{\mathbf{a}}_{.j}$, $\ddot{\mathbf{a}}_{i.}$ and $\ddot{\mathbf{a}}_{.j}$ are the ith rows and the jth columns of $\mathbf{A}^*\mathbf{A} \in \mathbb{H}^{n\times n}$ and $\mathbf{A}\mathbf{A}^* \in \mathbb{H}^{m\times m}$, respectively.

Recently, $\mathfrak{D}$-representations of the quaternion core-EP inverses were obtained in [1] as well.

Lemma 3.3. [1] *Suppose that* $\mathbf{A} \in \mathbb{H}^{n\times n}$, $\operatorname{Ind}(\mathbf{A}) = k$ *and* $\operatorname{rank}\left(\mathbf{A}^k\right) = s$. *Then* $\mathbf{A}^{\textcircled{\dagger}} = \left(a_{ij}^{\textcircled{\dagger},r}\right)$ *and* $\mathbf{A}_{\textcircled{\dagger}} = \left(a_{ij}^{\textcircled{\dagger},l}\right)$ *possess the determinantal representations, respectively,*

$$a_{ij}^{\textcircled{\dagger},r} = \frac{\sum_{\alpha \in I_{s,n}\{j\}} \operatorname{rdet}_j\left(\left(\mathbf{A}^{k+1}\left(\mathbf{A}^{k+1}\right)^*\right)_{j.}(\hat{\mathbf{a}}_{i.})\right)_\alpha^\alpha}{\sum_{\alpha \in I_{s,n}} \left|\mathbf{A}^{k+1}\left(\mathbf{A}^{k+1}\right)^*\right|_\alpha^\alpha}, \tag{9}$$

$$a_{ij}^{\textcircled{\dagger},l} = \frac{\sum_{\beta \in J_{s,n}\{i\}} \operatorname{cdet}_i\left(\left(\left(\mathbf{A}^{k+1}\right)^*\mathbf{A}^{k+1}\right)_{.i}(\check{\mathbf{a}}_{.j})\right)_\beta^\beta}{\sum_{\beta \in J_{s,n}} \left|\left(\mathbf{A}^{k+1}\right)^*\mathbf{A}^{k+1}\right|_\beta^\beta}, \tag{10}$$

where $\hat{\mathbf{a}}_{i.}$ is the ith row of $\hat{\mathbf{A}} = \mathbf{A}^k\left(\mathbf{A}^{k+1}\right)^*$ and $\check{\mathbf{a}}_{.j}$ is the jth column of $\check{\mathbf{A}} = \left(\mathbf{A}^{k+1}\right)^*\mathbf{A}^k$.

Theorem 3.4. *Let* $\mathbf{A} \in \mathbb{H}_s^{n\times n}$, $\operatorname{Ind}(\mathbf{A}) = k$ *and* $\operatorname{rank}\left(\mathbf{A}^k\right) = s_1$. *Then its MPCEP inverse* $\mathbf{A}^{\dagger,\textcircled{\dagger}} = \left(a_{ij}^{\dagger,\textcircled{\dagger}}\right)$ *is expressed by componentwise*

$$a_{ij}^{\dagger,\dagger} = \frac{\sum_{\alpha \in I_{s_1,n}\{j\}} \mathrm{rdet}_j \left(\left(\mathbf{A}^{k+1} \left(\mathbf{A}^{k+1} \right)^* \right)_{j.} \left(\mathbf{v}_{i.}^{(1)} \right) \right)_{\alpha}^{\alpha}}{\sum_{\beta \in J_{s,n}} |\mathbf{A}^* \mathbf{A}|_{\beta}^{\beta} \sum_{\alpha \in I_{s_1,n}} \left| \mathbf{A}^{k+1} \left(\mathbf{A}^{k+1} \right)^* \right|_{\alpha}^{\alpha}} \tag{11}$$

$$= \frac{\sum_{\beta \in J_{s,n}\{i\}} \mathrm{cdet}_i \left((\mathbf{A}^* \mathbf{A})_{.i} \left(\mathbf{u}_{j}^{(1)} \right) \right)_{\beta}^{\beta}}{\sum_{\beta \in J_{s,n}} |\mathbf{A}^* \mathbf{A}|_{\beta}^{\beta} \sum_{\alpha \in I_{s_1,n}} \left| \mathbf{A}^{k+1} \left(\mathbf{A}^{k+1} \right)^* \right|_{\alpha}^{\alpha}}, \tag{12}$$

where

$$\mathbf{v}_{i.}^{(1)} = \left[\sum_{\beta \in J_{s,n}\{i\}} \mathrm{cdet}_i \left((\mathbf{A}^* \mathbf{A})_{.i} (\tilde{\mathbf{a}}_{.l}) \right)_{\beta}^{\beta} \right] \in \mathbb{H}^{1 \times n}, \ l = 1, \dots, n, \tag{13}$$

$$\mathbf{u}_{j}^{(1)} = \left[\sum_{\alpha \in I_{s_1,n}\{j\}} \mathrm{rdet}_j \left(\left(\mathbf{A}^{k+1} \left(\mathbf{A}^{k+1} \right)^* \right)_{j.} (\tilde{\mathbf{a}}_{f.}) \right)_{\alpha}^{\alpha} \right] \in \mathbb{H}^{n \times 1}, \ f = 1, \dots, n,$$

and $\tilde{\mathbf{a}}_{.l}$ and $\tilde{\mathbf{a}}_{f.}$ are the lth column and the fth row of $\tilde{\mathbf{A}} = \mathbf{A}^* \mathbf{A}^{k+1} \left(\mathbf{A}^{k+1} \right)^*$.

Proof. By (2), we have

$$a_{ij}^{\dagger,\dagger} = \sum_{l=1}^{n} q_{il} a_{lj}^{\dagger,r}. \tag{14}$$

Using (7) and (9) for the determinantal representations of $\mathbf{Q}_A = \mathbf{A}^\dagger \mathbf{A} = \left(q_{ij} \right)$ and $\mathbf{A}^\dagger$, respectively, from (14) it follows

$$a_{ij}^{\dagger,\dagger} = \sum_{l=1}^{n} \sum_{f=1}^{n} \frac{\sum_{\beta \in J_{s,n}\{i\}} \mathrm{cdet}_i \left((\mathbf{A}^* \mathbf{A})_{.i} (\dot{\mathbf{a}}_{.f}) \right)_{\beta}^{\beta}}{\sum_{\beta \in J_{s,n}} |\mathbf{A}^* \mathbf{A}|_{\beta}^{\beta}} \times \frac{\sum_{\alpha \in I_{s_1,n}\{j\}} \mathrm{rdet}_j \left(\left(\mathbf{A}^{k+1} \left(\mathbf{A}^{k+1} \right)^* \right)_{j.} (\hat{\mathbf{a}}_{l.}) \right)_{\alpha}^{\alpha}}{\sum_{\alpha \in I_{s_1,n}} \left| \mathbf{A}^{k+1} \left(\mathbf{A}^{k+1} \right)^* \right|_{\alpha}^{\alpha}} =$$

$$= \sum_{l=1}^{n} \sum_{f=1}^{n} \frac{\sum_{\beta \in J_{s,n}\{i\}} \mathrm{cdet}_i \left((\mathbf{A}^* \mathbf{A})_{.i} (\mathbf{e}_{.f}) \right)_{\beta}^{\beta}}{\sum_{\beta \in J_{s,n}} |\mathbf{A}^* \mathbf{A}|_{\beta}^{\beta}} \tilde{a}_{fl} \frac{\sum_{\alpha \in I_{s_1,n}\{j\}} \mathrm{rdet}_j \left(\left(\mathbf{A}^{k+1} \left(\mathbf{A}^{k+1} \right)^* \right)_{j.} (\mathbf{e}_{l.}) \right)_{\alpha}^{\alpha}}{\sum_{\alpha \in I_{s_1,n}} \left| \mathbf{A}^{k+1} \left(\mathbf{A}^{k+1} \right)^* \right|_{\alpha}^{\alpha}},$$

where $\mathbf{e}_{.f}$ and $\mathbf{e}_{l.}$ are the fth column and the lth row of the unit matrix $\mathbf{I}_n$, $\hat{\mathbf{a}}_{l.}$ is the lth row of $\hat{\mathbf{A}} = \mathbf{A}^k \left(\mathbf{A}^{k+1} \right)^*$, and $\tilde{a}_{fl}$ is the (fl)th element of $\tilde{\mathbf{A}} = \mathbf{A}^* \mathbf{A}^{k+1} \left(\mathbf{A}^{k+1} \right)^*$.

If we denote by

$$v_{il}^{(1)} := \sum_{f=1}^{n} \sum_{\beta \in J_{s,n}\{i\}} \mathrm{cdet}_i \left((\mathbf{A}^* \mathbf{A})_{.i} (\mathbf{e}_{.f}) \right)_{\beta}^{\beta} \tilde{\mathbf{a}}_{fl} = \sum_{\beta \in J_{s,n}\{i\}} \mathrm{cdet}_i \left((\mathbf{A}^* \mathbf{A})_{.i} (\tilde{\mathbf{a}}_{.l}) \right)_{\beta}^{\beta}$$

the lth component of a row-vector $\mathbf{v}_{i.}^{(1)} = \left[v_{i1}^{(1)}, \dots, v_{in}^{(1)} \right]$, then

$$\sum_{l=1}^{n} v_{il}^{(1)} \sum_{\alpha \in I_{s_1,n}\{j\}} \mathrm{rdet}_j \left(\left(\mathbf{A}^{k+1} \left(\mathbf{A}^{k+1} \right)^* \right)_{j.} (\mathbf{e}_{l.}) \right)_{\alpha}^{\alpha}$$

$$= \sum_{\alpha \in I_{s_1,n}\{j\}} \mathrm{rdet}_j \left(\left(\mathbf{A}^{k+1} \left(\mathbf{A}^{k+1} \right)^* \right)_{j.} \left(\mathbf{v}_{i.}^{(1)} \right) \right)_{\alpha}^{\alpha}.$$

So, we have (11). By putting

$$u_{fj}^{(1)} := \sum_{l=1}^{n} \tilde{a}_{fl} \sum_{\alpha \in I_{s_1,n}\{j\}} \mathrm{rdet}_j\left(\left(\mathbf{A}^{k+1}\left(\mathbf{A}^{k+1}\right)^*\right)_{j.}(\mathbf{e}_{l.})\right)_\alpha^\alpha$$
$$= \sum_{\alpha \in I_{s_1,n}\{j\}} \mathrm{rdet}_j\left(\left(\mathbf{A}^{k+1}\left(\mathbf{A}^{k+1}\right)^*\right)_{j.}(\tilde{\mathbf{a}}_{f.})\right)_\alpha^\alpha$$

as the fth component of a column-vector $\mathbf{u}_j^{(1)} = \left[u_{1j}^{(1)}, \dots, u_{nj}^{(1)}\right]^T$, it follows

$$\sum_{f=1}^{n} \sum_{\beta \in J_{s,n}\{i\}} \mathrm{cdet}_i((\mathbf{A}^*\mathbf{A})_{.i}(\mathbf{e}_{.f}))_\beta^\beta u_{fj}^{(1)} = \sum_{\beta \in J_{s,n}\{i\}} \mathrm{cdet}_i\left((\mathbf{A}^*\mathbf{A})_{.i}\left(\mathbf{u}_j^{(1)}\right)\right)_\beta^\beta.$$

Hence, we obtain (12). □

Determinantal representations of a complex MPCEP inverse are obtained by substituting row-column determinants for usual determinants in (11)–(12).

Corollary 3.5. *Let* $\mathbf{A} \in \mathbb{C}_s^{n\times n}$, $\mathrm{Ind}(\mathbf{A}) = k$ *and* $\mathrm{rank}\left(\mathbf{A}^k\right) = s_1$. *Then its MPCEP inverse* $\mathbf{A}^{\dagger,\dagger} = \left(a_{ij}^{\dagger,\dagger}\right)$ *has the following determinantal representations*

$$a_{ij}^{\dagger,\dagger} = \frac{\sum_{\alpha \in I_{s_1,n}\{j\}} \left|\left(\mathbf{A}^{k+1}\left(\mathbf{A}^{k+1}\right)^*\right)_{j.}\left(\mathbf{v}_{i.}^{(1)}\right)\right|_\alpha^\alpha}{\sum_{\beta \in J_{s,n}} |\mathbf{A}^*\mathbf{A}|_\beta^\beta \sum_{\alpha \in I_{s_1,n}} \left|\mathbf{A}^{k+1}\left(\mathbf{A}^{k+1}\right)^*\right|_\alpha^\alpha}$$
$$= \frac{\sum_{\beta \in J_{s,n}\{i\}} \left|(\mathbf{A}^*\mathbf{A})_{.i}\left(\mathbf{u}_j^{(1)}\right)\right|_\beta^\beta}{\sum_{\beta \in J_{s,n}} |\mathbf{A}^*\mathbf{A}|_\beta^\beta \sum_{\alpha \in I_{s_1,n}} \left|\mathbf{A}^{k+1}\left(\mathbf{A}^{k+1}\right)^*\right|_\alpha^\alpha},$$

where

$$\mathbf{v}_{i.}^{(1)} = \left[\sum_{\beta \in J_{s,n}\{i\}} |(\mathbf{A}^*\mathbf{A})_{.i}(\tilde{\mathbf{a}}_{.l})|_\beta^\beta\right] \in \mathbb{C}^{1\times n},\ l = 1, \dots, n,$$
$$\mathbf{u}_{j}^{(1)} = \left[\sum_{\alpha \in I_{s_1,n}\{j\}} \left|\left(\mathbf{A}^{k+1}\left(\mathbf{A}^{k+1}\right)^*\right)_{j.}(\tilde{\mathbf{a}}_{f.})\right|_\alpha^\alpha\right] \in \mathbb{C}^{n\times 1},\ f = 1, \dots, n, \quad (15)$$

and $\tilde{\mathbf{a}}_{.l}$ *and* $\tilde{\mathbf{a}}_{f.}$ *are the lth column and the fth row of* $\tilde{\mathbf{A}} = \mathbf{A}^*\mathbf{A}^{k+1}\left(\mathbf{A}^{k+1}\right)^*$.

Theorem 3.6. *Let* $\mathbf{A} \in \mathbb{H}_s^{n\times n}$, $\mathrm{Ind}(\mathbf{A}) = k$ *and* $\mathrm{rank}\left(\mathbf{A}^k\right) = s_1$. *Then its CEPMP inverse* $\mathbf{A}^{\dagger,\dagger} = \left(a_{ij}^{\dagger,\dagger}\right)$ *has the following determinantal representations*

$$a_{ij}^{\dagger,\dagger} = \frac{\sum_{\alpha \in I_{s,n}\{j\}} \mathrm{rdet}_j \left((\mathbf{A}\mathbf{A}^*)_{j.} \left(\mathbf{v}_{i.}^{(2)} \right) \right)_\alpha^\alpha}{\sum_{\alpha \in I_{s,n}} |\mathbf{A}\mathbf{A}^*|_\alpha^\alpha \sum_{\beta \in J_{s_1,n}} \left| \left(\mathbf{A}^{k+1}\right)^* \mathbf{A}^{k+1} \right|_\beta^\beta} \tag{16}$$

$$= \frac{\sum_{\beta \in J_{s_1,n}\{i\}} \mathrm{cdet}_i \left(\left(\left(\mathbf{A}^{k+1}\right)^* \mathbf{A}^{k+1} \right)_{.i} \left(\mathbf{u}_{j}^{(2)} \right) \right)_\beta^\beta}{\sum_{\alpha \in I_{s,n}} |\mathbf{A}\mathbf{A}^*|_\alpha^\alpha \sum_{\beta \in J_{s_1,n}} \left| \left(\mathbf{A}^{k+1}\right)^* \mathbf{A}^{k+1} \right|_\beta^\beta}, \tag{17}$$

where

$$\mathbf{v}_{i.}^{(2)} = \left[\sum_{\beta \in J_{s_1,n}\{i\}} \mathrm{cdet}_i \left(\left(\left(\mathbf{A}^{k+1}\right)^* \mathbf{A}^{k+1} \right)_{.i} (\hat{\mathbf{a}}_{.l}) \right)_\beta^\beta \right] \in \mathbb{H}^{1\times n}, \ l = 1, \dots, n,$$
$$\mathbf{u}_{j}^{(2)} = \left[\sum_{\alpha \in I_{s,n}\{j\}} \mathrm{rdet}_j \left((\mathbf{A}\mathbf{A}^*)_{j.} (\hat{\mathbf{a}}_{f.}) \right)_\alpha^\alpha \right] \in \mathbb{H}^{n\times 1}, \ f = 1, \dots, n. \tag{18}$$

Here $\hat{\mathbf{a}}_{.l}$ and $\hat{\mathbf{a}}_{f.}$ are the lth column and the fth row of $\hat{\mathbf{A}} = \left(\mathbf{A}^{k+1}\right)^* \mathbf{A}^{k+1}\mathbf{A}^*$.

Proof. The proof is similar to the proof of Theorem 3.4 by using the representation (3) for the CEPMP inverse.

Corollary 3.7. *Let* $\mathbf{A} \in \mathbb{C}_s^{n\times n}$, $\mathrm{Ind}(\mathbf{A}) = k$ *and* $\mathrm{rank}\left(\mathbf{A}^k\right) = s_1$. *Then its CEPMP inverse* $\mathbf{A}^{\dagger,\dagger} = \left(a_{ij}^{\dagger,\dagger}\right)$ *has the following determinantal representations*

$$a_{ij}^{\dagger,\dagger} = \frac{\sum_{\alpha \in I_{s,n}\{j\}} \left| (\mathbf{A}\mathbf{A}^*)_{j.} \left(\mathbf{v}_{i.}^{(2)} \right) \right|_\alpha^\alpha}{\sum_{\alpha \in I_{s,n}} |\mathbf{A}\mathbf{A}^*|_\alpha^\alpha \sum_{\beta \in J_{s_1,n}} \left| \left(\mathbf{A}^{k+1}\right)^* \mathbf{A}^{k+1} \right|_\beta^\beta}$$
$$= \frac{\sum_{\beta \in J_{s_1,n}\{i\}} \mathrm{cdet}_i \left(\left(\left(\mathbf{A}^{k+1}\right)^* \mathbf{A}^{k+1} \right)_{.i} \left(\mathbf{u}_{j}^{(2)} \right) \right)_\beta^\beta}{\sum_{\alpha \in I_{s,n}} |\mathbf{A}\mathbf{A}^*|_\alpha^\alpha \sum_{\beta \in J_{s_1,n}} \left| \left(\mathbf{A}^{k+1}\right)^* \mathbf{A}^{k+1} \right|_\beta^\beta},$$

where

$$\mathbf{v}_{i.}^{(2)} = \left[\sum_{\beta \in J_{s_1,n}\{i\}} \left| \left(\left(\mathbf{A}^{k+1}\right)^* \mathbf{A}^{k+1} \right)_{.i} (\hat{\mathbf{a}}_{.l}) \right|_\beta^\beta \right] \in \mathbb{C}^{1\times n}, \ l = 1, \dots, n,$$
$$\mathbf{u}_{j}^{(2)} = \left[\sum_{\alpha \in I_{s,n}\{j\}} \left| (\mathbf{A}\mathbf{A}^*)_{j.} (\hat{\mathbf{a}}_{f.}) \right|_\alpha^\alpha \right] \in \mathbb{C}^{n\times 1}, \ f = 1, \dots, n. \tag{19}$$

Here $\hat{\mathbf{a}}_{.l}$ and $\hat{\mathbf{a}}_{f.}$ are the lth column and the fth row of $\hat{\mathbf{A}} = \left(\mathbf{A}^{k+1}\right)^* \mathbf{A}^{k+1}\mathbf{A}^*$.

Theorem 3.8. *Let* $\mathbf{A} \in \mathbb{H}_s^{n\times n}$, $\mathrm{Ind}(\mathbf{A}) = k$ *and* $\mathrm{rank}\left(\mathbf{A}^k\right) = s_1$. *Then its right MPCEPMP inverse* $\mathbf{A}^{\dagger,\dagger,\dagger,r} = \left(a_{ij}^{\dagger,\dagger,\dagger,r}\right)$ *has the following determinantal representations*

$$a_{ij}^{\dagger,\dagger,\dagger,r} = \frac{\sum_{\alpha \in I_{s,n}\{j\}} \operatorname{rdet}_j \left((\mathbf{AA}^*)_{j.} \left(\phi_{i.}^{(1)} \right) \right)_{\alpha}^{\alpha}}{\left(\sum_{\alpha \in I_{s,n}} |\mathbf{AA}^*|_{\alpha}^{\alpha} \right)^2 \sum_{\beta \in J_{s_1,n}} \left| \left(\mathbf{A}^{k+1} \right)^* \mathbf{A}^{k+1} \right|_{\beta}^{\beta}} = \tag{20}$$

$$= \frac{\sum_{\beta \in J_{s_1,n}\{i\}} \operatorname{cdet}_i \left(\left(\left(\mathbf{A}^{k+1} \right)^* \mathbf{A}^{k+1} \right)_{.i} \left(\psi_{.j}^{(1)} \right) \right)_{\beta}^{\beta}}{\left(\sum_{\beta \in I_{s,n}} |\mathbf{AA}^*|_{\beta}^{\beta} \right)^2 \sum_{\alpha \in J_{s_1,n}} \left| \left(\mathbf{A}^{k+1} \right)^* \mathbf{A}^{k+1} \right|_{\alpha}^{\alpha}}, \tag{21}$$

where

$$\phi_{i.}^{(1)} = \left[\sum_{\beta \in J_{s_1,n}\{i\}} \operatorname{cdet}_i \left(\left(\left(\mathbf{A}^{k+1} \right)^* \mathbf{A}^{k+1} \right)_{.i} (\hat{\mathbf{u}}_{.l}) \right)_{\beta}^{\beta} \right] \in \mathbb{H}^{1 \times n}, \; l = 1, \dots, n,$$

$$\psi_{.j}^{(1)} = \left[\sum_{\alpha \in I_{s,n}\{j\}} \operatorname{rdet}_j \left((\mathbf{AA}^*)_{j.} (\hat{\mathbf{u}}_{f.}) \right)_{\alpha}^{\alpha} \right] \in \mathbb{H}^{n \times 1}, \; f = 1, \dots, n.$$

Here $\hat{\mathbf{u}}_{.l}$ and $\hat{\mathbf{u}}_{f.}$ are the lth column and the fth row of $\hat{\mathbf{U}} = \mathbf{U}_2\mathbf{AA}^*$, and the matrix $\mathbf{U}_2$ is constructed from the columns (18).

Proof. Owing to (4), we have

$$a_{ij}^{\dagger,\dagger,\dagger,r} = \sum_{t=1}^{n} a_{it}^{\dagger,\dagger} p_{tj}. \tag{22}$$

Applying (8) for the determinantal representation of $\mathbf{P}_A = \mathbf{AA}^\dagger = \left(p_{ij} \right)$ and (17) for the determinantal representation of $\mathbf{A}^{\dagger,\dagger}$ in (22), we obtain

$$a_{ij}^{\dagger,\dagger,\dagger,r} = \sum_{t=1}^{n} \frac{\sum_{\beta \in J_{s_1,n}\{i\}} \operatorname{cdet}_i \left(\left(\left(\mathbf{A}^{k+1} \right)^* \mathbf{A}^{k+1} \right)_{.i} \left(\mathbf{u}_{.t}^{(2)} \right) \right)_{\beta}^{\beta}}{\sum_{\alpha \in I_{s,n}} |\mathbf{AA}^*|_{\alpha}^{\alpha} \sum_{\beta \in I_{s_1,n}} \left| \left(\mathbf{A}^{k+1} \right)^* \mathbf{A}^{k+1} \right|_{\beta}^{\beta}} \times \frac{\sum_{\alpha \in I_{s,n}\{j\}} \operatorname{rdet}_j \left((\mathbf{AA}^*)_{j.} (\ddot{\mathbf{a}}_{t.}) \right)_{\alpha}^{\alpha}}{\sum_{\alpha \in I_{s,n}} |\mathbf{AA}^*|_{\alpha}^{\alpha}} =$$

$$= \sum_{l=1}^{n} \sum_{f=1}^{n} \frac{\sum_{\beta \in J_{s_1,n}\{i\}} \operatorname{cdet}_i \left(\left(\left(\mathbf{A}^{k+1} \right)^* \mathbf{A}^{k+1} \right)_{.i} (\mathbf{e}_{.f}) \right)_{\beta}^{\beta}}{\sum_{\beta \in J_{s_1,n}} \left| \left(\mathbf{A}^{k+1} \right)^* \mathbf{A}^{k+1} \right|_{\beta}^{\beta}} \hat{u}_{fl} \times \frac{\sum_{\alpha \in I_{s,n}\{j\}} \operatorname{rdet}_j \left((\mathbf{AA}^*)_{j.} (\mathbf{e}_{l.}) \right)_{\alpha}^{\alpha}}{\left(\sum_{\alpha \in I_{s,n}} |\mathbf{AA}^*|_{\alpha}^{\alpha} \right)^2},$$

where $\mathbf{e}_{.f}$ and $\mathbf{e}_{l.}$ are the fth column and the lth row of the unit matrix $\mathbf{I}_n$, and $\hat{u}_{fl}$ is the (fl)th element of $\hat{\mathbf{U}} = \mathbf{U}_2\mathbf{AA}^*$. The matrix $\mathbf{U}_2 = \left[\mathbf{u}_{.1}^{(2)}, \dots, \mathbf{u}_{.n}^{(2)} \right]$ is constructed from the columns (18). If we denote by

$$\phi_{il}^{(1)} := \sum_{f=1}^{n} \sum_{\beta \in J_{s_1,n}\{i\}} \operatorname{cdet}_i \left(\left(\left(\mathbf{A}^{k+1} \right)^* \mathbf{A}^{k+1} \right)_{.i} (\mathbf{e}_{.f}) \right)_{\beta}^{\beta} \hat{u}_{fl}$$

$$= \sum_{\beta \in J_{s_1,n}\{i\}} \operatorname{cdet}_i \left(\left(\left(\mathbf{A}^{k+1} \right)^* \mathbf{A}^{k+1} \right)_{.i} (\hat{\mathbf{u}}_{.l}) \right)_{\beta}^{\beta}$$

the lth component of a row-vector $\phi_{i.}^{(1)} = \left[\phi_{i1}^{(1)}, \ldots, \phi_{in}^{(1)}\right]$, then

$$\sum_{l=1}^{n} \phi_{il}^{(1)} \sum_{\alpha \in I_{s,n}\{j\}} \operatorname{rdet}_j\left((\mathbf{AA}^*)_{j.}(\mathbf{e}_{l.})\right)_{\alpha}^{\alpha} = \sum_{\alpha \in I_{s,n}\{j\}} \operatorname{rdet}_j\left((\mathbf{AA}^*)_{j.}\left(\phi_{i.}^{(1)}\right)\right)_{\alpha}^{\alpha}.$$

Therefore, (20) holds.

By putting

$$\psi_{fj}^{(1)} := \sum_{l=1}^{n} \hat{a}_{fl} \sum_{\alpha \in I_{s,n}\{j\}} \operatorname{rdet}_j\left((\mathbf{AA}^*)_{j.}(\mathbf{e}_{l.})\right)_{\alpha}^{\alpha} = \sum_{\alpha \in I_{s,n}\{j\}} \operatorname{rdet}_j\left((\mathbf{AA}^*)_{j.}(\hat{\mathbf{u}}_{f.}\right)_{\alpha}^{\alpha}$$

as the fth component of a column-vector $\psi_{.j}^{(1)} = \left[\psi_{1j}^{(1)}, \ldots, \psi_{nj}^{(1)}\right]^T$, it follows

$$\sum_{f=1}^{n} \sum_{\beta \in J_{s_1,n}\{i\}} \operatorname{cdet}_i\left(\left(\left(\mathbf{A}^{k+1}\right)^* \mathbf{A}^{k+1}\right)_{.i}(\mathbf{e}_{f})\right)_{\beta}^{\beta} \psi_{fj}^{(1)} = \sum_{\beta \in J_{s_1,n}\{i\}} \operatorname{cdet}_i\left(\left(\left(\mathbf{A}^{k+1}\right)^* \mathbf{A}^{k+1}\right)_{.i}\left(\psi_{.j}^{(1)}\right)\right)_{\beta}^{\beta}.$$

Thus, Eq. (21) holds.

Corollary 3.9. *Let* $\mathbf{A} \in \mathbb{C}_s^{n\times n}$, $\operatorname{Ind}(\mathbf{A}) = k$ *and* $\operatorname{rank}\left(\mathbf{A}^k\right) = s_1$. *Then its right MPCEPMP inverse* $\mathbf{A}^{\dagger,\dagger,\dagger,r} = \left(a_{ij}^{\dagger,\dagger,\dagger,r}\right)$ *has the following determinantal representations*

$$a_{ij}^{\dagger,\dagger} = \frac{\sum_{\alpha \in I_{s,n}\{j\}} \left|(\mathbf{AA}^*)_{j.}\left(\phi_{i.}^{(1)}\right)\right|_{\alpha}^{\alpha}}{\left(\sum_{\alpha \in I_{s,n}} |\mathbf{AA}^*|_{\alpha}^{\alpha}\right)^2 \sum_{\beta \in J_{s_1,n}} \left|\left(\mathbf{A}^{k+1}\right)^* \mathbf{A}^{k+1}\right|_{\beta}^{\beta}}$$

$$= \frac{\sum_{\beta \in J_{s_1,n}\{i\}} \left|\left(\left(\mathbf{A}^{k+1}\right)^* \mathbf{A}^{k+1}\right)_{.i}\left(\psi_{.j}^{(1)}\right)\right|_{\beta}^{\beta}}{\left(\sum_{\beta \in I_{s,n}} |\mathbf{AA}^*|_{\beta}^{\beta}\right)^2 \sum_{\alpha \in J_{s_1,n}} \left|\left(\mathbf{A}^{k+1}\right)^* \mathbf{A}^{k+1}\right|_{\alpha}^{\alpha}},$$

where

$$\phi_{i.}^{(1)} = \left[\sum_{\beta \in J_{s_1,n}\{i\}} \left|\left(\left(\mathbf{A}^{k+1}\right)^* \mathbf{A}^{k+1}\right)_{.i}(\hat{\mathbf{u}}_{.l})\right|_{\beta}^{\beta}\right] \in \mathbb{C}^{1\times n}, \; l = 1, \ldots, n,$$

$$\psi_{.j}^{(1)} = \left[\sum_{\alpha \in I_{s,n}\{j\}} \left|(\mathbf{AA}^*)_{j.}(\hat{\mathbf{u}}_{f.})\right|_{\alpha}^{\alpha}\right] \in \mathbb{C}^{n\times 1}, \; f = 1, \ldots, n.$$

Here $\hat{\mathbf{u}}_{.l}$ *and* $\hat{\mathbf{u}}_{f.}$ *are the lth column and the fth row of* $\hat{\mathbf{U}} = \mathbf{U}_2\mathbf{AA}^*$, *and the matrix* $\mathbf{U}_2$ *is constructed from the columns* (19).

Theorem 3.10. *Let* $\mathbf{A} \in \mathbb{H}_s^{n\times n}$, $\operatorname{Ind}(\mathbf{A}) = k$ *and* $\operatorname{rank}\left(\mathbf{A}^k\right) = s_1$. *Then its left MPCEPMP inverse* $\mathbf{A}^{\dagger,\dagger,\dagger,l} = \left(a_{ij}^{\dagger,\dagger,\dagger,l}\right)$ *has the following determinantal representations*

$$a_{ij}^{\dagger,\dagger,\dagger,l} = \frac{\sum_{\alpha \in I_{s_1,n}\{j\}} \mathrm{rdet}_j \left(\left(\mathbf{A}^{k+1} \left(\mathbf{A}^{k+1} \right)^* \right)_{j.} \left(\phi_{i.}^{(2)} \right) \right)_{\alpha}^{\alpha}}{\left(\sum_{\beta \in J_{s,n}} |\mathbf{A}^* \mathbf{A}|_{\beta}^{\beta} \right)^2 \sum_{\alpha \in I_{s_1,n}} \left| \mathbf{A}^{k+1} \left(\mathbf{A}^{k+1} \right)^* \right|_{\alpha}^{\alpha}} = \tag{23}$$

$$= \frac{\sum_{\beta \in J_{s,n}\{i\}} \mathrm{cdet}_i \left((\mathbf{A}^* \mathbf{A})_{.i} \left(\psi_{.j}^{(2)} \right) \right)_{\beta}^{\beta}}{\left(\sum_{\beta \in J_{s,n}} |\mathbf{A}^* \mathbf{A}|_{\beta}^{\beta} \right)^2 \sum_{\alpha \in I_{s_1,n}} \left| \mathbf{A}^{k+1} \left(\mathbf{A}^{k+1} \right)^* \right|_{\alpha}^{\alpha}}, \tag{24}$$

where

$$\phi_{i.}^{(2)} = \left[\sum_{\beta \in J_{s,n}\{i\}} \mathrm{cdet}_i \left((\mathbf{A}^* \mathbf{A})_{.i} (\tilde{\mathbf{v}}_{.t}) \right)_{\beta}^{\beta} \right] \in \mathbb{H}^{1 \times n}, \quad t = 1, \dots, n,$$

$$\psi_{.j}^{(2)} = \left[\sum_{\alpha \in I_{s_1,n}\{j\}} \mathrm{rdet}_j \left(\left(\mathbf{A}^{k+1} \left(\mathbf{A}^{k+1} \right)^* \right)_{j.} (\tilde{\mathbf{v}}_{f.}) \right)_{\alpha}^{\alpha} \right] \in \mathbb{H}^{n \times 1}, \quad f = 1, \dots, n,$$

and $\tilde{\mathbf{v}}_{.l}$ *and* $\tilde{\mathbf{v}}_{f.}$ *are the lth column and the fth row of* $\tilde{\mathbf{V}} = \mathbf{A}^* \mathbf{A} \mathbf{V}_1$, *where the matrix* $\mathbf{V}_1$ *is determined from the rows* (13).

Proof. Due to (6),

$$a_{ij}^{\dagger,\dagger,\dagger,l} = \sum_{t=1}^{n} q_{it} a_{tj}^{\dagger,\dagger}. \tag{25}$$

Using (7) for the determinantal representation of $\mathbf{Q}_A = \mathbf{A}^\dagger \mathbf{A} = \left(q_{ij} \right)$ and (9) for the determinantal representation of $\mathbf{A}^{\dagger,\dagger}$ in (14), we obtain

$$a_{ij}^{\dagger,\dagger,\dagger,l} = \sum_{t=1}^{n} \frac{\sum_{\beta \in J_{s,n}\{i\}} \mathrm{cdet}_i \left((\mathbf{A}^* \mathbf{A})_{.i} (\dot{\mathbf{a}}_{.t}) \right)_{\beta}^{\beta}}{\sum_{\beta \in J_{s,n}} |\mathbf{A}^* \mathbf{A}|_{\beta}^{\beta}} \times \frac{\sum_{\alpha \in I_{s_1,n}\{j\}} \mathrm{rdet}_j \left(\left(\mathbf{A}^{k+1} \left(\mathbf{A}^{k+1} \right)^* \right)_{j.} \left(\mathbf{v}_{t.}^{(1)} \right) \right)_{\alpha}^{\alpha}}{\sum_{\beta \in J_{s,n}} |\mathbf{A}^* \mathbf{A}|_{\beta}^{\beta} \sum_{\alpha \in I_{s_1,n}} \left| \mathbf{A}^{k+1} \left(\mathbf{A}^{k+1} \right)^* \right|_{\alpha}^{\alpha}}$$

$$= \sum_{t=1}^{n} \sum_{f=1}^{n} \frac{\sum_{\beta \in J_{s,n}\{i\}} \mathrm{cdet}_i \left((\mathbf{A}^* \mathbf{A})_{.i} (\mathbf{e}_{.f}) \right)_{\beta}^{\beta}}{\left(\sum_{\beta \in J_{s,n}} |\mathbf{A}^* \mathbf{A}|_{\beta}^{\beta} \right)^2} \tilde{v}_{ft}$$

$$\times \frac{\sum_{\alpha \in I_{s_1,n}\{j\}} \mathrm{rdet}_j \left(\left(\mathbf{A}^{k+1} \left(\mathbf{A}^{k+1} \right)^* \right)_{j.} (\mathbf{e}_{t.}) \right)_{\alpha}^{\alpha}}{\sum_{\alpha \in I_{s_1,n}} \left| \mathbf{A}^{k+1} \left(\mathbf{A}^{k+1} \right)^* \right|_{\alpha}^{\alpha}},$$

where $\tilde{v}_{ft}$ is the (ft)th element of $\tilde{\mathbf{V}} = \mathbf{A}^* \mathbf{A} \mathbf{V}_1$ and the matrix $\mathbf{V}_1$ is constructed from the rows (13). If we put

$$\phi_{it}^{(2)} := \sum_{f=1}^{n} \sum_{\beta \in J_{s,n}\{i\}} \mathrm{cdet}_i \left((\mathbf{A}^* \mathbf{A})_{.i} (\mathbf{e}_{.f}) \right)_{\beta}^{\beta} \tilde{v}_{ft}$$

$$= \sum_{\beta \in J_{s,n}\{i\}} \mathrm{cdet}_i \left((\mathbf{A}^* \mathbf{A})_{.i} (\tilde{\mathbf{v}}_{.t}) \right)_{\beta}^{\beta}$$

as the lth component of a row-vector $\phi_{i.}^{(2)} = \left[\phi_{i1}^{(2)}, \dots, \phi_{in}^{(2)}\right]$, then

$$\sum_{t=1}^{n} \phi_{it}^{(2)} \sum_{\alpha \in I_{s_1,n}\{j\}} \mathrm{rdet}_j \left(\left(\mathbf{A}^{k+1} \left(\mathbf{A}^{k+1} \right)^* \right)_{j.} (\mathbf{e}_{t.}) \right)_\alpha^\alpha$$
$$= \sum_{\alpha \in I_{s_1,n}\{j\}} \mathrm{rdet}_j \left(\left(\mathbf{A}^{k+1} \left(\mathbf{A}^{k+1} \right)^* \right)_{j.} \left(\phi_{i.}^{(2)} \right) \right)_\alpha^\alpha,$$

then Eq. (23) holds. If we denote by

$$\psi_{fj}^{(2)} := \sum_{t=1}^{n} \tilde{u}_{ft} \sum_{\alpha \in I_{s_1,n}\{j\}} \mathrm{rdet}_j \left(\left(\mathbf{A}^{k+1} \left(\mathbf{A}^{k+1} \right)^* \right)_{j.} (\mathbf{e}_{t.}) \right)_\alpha^\alpha =$$
$$= \sum_{\alpha \in I_{s_1,n}\{j\}} \mathrm{rdet}_j \left(\left(\mathbf{A}^{k+1} \left(\mathbf{A}^{k+1} \right)^* \right)_{j.} (\tilde{\mathbf{u}}_{f.})_\alpha^\alpha \right.$$

the fth component of a column-vector $\psi_{.j}^{(2)} = \left[\psi_{1j}^{(2)}, \dots, \psi_{nj}^{(2)}\right]^T$, then

$$\sum_{f=1}^{n} \sum_{\beta \in J_{s,n}\{i\}} \mathrm{cdet}_i \left((\mathbf{A}^* \mathbf{A})_{.i} (\mathbf{e}_{.f}) \right)_\beta^\beta \psi_{fj}^{(2)} = \sum_{\beta \in J_{s,n}\{i\}} \mathrm{cdet}_i \left((\mathbf{A}^* \mathbf{A})_{.i} \left(\psi_{.j}^{(2)} \right) \right)_\beta^\beta.$$

Hence, we obtain (24). □

Corollary 3.11. *Let* $\mathbf{A} \in \mathbb{C}_s^{n \times n}$, $\mathrm{Ind}(\mathbf{A}) = k$ *and* $\mathrm{rank}\left(\mathbf{A}^k\right) = s_1$. *Then its left MPCEPMP inverse* $\mathbf{A}^{\dagger,\dagger,\dagger,l} = \left(a_{ij}^{\dagger,\dagger,\dagger,l}\right)$ *has the following determinantal representations*

$$a_{ij}^{\dagger,\dagger,\dagger,l} = \frac{\sum_{\alpha \in I_{s_1,n}\{j\}} \left| \left(\mathbf{A}^{k+1} \left(\mathbf{A}^{k+1} \right)^* \right)_{j.} \left(\phi_{i.}^{(2)} \right) \right|_\alpha^\alpha}{\left(\sum_{\beta \in J_{s,n}} |\mathbf{A}^* \mathbf{A}|_\beta^\beta \right)^2 \sum_{\alpha \in I_{s_1,n}} \left| \mathbf{A}^{k+1} \left(\mathbf{A}^{k+1} \right)^* \right|_\alpha^\alpha} =$$
$$= \frac{\sum_{\beta \in J_{s,n}\{i\}} \left| (\mathbf{A}^* \mathbf{A})_{.i} \left(\psi_{.j}^{(2)} \right) \right|_\beta^\beta}{\left(\sum_{\beta \in J_{s,n}} |\mathbf{A}^* \mathbf{A}|_\beta^\beta \right)^2 \sum_{\alpha \in I_{s_1,n}} \left| \mathbf{A}^{k+1} \left(\mathbf{A}^{k+1} \right)^* \right|_\alpha^\alpha},$$

where

$$\phi_{i.}^{(2)} = \left[\sum_{\beta \in J_{s,n}\{i\}} \left| (\mathbf{A}^* \mathbf{A})_{.i} (\tilde{\mathbf{v}}_{.t}) \right|_\beta^\beta \right] \in \mathbb{C}^{1 \times n}, \quad t = 1, \dots, n,$$
$$\psi_{.j}^{(2)} = \left[\sum_{\alpha \in I_{s_1,n}\{j\}} \left| \left(\mathbf{A}^{k+1} \left(\mathbf{A}^{k+1} \right)^* \right)_{j.} (\tilde{\mathbf{v}}_{f.}) \right|_\alpha^\alpha \right] \in \mathbb{C}^{n \times 1}, \quad f = 1, \dots, n,$$

and $\tilde{\mathbf{v}}_{.t}$ *and* $\tilde{\mathbf{v}}_{f.}$ *are the tth column and the fth row of* $\tilde{\mathbf{V}} = \mathbf{A}^* \mathbf{A} \mathbf{V}_1$, *where the matrix* $\mathbf{V}_1$ *is determined by* (13).

4. Conclusions

In this chapter, notions of the MPCEP and CEPMP inverses are extended to quaternion matrices, and the new right and left MPCEPMP inverses are introduced and their characterizations are explored. Their determinantal representations are obtained within the framework of the theory of noncommutative column-row determinants previously introduced by the author. Also, determinantal representations of these generalized inverses for complex matrices are derived by using regular determinants. The obtained determinantal representations give new direct methods of calculations of these generalized inverses.

Acknowledgements

The author thanks the Erwin Schrödinger Institute for Mathematics and Physics (ESI) at the University of Vienna for the support given by the Special Research Fellowship Programme for Ukrainian Scientists.

Author details

Ivan I. Kyrchei
Pidstryhach Institute for Applied Problems of Mechanics and Mathematics, NAS of Ukraine, Lviv, Ukraine

*Address all correspondence to: ivankyrchei26@gmail.com

References

[1] Kyrchei II. Determinantal representations of the quaternion core inverse and its generalizations. Advances in Applied Clifford Algebras. 2019;**29**(5): 104. DOI: 10.1007/s00006-019-1024-6

[2] Prasad KM, Mohana KS. Core-EP inverse. Linear and Multilinear Algebra. 2014;**62**(6):792-802. DOI: 10.1080/03081087.2013.791690

[3] Gao Y, Chen J. Pseudo core inverses in rings with involution. Communications in Algebra. 2018;**46**(1):38-50. DOI: 10.1080/00927872.2016.1260729

[4] Baksalary OM, Trenkler G. Core inverse of matrices. Linear and Multilinear Algebra. 2010;**58**(6):681-697. DOI: 10.1080/03081080902778222

[5] Ma H, Stanimirović PS. Characterizations, approximation and perturbations of the core-EP inverse. Applied Mathematics and Computation. 2019;**359**:404-417. DOI: 10.1080/03081087.2013.791690

[6] Wang H. Core-EP decomposition and its applications. Linear Algebra and its Applications. 2016;**508**:289-300. DOI: 10.1016/j.laa.2016.08.008

[7] Zhou MM, Chen JL, Li TT, Wan DG. Three limit representations of the core-EP inverse. Univerzitet u Nišu. 2018;**32**: 5887-5894. DOI: 10.2298/FIL1817887Z

[8] Gao Y, Chen J, Patricio P. Continuity of the core-EP inverse and its applications. Linear and Multilinear Algebra. 2021;**69**(3):557-571. DOI: 10.1080/03081087.2019.1608899

[9] Prasad KM, Raj MD. Bordering method to compute core-EP inverse. Special Matrices. 2018;**6**:193-200. DOI: 10.1515/spma-2018-0016

[10] Prasad KM, Raj MD, Vinay M. Iterative method to find core-EP inverse. Bulletin of Kerala Mathematics Association. 2018;**16**(1):139-152

[11] Kyrchei II. Determinantal representations of the core inverse and its generalizations with applications. Journal of Mathematics. 2019;**1631979**:13. DOI: 10.1155/2019/1631979

[12] Ferreyra DE, Levis FE, Thome N. Revisiting the core EP inverse and its extension to rectangular matrices. Quaestiones Mathematicae. 2018;**41**(2): 265-281. DOI: 10.2989/16073606.2017.1377779

[13] Mosić D, Djordjević DS. The gDMP inverse of Hilbert space operators. Journal of Spectral Theory. 2018;**8**(2): 555-573. DOI: 10.4171/JST/207

[14] Mosić D. Core-EP inverses in Banach algebras. Linear and Multilinear Algebra. 2021;**69**(16):2976-2989. DOI: 10.1080/03081087.2019.1701976

[15] Sahoo JK, Behera R, Stanimirović PS, Katsikis VN, Ma H. Core and core-EP inverses of tensors. Computational and Applied Mathematics. 2020;**39**:9. DOI: 10.1007/s40314-019-0983-5

[16] Chen JL, Mosić D, Xu SZ. On a new generalized inverse for Hilbert space operators. Quaestiones Mathematicae. 2020;**43**(9):1331-1348. DOI: 10.2989/16073606.2019.1619104

[17] Udwadia F, Schttle A. An alternative derivation of the quaternion equations of motion for rigid-body rotational dynamics. Journal of Applied Mechanics. 2010;**77**:044505.1-044505.4. DOI: 10.1115/1.4000917

[18] Gibbon JD. A quaternionic structure in the three-dimensional Euler and ideal magneto-hydrodynamics equation. Physica D: Nonlinear Phenomena. 2002; **166**:17-28. DOI: 10.1016/S0167-2789(02) 00434-7

[19] Gibbon JD, Holm DD, Kerr RM, Roulstone I. Quaternions and particle dynamics in the Euler fluid equations. Nonlinearity. 2006;**19**:1969-1983. DOI: 10.1088/0951-7715/19/8/011

[20] Adler SL. Quaternionic Quantum Mechanics and Quantum Fields. New York: Oxford University Press; 1995

[21] Jiang T, Chen L. Algebraic algorithms for least squares problem in quaternionic quantum theory. Computer Physics Communications. 2007;**176**: 481-485. DOI: 10.1016/j.cpc.2006.12.005

[22] Leo SD, Ducati G. Delay time in quaternionic quantum mechanics. Journal of Mathematical Physics. 2012;**53**:022102.8. DOI: 10.1063/ 1.3684747

[23] Took CC, Mandic DP. A quaternion widely linear adaptive filter. IEEE Transactions on Signal Processing. 2010; **58**:4427-4431. DOI: 10.1109/ TSP.2010.2048323

[24] Took CC, Mandic DP. Augmented second-order statistics of quaternion random signals. Signal Processing. 2011; **91**:214-224. DOI: 10.1016/j. sigpro.2010.06.024

[25] Le Bihan N, Sangwine SJ. Quaternion principal component analysis of color images. Proceedings ICIP. 2003;I-809. DOI: 10.1109/ICIP.2003.1247085

[26] Jia Z, Ng MK, Song GJ. Robust quaternion matrix completion with applications to image inpainting. Numerical Linear Algebra with Applications. 2019;**26**(4):e2245. DOI: 10.1002/nla.2245

[27] Jia Z, Ng MK, Song GJ. Lanczos method for large-scale quaternion singular value decomposition. Numerical Algorithms. 2019;**82**:699-717. DOI: 10.1007/s11075-018-0621-0

[28] Kyrchei II. Determinantal representations of the Moore–Penrose inverse over the quaternion skew field and corresponding Cramer's rules. Linear and Multilinear Algebra. 2011;**59**: 413-431. DOI: 10.1080/030810810035 86860

[29] Kyrchei II. Determinantal representations of the Drazin inverse over the quaternion skew field with applications to some matrix equations. Applied Mathematics and Computation. 2014;**238**:193-207. DOI: 10.1016/j. amc.2014.03.125

[30] Kyrchei II. Determinantal representations of the Drazin and W-weighted Drazin inverses over the quaternion skew field with applications. In: Griffin S, editor. Quaternions: Theory and Applications. New York: Nova Science Publishers; 2017. pp. 201-275

[31] Kyrchei II. Determinantal representations of the quaternion weighted Moore–Penrose inverse and its applications. In: Baswell AR, editor. Advances in Mathematics Research: Vol. 23. New York: Nova Science Publishers; 2017. pp. 35-96

[32] Kyrchei II. Determinantal representations of the weighted core-EP, DMP, MPD, and CMP inverses. J. Math. 2020;**9816038**: 12 p. DOI: 10.1155/2020/ 9816038

[33] Kyrchei II. Weighted quaternion core-EP, DMP, MPD, and CMP inverses and their determinantal representations.

Revista de La Real Academia Ciencias Exactas, Físicas y Naturales. Serie A. Matemáticas RACSAM. 2020;**114**:198. DOI: 10.1007/s13398-020-00930-3

[34] Aslaksen H. Quaternionic determinants. Mathematical Intelligence. 1996;**18**(3):57-65. DOI: 10.1007/BF03024312

[35] Cohen N, De Leo S. The quaternionic determinant. Electronic Journal of Linear Algebra. 2000;**7**:100-111. DOI: 10.13001/1081-3810.1050

[36] Zhang FZ. Quaternions and matrices of quaternions. Linear Algebra and its Applications. 1997;**251**:21-57. DOI: 10.1016/0024-3795(95)00543-9

[37] Kyrchei II. Cramer's rule for quaternionic systems of linear equations. Journal of Mathematical Sciences. 2008; **155**(6):839-858. DOI: 10.1007/s10958-008-9245-6

[38] Kyrchei II. The theory of the column and row determinants in a quaternion linear algebra. In: Baswell AR, editor. Advances in Mathematics Research: Vol. 15. New York: Nova Science Publishers; 2012. pp. 301-359

Section 2

Applications

Chapter 7

The COVID-19 DNA-RNA Genetic Code Analysis Using Double Stochastic and Block Circulant Jacket Matrix

Sung Kook Lee and Moon Ho Lee

Abstract

We present a COVID-19 DNA-RNA genetic code where $A = T = U = 31\%$ and $C = G = 19\%$, which has been developed from a base matrix $[C\ U; A\ G]$ where C, U, A, and G are RNA bases while C, U, A, and T are DNA bases that E. Chargaff found them complementary like $A = T = U = 30\%$, and $C = G = 20\%$ from his experimental results, which implied the structure of DNA double helix and its complementary combination. Unfortunately, they have not been solved mathematically yet. Therefore, in this paper, we present a simple solution by the information theory of a doubly stochastic matrix over the Shannon symmetric channel as well as prove it mathematically. Furthermore, we show that DNA-RNA genetic code is one kind of block circulant Jacket matrix. Moreover, general patterns by block circulant, upper-lower, and left-right scheme are presented, which are applied to the correct communication as well as means the healthy condition because it perfectly consists of 4 bases. Henceforth, we also provide abnormal patterns by block circulant, upper-lower, and left-right scheme, which cover the distorted signal as well as COVID-19.

Keywords: COVID-19 DNA-RNA, E. Chargaff, DNA-RNA genetic code, double stochastic matrix, symmetric channel, block circulant jacket matrix, general pattern, abnormal pattern

1. Introduction

In 1950, Chargaff's two rules [1] were presented. One is that the percentage of adenine is identical to that of thymine as well as the percentage of guanine is identical to that of cytosine, which gives a hint of the composition of the base pair for the double-strand DNA molecule. The other is that base complementarity is effective for each DNA strand, which gives an explanation for the overall characteristics of fundamental bases. To make an example of COVID-19 DNA, its four bases are satisfied with these two rules analogous to $A = T = 31\%$ and $C = G = 19\%$. In 1953, it was discovered that DNA has a double helix structure [2, 3], which results in an optimal and economical genetic code [4].

A RNA base matrix $[C\ U; A\ G]$ was based on stochastic matrices [5], which results in the genetic code [6, 7]. A symmetric capacity is calculated by applying the Markov process to these doubly stochastic matrices, which suggested the symmetry between Shannon [8] and RNA stochastic transition matrix $[C\ U; A\ G]$, which is defined as below. A square matrix of $\mathbf{P} = \left(p_{ij}\right)$ is stochastic, whose entries are positive as well as its sum in rows and columns is equal to one or constant. In other words, if the sum of all its elements in rows and columns is equal to one or invariable, it is double stochastic, which is able to describe the time-invariant binary symmetric channel. For the input x_n and the output x_{n+1}, two states e_0 and e_1 are able to depict Markov processes on an individual basis, which are indicated by two binary symbols "0" and "1", accordingly. The output signal is affected by the input signal whose information is fed into given a certain error probability. Assume that these channel probabilities α and β are less than a half, whose error probabilities have been kept steady over a time-variant channel for a wide variety of transmitted symbols such as

$$P\{\mathrm{x}_{n+1} = 1|\mathrm{x}_n = 0\} = p_{01} = \alpha, P\{\mathrm{x}_{n+1} = 0|\mathrm{x}_n = 1\} = p_{10} = \beta. \tag{1}$$

In addition, its Markov chain is homogeneous. $\mathbf{P}$ represents a 2×2 homogeneous probability transition matrix defined as

$$\mathbf{P} = \begin{bmatrix} p_{00} & p_{01} \\ p_{10} & p_{11} \end{bmatrix} = \begin{bmatrix} 1-\alpha & \alpha \\ \beta & 1-\beta \end{bmatrix} = \begin{bmatrix} 1-p & p \\ p & 1-p \end{bmatrix}_{p=0.5} = \frac{1}{2}\begin{bmatrix} 1 & 1 \\ 1 & 1 \end{bmatrix}, \tag{2}$$

whose two error probabilities are identical similarly to $\alpha = \beta = p$ over a binary symmetric channel. This paper proceeds as below. First of all, we derive the RNA stochastic entropy by applying it to the Shannon entropy in Section 2. Next, we make an estimate of the variance of RNA in Section 3. Then, the binary symmetric channel entropy is derived in Section 4. Henceforth, two user capacity is made an estimate of over symmetric interference channel in Section 5. Afterward, the construction scheme

Organism	Taxon	%A	%G	%C	%T	A / T	G / C	%GC	%AT
Maize	*Zea*	26.8	22.8	23.2	27.2	0.99	0.98	46.1	54.0
Octopus	*Octopus*	33.2	17.6	17.6	31.6	1.05	1.00	35.2	64.8
Chicken	*Gallus*	28.0	22.0	21.6	28.4	0.99	1.02	43.7	56.4
Rat	*Rattus*	28.6	21.4	20.5	28.4	1.01	1.00	42.9	57.0
Human	*Homo*	29.3	20.7	20.0	30.0	0.98	1.04	40.7	59.3
Grasshopper	*Orthoptera*	29.3	20.5	20.7	29.3	1.00	0.99	41.2	58.6
Sea urchin	*Echinoidea*	32.8	17.7	17.3	32.1	1.02	1.02	35.0	64.9
Wheat	*Triticum*	27.3	22.7	22.8	27.1	1.01	1.00	45.5	54.4
Yeast	*Saccharomyces*	31.3	18.7	17.1	32.9	0.95	1.09	35.8	64.4
E. coli	*Escherichia*	24.7	26.0	25.7	23.6	1.05	1.01	51.7	48.3
φX174	*PhiX174*	24.0	23.3	21.5	31.2	0.77	1.08	44.8	55.2
Covid-19	*SARS-CoV-2*	29.9	19.6	18.4	32.1	0.93	1.07	38.0	62.0

Table 1.
Ratio of bases [1, 9–11].

is proposed, which is enabled to create RNA genetic codes in Section 6. Later, a symmetric genetic Jacket block matrix is examined in Section 7. Hereupon, general patterns of block circulant symmetric genetic Jacket matrices are looked into in Section 8. In the end, this paper comes to a conclusion in Section 9.

Table 1 makes the description of the ratio of bases for several organisms [1, 9–11], which shows that the ratios are constant among the species.

2. Analytical approach to RNA stochastic entropy

In [1, 5, 12, 13], stochastic complementary RNA bases are given for the genetic code. On the assumption that $C = G = 19\%,\ A = T = U = 31\%$, $\mathbf{P}$ denotes the transition channel matrix expressed by

$$\mathbf{P} = \begin{bmatrix} C & U \\ A & G \end{bmatrix} = \begin{bmatrix} 0.19 & 0.31 \\ 0.31 & 0.19 \end{bmatrix}. \tag{3}$$

On the condition that the RNA base matrix $[C\,U; A\,G]$ for the Markov process described by two independent probabilities of its corresponding source varies from $0.19p$ to $0.31p$, the transition channel matrix $\mathbf{P}$ is defined by

$$\mathbf{P} = \begin{bmatrix} 0.19p & 1-0.19p \\ 1-0.19p & 0.19p \end{bmatrix} = \begin{bmatrix} 0.5 & 1-0.5 \\ 1-0.5 & 0.5 \end{bmatrix} = \begin{bmatrix} 0.5 & 0.5 \\ 0.5 & 0.5 \end{bmatrix}. \tag{4}$$

By comparison with Eq. (12), we have.

$$0.19p = 1 - 0.19p \tag{5}$$

where p is 2.631.

Applying in a similar fashion to the rest of (4),

$$\mathbf{P} = \begin{bmatrix} 0.31p & 1-0.31p \\ 1-0.31p & 0.31p \end{bmatrix} = \begin{bmatrix} 0.500 & 1-0.500 \\ 1-0.500 & 0.500 \end{bmatrix} = \begin{bmatrix} 0.5 & 0.5 \\ 0.5 & 0.5 \end{bmatrix}, \tag{6}$$

where $0.31p$ = 1-$0.31p$, where p is 1.613.

In order to make a double stochastic matrix by adding (6) to (4),

$$2\mathbf{P} = \begin{bmatrix} 0.5 & 0.5 \\ 0.5 & 0.5 \end{bmatrix} + \begin{bmatrix} 0.5 & 0.5 \\ 0.5 & 0.5 \end{bmatrix} = \begin{bmatrix} 1 & 1 \\ 1 & 1 \end{bmatrix}. \tag{7}$$

Applying in a similar way to (3),

$$2\mathbf{P} = 2\begin{bmatrix} C & U \\ A & G \end{bmatrix} = 2\begin{bmatrix} 0.19 & 0.31 \\ 0.31 & 0.19 \end{bmatrix} = \begin{bmatrix} 0.38 & 0.62 \\ 0.62 & 0.38 \end{bmatrix}. \tag{8}$$

If P is a random variable for source probability p corresponding to the first symbol event, we reach the entropy function [8] represented by

P	$-\log_2 p$	$-p\log_2 p$	$H_2(p)$
0.3800	1.3959	0.5305	0.9580
0.3900	1.3585	0.5298	0.9648
0.4000	1.3219	0.5288	0.9710
0.4100	1.2863	0.5274	0.9765
0.4200	1.2515	0.5256	0.9815
0.4300	1.2176	0.5236	0.9858
0.4400	1.1844	0.5211	0.9896
0.4500	1.1520	0.5184	0.9928
0.4600	1.1203	0.5153	0.9954
0.4700	1.0893	0.5120	0.9974
0.4800	1.0589	0.5083	0.9988
0.4900	1.0291	0.5043	0.9997
0.5000	1.0000	0.5000	1.0000
0.5100	0.9714	0.4954	0.9997
0.5200	0.9434	0.4906	0.9988
0.5300	0.9159	0.4854	0.9974
0.5400	0.8890	0.4800	0.9954
0.5500	0.8625	0.4744	0.9928
0.5600	0.8365	0.4684	0.9896
0.5700	0.8110	0.4623	0.9858
0.5800	0.7859	0.4558	0.9815
0.5900	0.7612	0.4491	0.9765
0.6000	0.7370	0.4422	0.9710
0.6100	0.7131	0.4350	0.9648
0.6200	0.6897	0.4276	0.9580

Table 2.
Shannon entropy for probability p.

$$H_2(P) = p\log_2\left(\frac{1}{p}\right) + (1-p)\log_2\left(\frac{1}{1-p}\right). \tag{9}$$

The last column of **Table 2** shows the result of Eq. (9). **Figure 1** portrays the curve of Shannon and RNA Entropy. Make a mental note to make sure that a vertical tangent can be drawn when $p = 0$ and $p = 1$ on account of the fact that

$$\frac{d}{dp}\left[p\log_2\left(\frac{1}{p}\right) + (1-p)\log_2\left(\frac{1}{1-p}\right)\right] = \left[\log_2\left(\frac{1}{p}\right) - 1 - \log_2\left(\frac{1}{1-p}\right) + 1\right]\log_2 e$$
$$= \log_2\left(\frac{1}{p}\right) - \log_2\left(\frac{1}{1-p}\right) = 0, \tag{10}$$

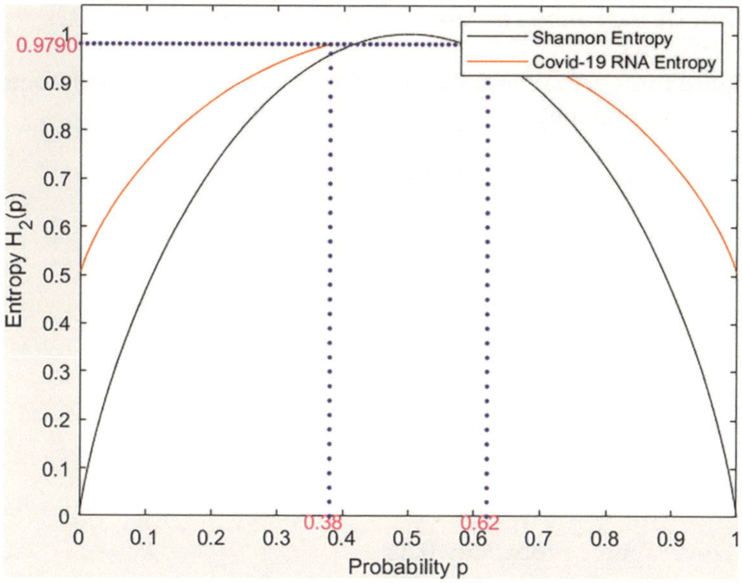

Figure 1.
Comparison between Shannon and RNA entropy for probability p.

which is maximized when p reaches a half because its derivative becomes 0. Therefore,

$$\log_2\left(\frac{1}{p}\right) - \log_2\left(\frac{1}{1-p}\right) = 0 \Rightarrow \left(\frac{1}{p}\right) - \left(\frac{1}{1-p}\right) = 0. \tag{11}$$

Then, we reach

$$p = 1 - p \Rightarrow p = \frac{1}{2}. \tag{12}$$

For the RNA base matrix $[C\ U; A\ G]$, its symmetric entropy is calculated as

$$H_2(P)_{RNA} = p\log_2\left(\frac{1}{p}\right) + (1-p)\log_2\left(\frac{1}{1-p}\right) = 0.9790, \tag{13}$$

when p is either 0.38 or 0.62. By the way, the Shannon entropy is calculated as

$$H_2(P)_{Shannon} = p\log_2\left(\frac{1}{p}\right) + (1-p)\log_2\left(\frac{1}{1-p}\right) = 1, \tag{14}$$

when p reaches a half.

Table 2 shows Shannon Entropy for probability p over a binary symmetric channel.

Figure 1 gives a comparison between Shannon and RNA Entropy for probability p under the RNA base matrix $[C\ U; A\ G]$.

3. Derivation of variance for the RNA base matrix $[C\ U; A\ G]$

The variance for RNA random variable X is denoted by $V(X)$ is the square of the mean, which is expressed by

$$E\{X\} = a = 0.5. \tag{15}$$

Therefore, for a random variable X, the variance is obtained such as

$$\begin{aligned} V(X) &= E\left\{(X-a)^2\right\} = E\{X^2\} - 2aE\{X\} + E\{a^2\} \\ &= E\{X^2\} - 2a^2 + a^2 = E\{X^2\} - a^2 = \sigma^2. \end{aligned} \tag{16}$$

Case I. Upper source probability 0.62

$$\sigma_{upper}^2 = (0.62)^2 - (0.5)^2 = 0.13. \tag{17}$$

Case II. Lower source probability 0.38

$$\sigma_{lower}^2 = (0.5)^2 - (0.38)^2 = 0.10. \tag{18}$$

If X_1 and X_2 are the independent random variables, on an individual basis, its expectation and variance are

$$E\{X_1\} = a_1, \quad V\{X_1\} = \sigma_1^2. \tag{19}$$

$$E\{X_2\} = a_2, \quad V\{X_2\} = \sigma_2^2. \tag{20}$$

Therefore, we reach

$$E\{(X_1 - a_1)(X_2 - a_2)\} = E\{(X_1 - a_1)\}E\{(X_2 - a_2)\} = 0. \tag{21}$$

Assuming that X_1 and X_2 are independent random variables, the sum of its variances is calculated as

$$\begin{aligned} V\{X_1 + X_2\} &= E\left\{(X_1 + X_2 - a_1 - a_2)^2\right\} \\ &= E\left\{(X_1 - a_1)^2\right\} + 2E\{(X_1 - a_1)(X_2 - a_2)\} + E\left\{(X_2 - a_2)^2\right\} \\ &= V\{X_1\} + V\{X_2\} = \sigma_1^2 + \sigma_2^2 = 0.13 + 0.10 = 0.23, \end{aligned} \tag{22}$$

which is approximately 23% corresponding to the difference between $A = U$ and $C = G$. It means that RNA entropy cannot reach the Shannon entropy because the probabilities of its bases are 23% away from a half that is exactly identical to the sum of its variances.

4. RNA complement base matrix $[C\ U; A\ G]$ for symmetric noise immune-free channel

If over a noise immune-free binary symmetric channel the bases of RNA genetic code $[C\ U; A\ G]$ are complementary such as $C = U$ and $A = G$, the conditional

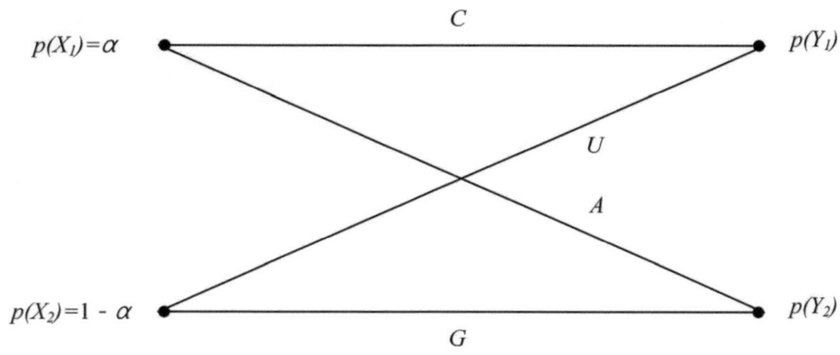

Figure 2.
Complementary bases of RNA genetic code [C U; A G] over noise immune-free binary symmetric channel.

probability $P(b_j|a_i) = P_{i,j}$ makes description of this channel, whose maximum amount of information can be transmitted as depicted in **Figure 2**. On the assumption that C and G are one's complement of its corresponding error probability as well as A and U are interference signals, the matrix [8] for this channel is made description of by

$$[p(X)]_{1\times 2}[P]_{2\times 2} = [\alpha \;\; 1-\alpha]\begin{bmatrix} C & U \\ A & G \end{bmatrix} = [p(Y)]_{1\times 2} = [p(Y_1) \quad p(Y_2)]. \tag{23}$$

Under the condition that p and 1-p are the selection probability $(\alpha = 0)$ and $(\alpha = 1)$ over the uniform channel on an individual basis, the mutual information is defined by

$$I(X;Y) = H(Y) - H(Y|X). \tag{24}$$

From Eq. (23), we are confronted with

$$[\alpha \quad 1-\alpha]\begin{bmatrix} -C\log_2 C & -U\log_2 U \\ -A\log_2 A & -G\log_2 G \end{bmatrix} = [\alpha \quad 1-\alpha]\begin{bmatrix} -U\log_2 U & -C\log_2 C \\ G\log_2 G & -A\log_2 A \end{bmatrix}, \tag{25}$$

where

$$\begin{aligned} H(Y|X) &= -\alpha C\log_2 C - \alpha A\log_2 A - (1-\alpha)U\log_2 U - (1-\alpha)G\log_2 G \\ &= -U\log_2 U - G\log_2 G = -C\log_2 C - A\log_2 A = 0.9790, \end{aligned} \tag{26}$$

where $A = U = 0.31$ and $C = G = 0.19$.
Therefore, its capacity is derived as

$$C_{RNA} = \max I(X;Y)|_{p=0.38 \text{ or } 0.62} = H(Y) - H(Y|X) = 1 - 0.9790 = 0.021, \tag{27}$$

i.e. $H(Y) = -p\log_2 p - (1-p)\log_2(1-p) = -0.38\log_2 0.38 - 0.62\log_2 0.62 = 1.$
while Shannon capacity is derived as

$$C_{Shannon} = \max I(X;Y)|_{p=0.5} = H(Y) - H(Y|X) = 1 - 1 = 0. \tag{28}$$

In **Figure 3**, we compare Shannon and RNA capacity for probability p. As forementioned in Section 3, if only if under the ideal circumstance, Shannon capacity can

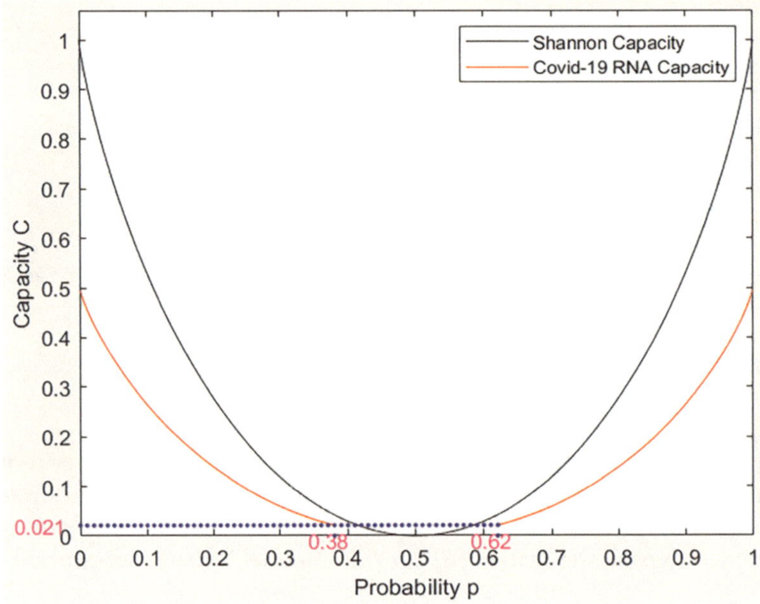

Figure 3.
Shannon and RNA capacity vary with probability p.

be reached. In other words, the difference between Shannon and RNA capacity exists, which is identical to the sum of variances of RNA base random variables because they are unable to become a half over a symmetric channel.

5. Two user capacity over symmetric interference channel

Figure 4 makes the description of the environment of the binary symmetric channel with the RNA base matrix $[C\ U; A\ G]$ as well as that of the symmetric interference channel for two users where two independent messages W_1 and W_2 with the common message set W_i are transmitted. Assume that $C = G = 19\%$ and $A = U = 31\%$ where $C = H^{11}$ is the direct signal and its corresponding interference signal is $U = H^{12}$ for Y_1. Analogously, the direct signal for the second user Y_2 is $G = H^{22}$ and its corresponding interference signal is $A = H^{21}$.

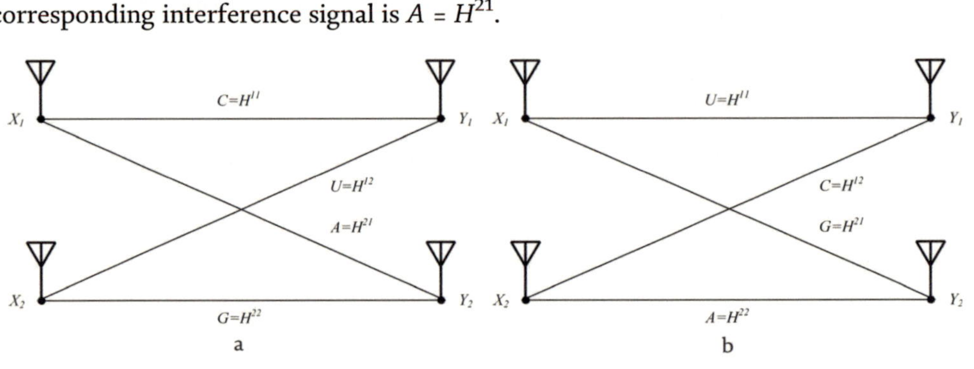

Figure 4.
Two-user symmetric Interference Channel. (a) Strong Interference Channel. (b) Weak Interference Channel.

$$H^{11} = H^{12} = h_d\sqrt{P_{SNR}},$$
$$H^{12} = H^{21} = h_c\sqrt{P_{SNR}}.$$

The relationship between the input and output for two user symmetric channel is described as follows [14],

$$Y_1 = h_d\sqrt{P_{SNR}}X_1 + h_c\sqrt{P_{SNR}^{\alpha}}X_2 + Z_1, \tag{29}$$

$$Y_2 = h_c\sqrt{P_{SNR}^{\alpha}}X_1 + h_d\sqrt{P_{SNR}}X_2 + Z_2, \tag{30}$$

where the powers of input symbols X_1, $X_{2,}$ and additive white Gaussian noise (AWGN) terms Z_1 and Z_2 are normalized to unity. Analogous to the definition of the degree of freedom (*DoF*), the total *GDoF* metric $d(\alpha)$ is defined as

$$d(\alpha) = \lim_{P_{SNR}\to\infty} \frac{C(P_{SNR}, \alpha)}{\log{(P_{SNR})}}, \tag{31}$$

where C (P_{SNR}, α) is the sum-capacity parameterized by P_{SNR} and α. Here α is the ratio (on the decibel scale) of cross channel strength compared to straight channel strength and P_{SNR} indicates the ratio (on the decibel scale) of signal to the noise. Importantly, in order to find the achievable *DoF*, take the limit of Eq. (31) by letting P_{SNR} go to infinity. Make a mental note of the *DoF* metric resembling to that at the point α =1. Thus, the *GDoF* curve gives a significant hint for optimal interference management strategies, which has been made use of most successfully to estimate the capacity of two-user interference channel to contain a constant gap in [14]. To take an example, for RNA genetic code, assuming that its bases $C = G = 19\%$ and $A = T = U = 31\%$, this symmetric interference channel for two users can be analyzed in strong and weak interference region as below. The noise immune channel is described as below where X_1 and X_2 denote the input symbols while Y_1 and Y_2 denote the output symbols

$$Y_1 = CX_1 + UX_2, \tag{32}$$

$$Y_2 = GX_1 + AX_2. \tag{33}$$

Case 1. Strong Interference region.

Figure 4 (a) makes the description of the channel in a strong interference regime, where its receivers have to try to decode the interfering signal in order to recover its desired signal. The general condition for a strong interference signal is represented by,

$$C < A, U > G. \tag{34}$$

Regretfully, it is still challenging to propose the scheme achieving a symmetric rate as well as being upper-bounded unlike in the weak interference region.

Case 2. Weak Interference region.

Figure 4 (b) makes the description of the channel in a very weak interference regime, where its receivers do not need to try to decode any portion of the interference signal by regarding it as noise. This scheme is enabled to achieve a symmetric rate per user as below [14],

$$R = \min\left\{\frac{1}{2}\log{(1 + INR + SNR)} + \frac{1}{2}\log\left(2 + \frac{SNR}{INR}\right) - 1,\ \log\left(1 + INR + \frac{SNR}{INR}\right) - 1\right\}. \tag{35}$$

The upper bound on the symmetric capacity is,

$$C_{Sym} \leq \min\left(\frac{1}{2}\log\left(1+SNR\right)+\frac{1}{2}\log\left(1+\frac{SNR}{1+INR}\right), \log\left(1+INR+\frac{SNR}{1+INR}\right)\right). \tag{36}$$

Letting $A = T = U = 31\%$, $C = G = 19\%$, i.e. $INR = 31$ and $SNR = 19$, we are confronted with the symmetric achievable rate such as

$$\begin{aligned} R = &\min\left\{\frac{1}{2}\log_2(1+31+19)+\frac{1}{2}\log_2\left(2+\frac{19}{31}\right)-1, \log_2\left(1+31+\frac{19}{31}\right)-1\right\} \\ &= \min\{2.83+0.69-1, 5.02-1\} = \min\{2.53, 4.02\} = 2.52. \end{aligned} \tag{37}$$

Analogously, the symmetric capacity is made the description of by

$$\begin{aligned} C_{sym} \leq &\min\left\{\frac{1}{2}\log_2(1+19)+\frac{1}{2}\log_2\left(1+\frac{19}{31}\right), \log_2\left(1+31+\frac{19}{31}\right)\right\} \\ &\leq \min\{2.16+0.34, 5.02\} \leq \min\{2.50, 5.02\} = 2.50. \end{aligned} \tag{38}$$

Following the above steps, in a weak interference regime, by treating interference as noise, the symmetric capacity is close to its achievable capacity such as

$$C_{sym} = R. \tag{39}$$

Figure 5 makes the description of the weak and strong interference region where the leftmost indicates a very weak interference region while the rightmost suggests a very strong interference region.

Analysis:

In 1948, Shannon proposed the code generation method by exploiting the random codebook in point-to-point communication with inverse Gaussian distribution (Gaussian distribution variance towards infinity is called inverse Gaussian) to achieve the channel capacity, which is described as follows [8],

$$C = \frac{1}{2}\log_2\left(1+\frac{S}{N}\right), \tag{40}$$

where the signal power is S and the noise power is N.

The point-to-point channel capacity is

$$C_{AWGN} = \log_2\left(1+\frac{S}{N}\right), \tag{41}$$

where the signal power is S and the noise power is N.

From Eq. (31), the degree of freedom is [14].

$$DoF = \lim_{x\to\infty}\left(\frac{1+\frac{S}{N}}{1+\frac{S}{N}}\right) = 1, \tag{42}$$

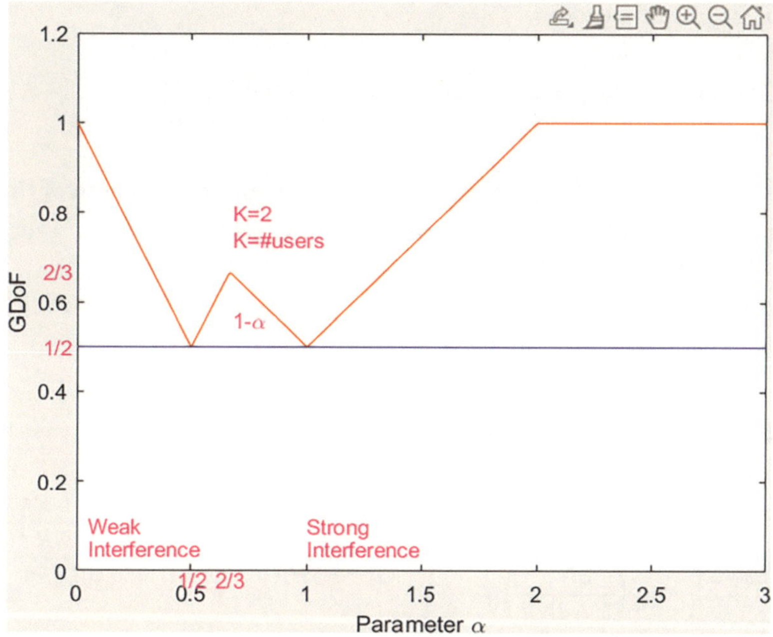

Figure 5.
Generalized degree of freedom for Gaussian Channel (W curve).

And the achievable rate is orthogonalized as

$$\sum_{i=1}^{K} R_i = \log_2\left(1 + \frac{\sum_{i=1}^{K} P_i}{N}\right), \tag{43}$$

where K means the number of users.
For two users,

$$2R = \log_2\left(1 + 2\frac{P}{N}\right) = \log_2(1 + 2SNR). \tag{44}$$

Therefore, the achievable rate is,

$$R = \frac{1}{2}\log_2(1 + 2SNR). \tag{45}$$

SNR = 19 and SNR = 31 case:

$$\text{The capacity}: C = \frac{1}{2}\log_2\left(1 + \frac{19}{31}\right) = \frac{1}{2}\log_2(1 + 0.61) = 0.34 \tag{46}$$

$$\begin{aligned} & 2R = \log_2\left(1 + 2\left(\frac{19}{31}\right)\right) \\ \text{Achievable rate}: \; & 2R = \log_2(2.22) \\ & 2R = 1.15 \\ & R = 0.57 \end{aligned} \tag{47}$$

And the degree of freedom,

$$DoF = \lim_{SNR\to\infty}\left(\frac{R}{\log_2(2SNR)}\right) \approx \frac{1}{2}\left(\frac{\log_2(1+2SNR)}{\log_2(2SNR)}\right) \approx \frac{1}{2}. \tag{48}$$

On the condition that the ratio $\alpha = \frac{\log_2 INR}{\log_2 SNR}$ is fixed and the strength of the signal is much larger than that of interference and noise, it is able to treat interference as noise. Therefore, the achievable rate is represented by

$$R = \log_2\left(1+\frac{SNR}{1+INR}\right). \tag{49}$$

From Eq. (49), the *DoF* is represented by [14].

$$DoF = \lim_{SNR\to\infty}\left(\frac{R}{\log_2\left(\frac{SNR}{1+INR}\right)}\right) = \left(\frac{\log_2\left(\frac{SNR}{1+INR}\right)}{\log_2(SNR)}\right) \approx \left(\frac{\log_2\left(\frac{SNR}{INR}\right)}{\log_2(SNR)}\right) \tag{50}$$
$$= \left(\frac{\log_2(SNR)-\log_2(INR)}{\log_2(SNR)}\right) = 1-\left(\frac{\log_2(INR)}{\log_2(SNR)}\right) = (1-\alpha).$$

In the conventional binary symmetric channel, p is a random variable and a large amount of resources are used up to make an estimate of p corresponding to the given channel. By the way, p can be determined deterministically for the RNA base matrix $[C\ U; A\ G]$, which is either 0.38 or 0.62. Because the specific value of p is given, the channel estimation should be investigated. The reason why the specific numerical values are selected is that for the RNA model, its maximum channel capacity is maintained even if p is determined deterministically, the variance of signal is not large, and a generalized *DoF*'s point of view shows a reasonable performance in the W curve. In the actual implementation, the receiver has to be satisfied with the 1-α = p shown in **Figure 2**. Under this circumstance, signal strength and the interference intensity are important to analyze the given channel where strong interference environment and weak interference environment are classified according to α. To take an example, if α = 1-p = 0.38, we need to analyze the strong interference channel. If α = 1-p = 0.62, we need to analyze the weak interference channel. This p estimation is able to minimize performance degradation in the binary symmetric channel while significantly reducing computational complexity. The *GDoF* curve of two user interference symmetric channel in **Figure 5** is the highly recognizable "W" curve shown that it greatly improves understanding of interference channel by identifying two regimes. From the abovementioned example, over the symmetric channel, when α = 0.62, the signal is relatively stronger than interference. By the way, when α = 0.38, signal is relatively weaker than interference.

6. RNA genetic code constructed by block circulant jacket matrix

A block circulant Jacket matrix (BCJM) is defined by [7, 12, 13, 15].

$$\mathbf{C}_4 = \begin{pmatrix} \mathbf{C}_0 & \mathbf{C}_1 \\ \mathbf{C}_1 & \mathbf{C}_0 \end{pmatrix} = \begin{pmatrix} 1 & 1 & a & -a \\ 1 & -1 & -1/a & -1/a \\ a & -a & 1 & 1 \\ -1/a & -1/a & 1 & -1 \end{pmatrix}_{a=1}$$
$$= \begin{pmatrix} 1 & 1 & 0 & 0 \\ 1 & -1 & 0 & 0 \\ 0 & 0 & 1 & 1 \\ 0 & 0 & 1 & -1 \end{pmatrix} + \begin{pmatrix} 0 & 0 & 1 & -1 \\ 0 & 0 & -1 & -1 \\ 1 & -1 & 0 & 0 \\ -1 & -1 & 0 & 0 \end{pmatrix} = \left(\begin{array}{cc|cc} 1 & 1 & 1 & -1 \\ 1 & -1 & -1 & -1 \\ \hline 1 & -1 & 1 & 1 \\ -1 & -1 & 1 & -1 \end{array}\right), \tag{51}$$

where $\mathbf{C}_0$ and $\mathbf{C}_1$ are the Hadamard matrix.

The circulant submatrices are 2×2 matrices, whose entries are moved by block diagonal cyclic shifts. These submatrices are block circulant Jacket matrices. The BCJM $\mathbf{C}_4$ is defined by

$$\underbrace{\mathbf{C}_4 \triangleq \mathbf{I}_0 \otimes \mathbf{C}'_0 + \mathbf{I}_1 \otimes \mathbf{C}_1,} \tag{52}$$

where $\mathbf{I}_0 = \begin{pmatrix} 1 & 0 \\ 0 & 1 \end{pmatrix}, \mathbf{I}_1 = \begin{pmatrix} 0 & 1 \\ 1 & 0 \end{pmatrix}, \mathbf{C}'_0 = \begin{pmatrix} 1 & 1 \\ 1 & -1 \end{pmatrix}$, and $\mathbf{C}_1 = \begin{pmatrix} 1 & -1 \\ -1 & -1 \end{pmatrix}$, while $\otimes$ is the Kronecker product.

From Eq. (52), the genetic matrix $[C\ U; A\ G]^3$ generates RNA sequences such as [12, 13].

$$\mathbf{P}^1 = \begin{pmatrix} C & U \\ A & G \end{pmatrix},\ \mathbf{P}^2 = \begin{pmatrix} C & U \\ A & G \end{pmatrix} \otimes \begin{pmatrix} C & U \\ A & G \end{pmatrix},\ \mathbf{P}^3 = \begin{pmatrix} C & U \\ A & G \end{pmatrix}^2 \otimes \begin{pmatrix} C & U \\ A & G \end{pmatrix}, \tag{53}$$

where $\otimes$ denotes the Kronecker product. RNA consists of the sequence of 4 bases where *C, U, A,* and *G* indicate cytosine, uracil, adenine, and guanine, on an individual basis.

According to the theory of noise-immunity coding, for 64 triplets, by comparing them with strong roots and weak roots, it is able to construct a mosaic gene matrix $[C\ U; A\ G]^3$. If any triplet belongs to one of the strong roots, it is substituted for 1. In an analogous fashion, if any triplet is included with one of the weak roots, it is replaced with -1. Here, the strong roots are $(CC, CU, CG, AC, UC, GC, GU, GG)$ and $(CA, AA, AU, AG, UA, UU, UG, GA)$ are the weak roots, which results in the singular Rademacher matrix $\mathbf{R}_8$ is in **Table 3** [6, 16].

A novel encoding scheme is proposed as

$$\mathbf{R}_8 = \left(\begin{array}{cccc|cccc} 1 & 1 & 1 & 1 & 1 & 1 & -1 & -1 \\ 1 & 1 & 1 & 1 & 1 & 1 & -1 & -1 \\ -1 & -1 & 1 & 1 & -1 & -1 & -1 & -1 \\ -1 & -1 & 1 & 1 & -1 & -1 & -1 & -1 \\ \hline 1 & 1 & -1 & -1 & 1 & 1 & 1 & 1 \\ 1 & 1 & -1 & -1 & 1 & 1 & 1 & 1 \\ -1 & -1 & -1 & -1 & -1 & -1 & 1 & 1 \\ -1 & -1 & -1 & -1 & -1 & -1 & 1 & 1 \end{array}\right). \tag{54}$$

	000 (0)	001 (1)	010 (2)	011 (3)	100 (4)	101 (5)	110 (6)	111 (7)
000 (0)	CCC 000	CCU 001	CUC 010	CUU 011	UCC 100	UCU 101	UUC 110	UUU 111
001 (1)	CCA 001	CCG 000	CUA 011	CUG 010	UCA 101	UCG 100	UUA 111	UUG 110
010 (2)	CAC 010	CAU 011	CGC 000	CGU 001	UAC 110	UAU 111	UGC 100	UGU 101
011 (3)	CAA 011	CAG 010	CGA 001	CGG 000	UAA 111	UAG 110	UGA 101	UGG 100
100 (4)	ACC 100	ACU 101	AUC 110	AUU 111	GCC 000	GCU 001	GUC 010	GUU 011
101 (5)	ACA 101	ACG 100	AUA 111	AUG 110	GCA 001	GCG 000	GUA 011	GUG 010
110 (6)	AAC 110	AAU 111	AGC 100	AGU 101	GAC 010	GAU 011	GGC 000	GGU 001
111 (7)	AAA 111	AAG 110	AGA 101	AGG 100	GAA 011	GAG 010	GGA 001	GGG 000

Table 3.
[C U;A G]^3 code [6, 16].

The Eq. (54) gives a hint of the DNA double helix.
Make a mental note to ensure that

$$\underbrace{\mathbf{R}_8 \triangleq \mathbf{I}_0 \otimes \mathbf{C}_0 \otimes \mathbf{P}_2 + \mathbf{I}_1 \otimes \mathbf{C}_1 \otimes \mathbf{P}_2,} \tag{55}$$

where $\mathbf{I}_0 = \begin{pmatrix} 1 & 0 \\ 0 & 1 \end{pmatrix}, \mathbf{I}_1 = \begin{pmatrix} 0 & 1 \\ 1 & 0 \end{pmatrix}, \mathbf{C}_0 = \begin{pmatrix} 1 & 1 \\ -1 & 1 \end{pmatrix}, \mathbf{C}_1 = \begin{pmatrix} 1 & -1 \\ -1 & -1 \end{pmatrix}$, and $\mathbf{P}_2$ is the double stochastic permutation matrix represented by $\mathbf{P}_2 = \begin{pmatrix} 1 & 1 \\ 1 & 1 \end{pmatrix}$. Eq. (54) has a series of redundant rows which just repeat and are able to be canceled. From the Rademacher matrix $\mathbf{R}_8$, one version of its mosaic gene matrices can be reached as

$$\mathbf{R}'_8 = \begin{pmatrix} 1 & 1 & 1 & 1 & 1 & 1 & -1 & -1 \\ -1 & -1 & 1 & 1 & -1 & -1 & -1 & -1 \\ 1 & 1 & -1 & -1 & 1 & 1 & 1 & 1 \\ -1 & -1 & -1 & -1 & -1 & -1 & 1 & 1 \end{pmatrix}. \tag{56}$$

Furthermore, by canceling the repeated column from Eq. (56) by means of CRISPR, another version of the mosaic gene matrices can be reached as Eq. (57), which is a singular RNA matrix.

$$\mathbf{R}''_4 = \left(\begin{array}{cc:cc} 1 & 1 & 1 & -1 \\ -1 & 1 & -1 & -1 \\ \hdashline 1 & -1 & 1 & 1 \\ -1 & -1 & -1 & 1 \end{array}\right) = \begin{pmatrix} C_0 & C_1 \\ C_1 & C_0 \end{pmatrix}, \tag{57}$$

where $\mathbf{C}_0 = \begin{pmatrix} 1 & 1 \\ -1 & 1 \end{pmatrix}$ and $\mathbf{C}_1 = \begin{pmatrix} 1 & -1 \\ -1 & -1 \end{pmatrix}$. These matrices are able to be expanded into the DNA double helix or the RNA single strand, which indicates the process by that DNA replicates its genetic information for itself, which is transcribed into RNA and used to synthesize protein for its translation. Therefore,

$$\underbrace{\mathbf{R}_4'' \triangleq \mathbf{I}_0 \otimes \mathbf{C}_0 + \mathbf{I}_1 \otimes \mathbf{C}_1,} \tag{58}$$

where $\mathbf{C}_0$ has eigenvalues such that $\lambda_1^{(1)} = 1 + i$ and $\lambda_2^{(1)} = 1 - i$, and their eigenvectors $\varsigma_1 = (1 \quad -i)^T$ and $\varsigma_2 = (1 \quad i)^T$, correspondingly. In addition, $\mathbf{C}_1$ has eigenvalues such that $\lambda_1^{(2)} = \sqrt{2}$ and $\lambda_2^{(2)} = -\sqrt{2}$ where their eigenvectors $\varsigma_1 = (-1 + \sqrt{2} \quad 1)^T$ and $\varsigma_1 = (-1 - \sqrt{2} \quad 1)^T$ on an individual basis [3, 17]. Then,

$$\mathbf{R}_4'' \otimes \mathbf{P}_2 \Rightarrow \mathbf{R}_8 = \mathbf{R}_{4\times 2^k}, \tag{59}$$

where $k = 1$.

7. Symmetric genetic jacket block matrix

It is demonstrated that the genomatrices are constructed based on the kernel $[C\,A; U\,G]$ and the mosaic genomatrices $[C\,A; U\,G]^3$ are built by a series of Kronecker products, which are expanded by permuting the 4 bases C, A, U, and G on their locations in the matrix.

7.1 Permutation scheme from upper to lower

Following this scheme, we are confronted with 24 variants of genomatrices, which distinguish them from each other by replacing their subsets by the kernel $[C\,A; U\,G]$. To take an analogous instance, by applying the upper-low scheme to $[C\,A;U\,G]$, the standard genetic code is expanded into $[U \quad C \quad A \quad G]^T \otimes [U \quad C \quad A \quad G] \otimes [U \quad C \quad A \quad G]^T$, where ^{T} is the transpose. Analogous to Eq. (56), one version of variants of genomatrices is constructed as

$$\begin{bmatrix} -1 & -1 & -1 & -1 & 1 & 1 & 1 & 1 & -1 & -1 & -1 & -1 & 1 & 1 & 1 & 1 \\ 1 & 1 & 1 & 1 & 1 & 1 & 1 & 1 & 1 & 1 & 1 & 1 & 1 & 1 & 1 & 1 \\ -1 & -1 & -1 & -1 & -1 & -1 & -1 & -1 & -1 & -1 & -1 & -1 & -1 & -1 & -1 & -1 \\ -1 & -1 & -1 & -1 & 1 & 1 & 1 & 1 & -1 & -1 & -1 & -1 & 1 & 1 & 1 & 1 \end{bmatrix}^T = \left(\underbrace{[1\ 0] \otimes \left(\begin{bmatrix} 1 \\ 1 \end{bmatrix} \otimes \begin{bmatrix} -1 & 1 \\ 1 & 1 \end{bmatrix} \right)} + \underbrace{[0\ 1] \otimes \left(\begin{bmatrix} 1 \\ 1 \end{bmatrix} \otimes \begin{bmatrix} -1 & -1 \\ -1 & 1 \end{bmatrix} \right)} \right) \otimes \begin{bmatrix} 1 \\ 1 \\ 1 \\ 1 \end{bmatrix}. \tag{60}$$

Eq. (60) is also another version of variants of genomatrices by a series of Kronecker product on $[1\,1\,1\,1]^T$, which is expanded into Eq. (61) indicating the process transcribing from $\mathbf{R}_8$ DNA to $\mathbf{R}_4''$ RNA.

$$\mathbf{R}_4'' = \left[\begin{array}{cc|cc} -1 & +1 & -1 & -1 \\ +1 & +1 & -1 & +1 \\ \hline -1 & +1 & -1 & -1 \\ +1 & +1 & -1 & +1 \end{array}\right] = \underbrace{[1\ 0]\otimes\left(\begin{bmatrix}1\\1\end{bmatrix}\otimes\begin{bmatrix}-1 & 1\\1 & 1\end{bmatrix}\right)} + \underbrace{[0\ 1]\otimes\left(\begin{bmatrix}1\\1\end{bmatrix}\otimes\begin{bmatrix}-1 & -1\\-1 & 1\end{bmatrix}\right)}. \quad (61)$$

Example 7.1. If $A = U$, $C = G$, we are confronted with six versions of variants of the genomatrices constructed by a series of Kronecker product of the kernel $[C\,A; U\,G]$.

$$\begin{bmatrix} A & C \\ U & G \end{bmatrix} = \begin{bmatrix} -1 & 1 & -1 & 1 \\ -1 & -1 & 1 & 1 \\ -1 & 1 & -1 & 1 \\ -1 & -1 & 1 & 1 \end{bmatrix} = [1\ \ 0]\otimes\begin{bmatrix}1\\1\end{bmatrix}\otimes\begin{bmatrix}-1 & 1\\-1 & -1\end{bmatrix} + [0\ \ 1]\otimes\begin{bmatrix}1\\1\end{bmatrix}\otimes\begin{bmatrix}-1 & 1\\1 & 1\end{bmatrix}, \quad (62)$$

which is expanded into Eq. (63) and Eq. (64). These are other versions of variants of genomatrices.

$$\begin{bmatrix} A & G \\ U & C \end{bmatrix} = \begin{bmatrix} -1 & -1 & -1 & 1 \\ -1 & 1 & 1 & 1 \\ -1 & -1 & -1 & 1 \\ -1 & 1 & 1 & 1 \end{bmatrix} = [1\ \ 0]\otimes\begin{bmatrix}1\\1\end{bmatrix}\otimes\begin{bmatrix}-1 & -1\\-1 & 1\end{bmatrix} + [0\ \ 1]\otimes\begin{bmatrix}1\\1\end{bmatrix}\otimes\begin{bmatrix}-1 & 1\\1 & 1\end{bmatrix}, \quad (63)$$

$$\begin{bmatrix} G & U \\ C & A \end{bmatrix} = \begin{bmatrix} 1 & 1 & -1 & -1 \\ 1 & -1 & 1 & -1 \\ 1 & 1 & -1 & -1 \\ 1 & -1 & 1 & -1 \end{bmatrix} = [1\ \ 0]\otimes\begin{bmatrix}1\\1\end{bmatrix}\otimes\begin{bmatrix}1 & 1\\1 & -1\end{bmatrix} + [0\ \ 1]\otimes\begin{bmatrix}1\\1\end{bmatrix}\otimes\begin{bmatrix}-1 & -1\\1 & -1\end{bmatrix}, \quad (64)$$

$$\begin{bmatrix} C & U \\ G & A \end{bmatrix} = \begin{bmatrix} 1 & 1 & 1 & -1 \\ 1 & -1 & -1 & -1 \\ 1 & 1 & 1 & -1 \\ 1 & -1 & -1 & -1 \end{bmatrix} = [1\ \ 0]\otimes\begin{bmatrix}1\\1\end{bmatrix}\otimes\begin{bmatrix}1 & 1\\1 & -1\end{bmatrix} + [0\ \ 1]\otimes\begin{bmatrix}1\\1\end{bmatrix}\otimes\begin{bmatrix}1 & -1\\-1 & -1\end{bmatrix}, \quad (65)$$

$$\begin{bmatrix} C & A \\ G & U \end{bmatrix} = \begin{bmatrix} 1 & -1 & 1 & -1 \\ 1 & 1 & -1 & -1 \\ 1 & -1 & 1 & -1 \\ 1 & 1 & -1 & -1 \end{bmatrix}, \tag{66}$$
$$= [1 \quad 0] \otimes \begin{bmatrix} 1 \\ 1 \end{bmatrix} \otimes \begin{bmatrix} 1 & -1 \\ 1 & 1 \end{bmatrix} + [0 \quad 1] \otimes \begin{bmatrix} 1 \\ 1 \end{bmatrix} \otimes \begin{bmatrix} 1 & -1 \\ -1 & -1 \end{bmatrix}$$

$$\begin{bmatrix} G & A \\ C & U \end{bmatrix} = \begin{bmatrix} 1 & -1 & -1 & -1 \\ 1 & 1 & 1 & -1 \\ 1 & -1 & -1 & -1 \\ 1 & 1 & 1 & -1 \end{bmatrix}. \tag{67}$$
$$= [1 \quad 0] \otimes \begin{bmatrix} 1 \\ 1 \end{bmatrix} \otimes \begin{bmatrix} 1 & -1 \\ 1 & 1 \end{bmatrix} + [0 \quad 1] \otimes \begin{bmatrix} 1 \\ 1 \end{bmatrix} \otimes \begin{bmatrix} -1 & -1 \\ 1 & -1 \end{bmatrix}$$

Eq. (62–67) are six versions of variants of genomatrices, which indicate six half pairs expanded from symmetric RNA genetic matrices by an upper-lower scheme. In other words, they are constructed by rotating the block in the direction from upper to low or vice versa.

7.2 Permutation scheme from left to right

Following this scheme, we are confronted with 6 variants of genomatrices, which distinguish them from each other with the kernel $[C\,A; U\,G]$. To take an analogous instance, by applying the left-right scheme to $[C\,A; U\,G]$, the standard genetic code is expanded into $\mathbf{R}_8$

$$\left(\begin{array}{cccc|cccc} 1 & 1 & 1 & 1 & 1 & 1 & 1 & 1 \\ 1 & 1 & 1 & 1 & 1 & 1 & 1 & 1 \\ 1 & 1 & -1 & -1 & 1 & 1 & -1 & -1 \\ 1 & 1 & -1 & -1 & 1 & 1 & -1 & -1 \\ \hline 1 & 1 & -1 & -1 & 1 & 1 & -1 & -1 \\ 1 & 1 & -1 & -1 & 1 & 1 & -1 & -1 \\ -1 & -1 & -1 & -1 & -1 & -1 & -1 & -1 \\ -1 & -1 & -1 & -1 & -1 & -1 & -1 & -1 \end{array}\right) \tag{68}$$
$$= \begin{bmatrix} \underbrace{\begin{pmatrix} 1 \\ 0 \end{pmatrix} \otimes (1 \quad 1) \otimes \begin{pmatrix} 1 & 1 \\ 1 & -1 \end{pmatrix}} \\ + \underbrace{\begin{pmatrix} 0 \\ 1 \end{pmatrix} \otimes (1 \quad 1) \otimes \begin{pmatrix} 1 & -1 \\ -1 & -1 \end{pmatrix}} \end{bmatrix} \otimes \begin{pmatrix} 1 & 1 \\ 1 & 1 \end{pmatrix}.$$

Eq. (68) is also another version of variants of genomatrices by a series of Kronecker product on [1 1;1 1], which is expanded into Eq. (69) indicating the process transcribing from $\mathbf{R}_8$ DNA to $\mathbf{R}_4''$ RNA.

$$\mathbf{R}_4'' = \left[\begin{array}{cc|cc} +1 & +1 & +1 & +1 \\ +1 & -1 & +1 & -1 \\ \hline +1 & -1 & +1 & -1 \\ -1 & -1 & -1 & -1 \end{array}\right] = \underbrace{\begin{pmatrix} 1 \\ 0 \end{pmatrix} \otimes \begin{pmatrix} 1 & 1 \end{pmatrix} \otimes \begin{pmatrix} 1 & 1 \\ 1 & -1 \end{pmatrix}} + \underbrace{\begin{pmatrix} 0 \\ 1 \end{pmatrix} \otimes \begin{pmatrix} 1 & 1 \end{pmatrix} \otimes \begin{pmatrix} 1 & -1 \\ -1 & -1 \end{pmatrix}}. \tag{69}$$

Example 7.2. If $A = U$, $C = G$, we are confronted with six versions of variants of the genomatrices constructed by a series of Kronecker product of the kernel $[C\,A; U\,G]$.

$$\begin{pmatrix} C & G \\ U & A \end{pmatrix} = \begin{pmatrix} 1 & 1 & 1 & 1 \\ 1 & -1 & 1 & -1 \\ 1 & -1 & 1 & -1 \\ -1 & -1 & -1 & -1 \end{pmatrix}$$

$$= \left[\underbrace{\begin{pmatrix} 1 \\ 0 \end{pmatrix} \otimes \begin{pmatrix} 1 & 1 \end{pmatrix} \otimes \begin{pmatrix} 1 & 1 \\ 1 & -1 \end{pmatrix}} + \underbrace{\begin{pmatrix} 0 \\ 1 \end{pmatrix} \otimes \begin{pmatrix} 1 & 1 \end{pmatrix} \otimes \begin{pmatrix} 1 & -1 \\ -1 & -1 \end{pmatrix}}\right], \tag{70}$$

which is expanded into Eq. (71) and Eq. (72). These are other versions of variants of genomatrices.

$$\begin{bmatrix} G & C \\ U & A \end{bmatrix} = \begin{bmatrix} 1 & 1 & 1 & 1 \\ 1 & -1 & 1 & -1 \\ -1 & 1 & -1 & 1 \\ -1 & -1 & -1 & -1 \end{bmatrix}, \tag{71}$$

$$= \begin{bmatrix} 1 \\ 0 \end{bmatrix} \otimes \begin{bmatrix} 1 & 1 \end{bmatrix} \otimes \begin{bmatrix} 1 & 1 \\ 1 & -1 \end{bmatrix} + \begin{bmatrix} 0 \\ 1 \end{bmatrix} \otimes \begin{bmatrix} 1 & 1 \end{bmatrix} \otimes \begin{bmatrix} 1 & -1 \\ -1 & -1 \end{bmatrix}$$

$$\begin{bmatrix} U & A \\ C & G \end{bmatrix} = \begin{bmatrix} -1 & -1 & -1 & -1 \\ 1 & -1 & 1 & -1 \\ 1 & -1 & 1 & -1 \\ 1 & 1 & 1 & 1 \end{bmatrix}, \tag{72}$$

$$= \begin{bmatrix} 1 \\ 0 \end{bmatrix} \otimes \begin{bmatrix} 1 & 1 \end{bmatrix} \otimes \begin{bmatrix} -1 & -1 \\ 1 & -1 \end{bmatrix} + \begin{bmatrix} 0 \\ 1 \end{bmatrix} \otimes \begin{bmatrix} 1 & 1 \end{bmatrix} \otimes \begin{bmatrix} 1 & -1 \\ 1 & 1 \end{bmatrix}$$

$$\begin{bmatrix} A & U \\ G & C \end{bmatrix} = \begin{bmatrix} -1 & -1 & -1 & -1 \\ -1 & 1 & -1 & 1 \\ -1 & 1 & -1 & 1 \\ 1 & 1 & 1 & 1 \end{bmatrix} = \begin{bmatrix} 1 \\ 0 \end{bmatrix} \otimes [1 \quad 1] \otimes \begin{bmatrix} -1 & -1 \\ -1 & 1 \end{bmatrix} + \begin{bmatrix} 0 \\ 1 \end{bmatrix} \otimes [1 \quad 1] \otimes \begin{bmatrix} -1 & 1 \\ 1 & 1 \end{bmatrix}, \tag{73}$$

$$\begin{bmatrix} G & C \\ A & U \end{bmatrix} = \begin{bmatrix} 1 & 1 & 1 & 1 \\ -1 & 1 & -1 & 1 \\ -1 & 1 & -1 & 1 \\ -1 & -1 & -1 & -1 \end{bmatrix} = \begin{bmatrix} 1 \\ 0 \end{bmatrix} \otimes [1 \quad 1] \otimes \begin{bmatrix} 1 & 1 \\ -1 & 1 \end{bmatrix} + \begin{bmatrix} 0 \\ 1 \end{bmatrix} \otimes [1 \quad 1] \otimes \begin{bmatrix} -1 & 1 \\ -1 & -1 \end{bmatrix}, \tag{74}$$

$$\begin{bmatrix} C & G \\ A & U \end{bmatrix} = \begin{bmatrix} 1 & 1 & 1 & 1 \\ -1 & 1 & -1 & 1 \\ 1 & -1 & 1 & -1 \\ -1 & -1 & -1 & -1 \end{bmatrix} = \begin{bmatrix} 1 \\ 0 \end{bmatrix} \otimes [1 \quad 1] \otimes \begin{bmatrix} 1 & 1 \\ -1 & 1 \end{bmatrix} + \begin{bmatrix} 0 \\ 1 \end{bmatrix} \otimes [1 \quad 1] \otimes \begin{bmatrix} 1 & -1 \\ -1 & -1 \end{bmatrix}. \tag{75}$$

Eqs. (70)-(75) are 6 versions of variants of genomatrices, which indicate six half pairs expanded from symmetric RNA genetic matrices by the left-right scheme. In other words, they are constructed by rotating the block in the direction from upper to low or vice versa.

7.3 Block Circulant jacket matrix

Construct a block matrix $[C]_N$ by Jacket matrices $[C_0]_p$ and $[C_1]_p$ such as $[C]_N = \begin{pmatrix} C_0 & C_1 \\ C_1 & C_0 \end{pmatrix}$ where its order N is $2p$. This matrix is called block circulant if only if $C_0 C_1^{RT} + C_1^{RT} C_0 = [0]_N$, where ^{RT} is the reciprocal transpose. In other words, $[C]_N$ is a block circulant Jacket matrix (BCJM) [12, 13, 15, 18]. From the fact that $C_0 C_0^{RT} = p[I]_p$ and $C_1 C_1^{RT} = p[I]_p$, C_0 and C_1 are Jacket matrices. Look back on the fact that $[C]_N$ is a Jacket matrix if only if $[C][C]^{RT} = NI_N$, where ^{RT} is the reciprocal transpose. Therefore, $[C]$ is a Jacket matrix if only if

$$[C][C]^{RT} = \begin{pmatrix} C_0 & C_1 \\ C_1 & C_0 \end{pmatrix} \begin{pmatrix} C_0 & C_1 \\ C_1 & C_0 \end{pmatrix}^{RT} = \begin{pmatrix} 2p[I]_p & C_0 C_1^{RT} + C_1^{RT} C_0 \\ C_0 C_1^{RT} + C_1^{RT} C_0 & 2p[I]_p \end{pmatrix} = NI_N, \tag{76}$$

where ^{RT} is the reciprocal transpose. Therefore, Eq. (76) results in plenty of BCJMs.

Example 7.3. Two 2 × 2 matrices are given such as

$$C_0 = \begin{pmatrix} 1 & 1 \\ 1 & -1 \end{pmatrix}, C_1 = \begin{pmatrix} a & -a \\ -1/a & -1/a \end{pmatrix}.$$

It is easy to know that $C_0 C_0^{RT} = 2[I]_2$ and $C_1 C_1^{RT} = 2[I]_2$ are satisfied. Therefore, C_0 and C_1 are Jacket matrices.

Moreover,

$$C_0 C_1^{RT} + C_1^{RT} C_0 = \begin{pmatrix} 1 & 1 \\ 1 & -1 \end{pmatrix}\begin{pmatrix} -1/a & -a \\ -1/a & -a \end{pmatrix} + \begin{pmatrix} a & -a \\ -1/a & -1/a \end{pmatrix}\begin{pmatrix} 1 & 1 \\ 1 & -1 \end{pmatrix} = [0]_2. \quad (77)$$

8. General pattern of block circulant symmetric genetic jacket matrix

We present 24(=4 × $_4\mathbf{C}_2$) DNA classes of genomatrices with their own characteristics. The main kernel of Eq. (78) is

$$\underbrace{E}_{\text{Position}} \otimes \underbrace{\{(I_0 \otimes A) + (I_1 \otimes B)\}}_{\text{Main Body Kernel}} \otimes \underbrace{F}_{\text{Extending}}. \quad (78)$$

Eq. (58) is an RNA pattern by the main kernel. By applying an upper-lower or left-right scheme to the genetic matrix, the position matrix **E** creates the patterns analogous to Eq. (61, 69). Analogously, by applying the upper-lower and left-right scheme to the genetic matrix, the extending matrix F creates the patterns analogous to Eq. (60, 68).

South Korea's national flag stands for different symbols of trigrams and Yin-Yang located in its middle, which is analogous to that of **Figure 6**. We present 24 versions of variants of genomatrices, which distinguish from each other by replacing their subsets with the kernel shown in **Figure 6** like its left-hand side $\begin{bmatrix} 1 \\ 0 \end{bmatrix} \otimes [1 \quad 1]$, its right-hand

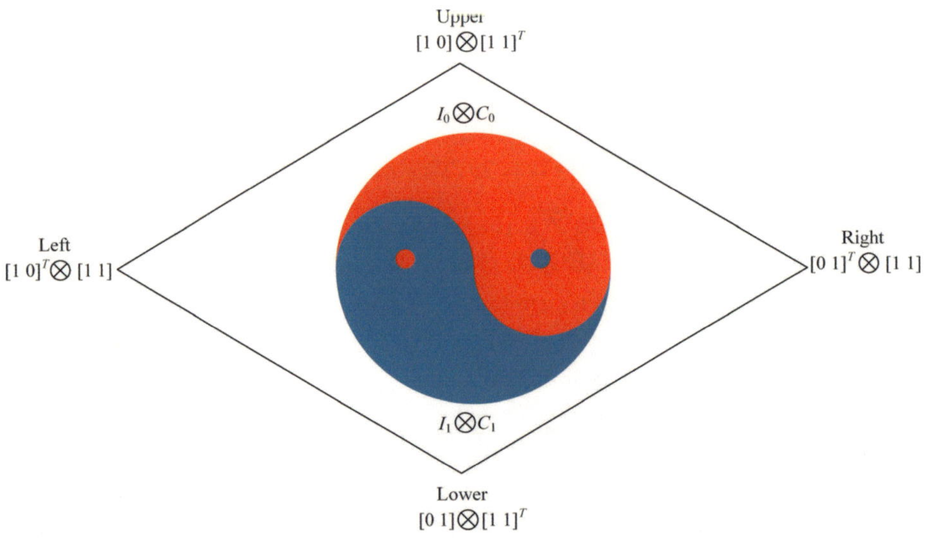

Figure 6.
General pattern by block circulant, upper-lower, and left–right scheme: Normal case.

side $\begin{bmatrix} 0 \\ 1 \end{bmatrix} \otimes [1 \quad 1]$, its upper position $(1 \quad 0) \otimes \begin{pmatrix} 1 \\ 1 \end{pmatrix}$, its lower position $(0 \quad 1) \otimes \begin{pmatrix} 1 \\ 1 \end{pmatrix}$, and its center part $I_0 \otimes C_0 + I_1 \otimes C_1$, on an individual basis.

From the fact that $(1 \quad 0) \otimes \begin{pmatrix} 1 \\ 1 \end{pmatrix} \leftrightarrow (0 \quad 1) \otimes \begin{pmatrix} 1 \\ 1 \end{pmatrix}$ and $\begin{pmatrix} 1 \\ 0 \end{pmatrix} \otimes (1 \quad 1) \leftrightarrow \begin{pmatrix} 0 \\ 1 \end{pmatrix} \otimes (1 \quad 1)$, upper symmetric genetic matrices are complementary with lower ones while left ones are complementary with right ones.

In addition, the pattern is created by block circulant, upper-lower, and left–right scheme on the ½ symmetric block, which are analyzed in three cases.

Case 1. Block circulant scheme

$$\begin{bmatrix} C & U \\ A & G \end{bmatrix} = \begin{bmatrix} 1 & 1 & 1 & -1 \\ -1 & 1 & -1 & -1 \\ 1 & -1 & 1 & 1 \\ -1 & -1 & 1 & -1 \end{bmatrix} = \begin{bmatrix} 1 & 0 \\ 0 & A^{diag} \end{bmatrix} \otimes \begin{bmatrix} 1 & 1 \\ -1 & 1 \end{bmatrix} + \begin{bmatrix} 0 & 1 \\ 1 & 0 \end{bmatrix} \otimes \begin{bmatrix} 1 & -1 \\ -1 & -1 \end{bmatrix}. \tag{79}$$

$$\begin{bmatrix} U & C \\ G & A \end{bmatrix} = \begin{bmatrix} -1 & 1 & 1 & 1 \\ -1 & -1 & 1 & -1 \\ 1 & 1 & -1 & 1 \\ -1 & 1 & -1 & -1 \end{bmatrix} = \begin{bmatrix} 1 & 0 \\ 0 & 1 \end{bmatrix} \otimes \begin{bmatrix} -1 & 1 \\ -1 & -1 \end{bmatrix} + \begin{bmatrix} 0 & 1 \\ A^{Anti-diag} & 0 \end{bmatrix} \otimes \begin{bmatrix} 1 & 1 \\ 1 & -1 \end{bmatrix}. \tag{80}$$

Case 2. Upper-lower scheme

$$\begin{bmatrix} U & G \\ A & C \end{bmatrix} = \begin{bmatrix} -1 & -1 & 1 & 1 \\ -1 & 1 & -1 & 1 \\ -1 & -1 & 1 & -1 \\ -1 & 1 & 1 & 1 \end{bmatrix} = [1 \quad 0] \otimes \begin{bmatrix} 1 \\ 1 \end{bmatrix} \otimes \begin{bmatrix} -1 & -1 \\ -1 & 1 \end{bmatrix} + [0 \quad A^{Upper}] \otimes \begin{bmatrix} 1 \\ 1 \end{bmatrix} \otimes \begin{bmatrix} 1 & 1 \\ -1 & 1 \end{bmatrix}. \tag{81}$$

$$\begin{bmatrix} U & C \\ A & G \end{bmatrix} = \begin{bmatrix} -1 & 1 & 1 & 1 \\ -1 & -1 & 1 & -1 \\ -1 & 1 & 1 & 1 \\ -1 & -1 & -1 & 1 \end{bmatrix} = [1 \quad 0] \otimes \begin{bmatrix} 1 \\ 1 \end{bmatrix} \otimes \begin{bmatrix} -1 & 1 \\ -1 & -1 \end{bmatrix} + [0 \quad A^{Lower}] \otimes \begin{bmatrix} 1 \\ 1 \end{bmatrix} \otimes \begin{bmatrix} 1 & 1 \\ 1 & -1 \end{bmatrix}. \tag{82}$$

Case 3. Left-right scheme

$$\begin{bmatrix} A & U \\ C & G \end{bmatrix} = \begin{bmatrix} -1 & -1 & -1 & -1 \\ 1 & -1 & 1 & -1 \\ 1 & 1 & -1 & 1 \\ 1 & -1 & 1 & 1 \end{bmatrix} = \begin{bmatrix} 1 \\ 0 \end{bmatrix} \otimes [1 \quad 1] \otimes \begin{bmatrix} -1 & -1 \\ 1 & -1 \end{bmatrix} + \begin{bmatrix} 0 \\ A^{Left} \end{bmatrix} \otimes [1 \quad 1] \otimes \begin{bmatrix} 1 & 1 \\ 1 & -1 \end{bmatrix}. \tag{83}$$

$$\begin{bmatrix} U & A \\ G & C \end{bmatrix} = \begin{bmatrix} -1 & -1 & -1 & -1 \\ -1 & 1 & -1 & 1 \\ 1 & -1 & 1 & 1 \\ 1 & 1 & -1 & 1 \end{bmatrix} = \begin{bmatrix} 1 \\ 0 \end{bmatrix} \otimes [1 \quad 1] \otimes \begin{bmatrix} -1 & -1 \\ -1 & 1 \end{bmatrix} + \begin{bmatrix} 0 \\ A^{Right} \end{bmatrix} \otimes [1 \quad 1] \otimes \begin{bmatrix} 1 & -1 \\ 1 & 1 \end{bmatrix}. \tag{84}$$

Eq. (79) is a block circulant while Eq. (80) is not. Meanwhile, one part of Eq. (81, 82) is upper-lower symmetric while the other is not. By the way, one part of Eq. (83, 84) is left–right symmetric while the other part is not. **Figure 7** shows a certain pattern constructed by a series of the product of $[C\,A; U\,G]$ as well as a distorted pattern in comparison with that in **Figure 6**. Therefore, these are called sickness pattern, which can cover COVID-19.

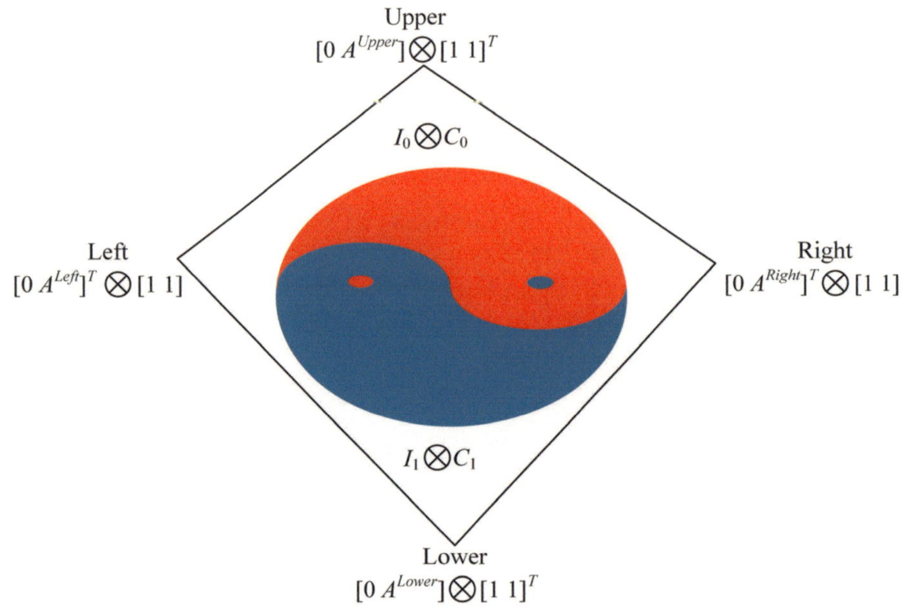

Figure 7.
Abnormal pattern by block circulant, upper-lower, and left–right scheme.

To take an analogous instance,

$$\begin{pmatrix} C & U \\ A & G \end{pmatrix} \Rightarrow \begin{pmatrix} A & B \\ C & D \end{pmatrix}, \tag{85}$$

Make a mental note to ensure.
Case 1. $A \neq D,\ B = C$ and $A = D,\ B \neq C$.
Case 2. $A = C,\ B \neq D$ and $A \neq C,\ B = D$.
Case 3: $A = B,\ C \neq D$ and $A \neq B,\ C = D$.
From the aforementioned processes, we are confronted with six half symmetric blocks such as $\begin{pmatrix} C & U \\ A & G \end{pmatrix}, \begin{pmatrix} U & C \\ G & A \end{pmatrix}, \begin{pmatrix} U & G \\ A & C \end{pmatrix}, \begin{pmatrix} U & C \\ A & G \end{pmatrix}, \begin{pmatrix} A & U \\ C & G \end{pmatrix}$, and $\begin{pmatrix} U & A \\ G & C \end{pmatrix}$.

9. Conclusion

We show the experimental results of $C = G = 19\%$ and $A = U = T = 31\%$ for the COVID-19 with the RNA base matrix $[C\ U; A\ G]$, which are expanded into our mathematical proof based on the information theory of doubly stochastic matrix. RNA entropy cannot reach the Shannon entropy because the probabilities of its bases are 23% away from a half that is exactly identical to the sum of its variances. In other words, there is a difference between Shannon capacity and RNA capacity, which is identical to the sum of variances of RNA base random variables because they are unable to become a half over a symmetric channel. We present a straightforward way of laying out a mathematical basis for double helix DNA in the process of reverse transcription from RNA to DNA, which is straightforward and explicit by decomposing a DNA matrix into sparse matrices which have non-redundant columns and rows. And we introduce a general pattern by block circulant, upper-lower, and left–right scheme, which is applied to the correct communication as well as means the healthy condition because it perfectly consists of 4 bases. Furthermore, we introduce an abnormal pattern by block circulant, upper-lower, and left–right scheme, which covers the distorted signal as well as COVID-19. The Equation 57, RNA matrix is the same as the Reference [12] USA patent MIMO Comm. definition 3.1 matrix.

Conflict of interest

The authors declare no conflict of interest.

Author details

Sung Kook Lee[1] and Moon Ho Lee[2*]

1 Jeju International School, Jeju, Republic of Korea

2 Department of Electronics, Jeonbuk National University, Jeonju, Republic of Korea

*Address all correspondence to: moonho@jbnu.ac.kr

References

[1] Chargaff E, Zamenhof S, Green C. Human desoxypentose nucleic acid: Composition of human desoxypentose nucleic acid. Nature. 1950;**165**:756-757. DOI: 10.1038/165756b0

[2] Watson J, Crick F. Molecular structure of nucleic acids: A structure for deoxyribose nucleic acid. Nature. 1953; **171**:737-738. DOI: 10.1038/171737a0

[3] Temin HM. Nature of the provirus of Rous sarcoma. National Cancer Institute Monograph. 1964;**17**:557-570

[4] Lee MH, Lee SK, Cho KM. A Life Ecosystem Management With DNA Base Complementarity. Moscow: Proceedings of the International Conference of Artificial Intelligence, Medical Engineering, Education (AIMEE 2018); 6–8 October 2018; Springer Nature; 2020

[5] Papoulis A, Pillai SU. Probability, Random Variables and Stochastic Process. 4th ed. Boston: McGraw Hill; 2002

[6] He M, Petoukhov S. Mathematics of Bioinformatics: Theory, Practice, and Applications. 1st ed. New Jersey: John Wiley & Sons; 2010. DOI: 10.1002/9780470904640

[7] Lee SK, Park DC, Lee MH. RNA genetic 8 by 8 matrix construction from the block circulant Jacket matrix. Springer Nature: Proceedings of Symmetry Festival 2016; 18-22 July 2016, Vienna, Cham; 2017

[8] Shannon CE. A Mathematical Theory of Communication. The Bell System Technical Journal. 1948;**27**:31-423-623-656. DOI: 10.1002/j.1538-7305.1948.tb01338.x\

[9] Azgari C, Kilinc Z, Turhan B, Circi D, Adebali O. The mutation profile of SARS-CoV-2 is primarily shaped by the host antiviral defense. Viruses. 2021; **13**(3):394. DOI: 10.3390/v13030394

[10] Berkhout B, Hemert VF. On the biased nucleotide composition of the human coronavirus RNA genome. Virus Research. 2015;**202**:41-47. DOI: 10.1016/j.virusres.2014.11.031

[11] Xia X. Extreme Genomic CpG Deficiency in SARS-CoV-2 and Evasion of Host Antiviral Defense. Molecular Biology and Evolution. 2020;**37**(9):2699-2705. DOI: 10.1093/molbev/msaa094

[12] Lee MH, Hai H, Zhang XD. MIMO Communication Method and System using the Block Circulant Jacket Matrix. United States Patent US 009356671B1 [Internet]. 31 May 2016. Available from: https://patentimages.storage.googleapis.com/cb/46/34/4acf23e5a9b6e1/US9356671.pdf [Accessed: 12 December 2021]

[13] Lee MH. Jacket Matrices: Construction and Its Application for Fast Cooperative Wireless Signal Processing. 1st ed. Germany, Saarbrucken: LAP LAMBERT Academic Publishing; 2012

[14] Tse D, Viswanath P. Fundamentals of Wireless Communication. 1st ed. New York: Cambridge University Press; 2005. DOI: 10.1017/CBO9780511807213

[15] Wikipedia, the free encyclopedia. Jacket Matrix [Internet]. 1999. Available from: https://en.wikipedia.org/wiki/Jacket_matrix [Accessed: 12 December 2021]

[16] Rumer YB. Translation of 'Systematization of Codons in the Genetic Code [II]' by Yu. B. Rumer (1968). Royal Society. 2016;**374**:2063. DOI: 10.1098/rsta.2015.0447

[17] Lee MH, Hai H, Lee SK, Petoukhov SV. A Mathematical Proof of Double Helix DNA to Reverse Transcription RNA for Bioinformatics. Moscow: Proceedings of the 1st International Conference of Artificial Intelligence, Medical Engineering, and Education (AIMEE 2017); 21–23 August 2017; Springer Nature; 2018

[18] Chen Z, Lee MH, Zeng G. Fast cocyclic Jacket transform. IEEE trans. on Signal Processing. 2008;**56**(5):2143-2148. DOI: 10.1109/TSP.2007.912895

Chapter 8

Joint EigenValue Decomposition for Quantum Information Theory and Processing

Gilles Burel, Hugo Pillin, Paul Baird, El-Houssaïn Baghious and Roland Gautier

Abstract

The interest in quantum information processing has given rise to the development of programming languages and tools that facilitate the design and simulation of quantum circuits. However, since the quantum theory is fundamentally based on linear algebra, these high-level languages partially hide the underlying structure of quantum systems. We show that in certain cases of practical interest, keeping a handle on the matrix representation of the quantum systems is a fruitful approach because it allows the use of powerful tools of linear algebra to better understand their behavior and to better implement simulation programs. We especially focus on the Joint EigenValue Decomposition (JEVD). After giving a theoretical description of this method, which aims at finding a common basis of eigenvectors of a set of matrices, we show how it can easily be implemented on a Matrix-oriented programming language, such as Matlab (or, equivalently, Octave). Then, through two examples taken from the quantum information domain (quantum search based on a quantum walk and quantum coding), we show that JEVD is a powerful tool both for elaborating new theoretical developments and for simulation.

Keywords: quantum information, quantum coding, quantum walk, quantum search, joint eigenspaces, joint eigenvalues, joint eigenvectors

1. Introduction

The field of quantum information is experiencing a resurgence of interest due to the recent implementation of secure transmission systems [1] based on the teleportation of quantum states in metropolitan networks and in the context of satellite transmissions, further underscored by the development of quantum computers. A new path for intercontinental quantum communication opened up in 2017 when a source onboard a Chinese satellite made it possible to distribute entangled photons between two ground stations, separated by more than 1000 km [2, 3]. Experiments using optical fibers [4] and terrestrial free-space channels [5] have also proved that the use of quantum entanglement can be achieved over large distances.

Quantum programming languages, such as Q# [6] have been developed to facilitate the design and simulation of quantum circuits. The underlying quantum theory is quite complex and often counter-intuitive due to the fact that it relies on linear algebra and tensor products—for instance, the state of a set of three independent qubits (quantum bits) is not described by a 3-dimensional vector, as would be the case for classical bits, but by a 2^3-dimensional vector which lives in a Hilbert space constructed by tensor products of lower-dimensional spaces. Therefore, these programming languages are helpful for people who do not need to bother with the underlying theory.

However, since the quantum theory is fundamentally based on linear algebra, there are cases of practical interest for the researcher in which keeping a handle on the matrix representation of the quantum systems is a fruitful approach because it allows the use of powerful tools of linear algebra to better understand their behavior and to better implement simulation programs.

In this chapter, our objective is to illustrate how the concept of Joint EigenValue Decomposition (JEVD) can provide interesting results in the domain of quantum information. The chapter is organized as follows. In Section 2, we give some mathematical background and in Section 3, we provide basic elements to understand quantum information. Then, in Section 4, we show an example of the application of JEVD to quantum coding, more precisely we propose an algorithm, based on JEVD, to identify a quantum encoder matrix from a collection of given Pauli errors. Finally, in Section 5, we show that JEVD is a powerful tool for the analysis of a quantum walk search. More precisely, we prove that, while the quantum walk operates in a huge state space, there exists a small subspace that captures all the essential elements of the quantum walk, and this subspace can be determined thanks to JEVD.

2. Mathematical background

2.1 Matrices and notations

We note U^T the transpose of a matrix U and U^* the transpose conjugate of U.

H is the normalized Hadamard 2×2 matrix and H_N the $N \times N$ Hadamard matrix obtained by the Kronecker product (defined in the next subsection):

$$H = \frac{1}{\sqrt{2}}\begin{pmatrix} 1 & 1 \\ 1 & -1 \end{pmatrix} \quad \text{and} \quad H_N = H^{\otimes n} \quad (N = 2^n) \tag{1}$$

I_N is the $N \times N$ identity matrix (which will sometimes be noted I when its dimension is implicit).

In the domain of quantum information processing, we mainly have unitary matrices. A square matrix U is unitary [7] if $U^*U = UU^* = I$. The columns of a unitary matrix are orthonormal and its eigenvalues are of norm 1. If the unitary matrix is real, its eigenvalues come by conjugate pairs.

We call “shuffle matrix” the permutation matrix $P_{a,b}$ which represents the permutation obtained when one writes elements row by row in an $a \times b$ matrix and reads them column by column. For instance, set $a = 2$ and $b = 3$. If one writes the elements $1, 2, 3, 4, 5, 6$ row by row in a 2×3 matrix and reads them column by column, the order becomes $1, 4, 2, 5, 3, 6$. Then the shuffle matrix is the permutation matrix such that $(1\ 4\ 2\ 5\ 3\ 6) = (1\ \ 2\ \ 3\ \ 4\ \ 5\ \ 6)P_{2,3}$. The inverse of $P_{a,b}$ is $P_{b,a} = (P_{a,b})^T$.

G_n is the $n \times n$ Grover diffusion matrix defined by [8]:

$$G_n = -I_n + 2\theta_n\theta_n^T \tag{2}$$

where θ_n the $(n \times 1)$ vector is defined by $\theta_n = [1 \quad 1 \quad \cdots \quad 1]^T/\sqrt{n}$. It is easy to see that $G_n\theta_n = \theta_n$. Therefore, θ_n is an eigenvector of G_n with eigenvalue $+1$. We can also see that for any vector v orthogonal to θ_n we have $G_n v = -v$. It follows that G_n has two eigenvalues, -1 and $+1$, and the dimensions of the associated eigenspaces are $n-1$ and 1.

2.2 Kronecker product

The Kronecker product, denoted by $\otimes$, is a bilinear operation on two matrices. If A is a $k \times l$ matrix and B is a $m \times n$ matrix, then the Kronecker product is the $km \times ln$ block matrix C below:

$$C = A \otimes B = \begin{pmatrix} a_{11}B & \cdots & a_{1l}B \\ \vdots & \ddots & \vdots \\ a_{k1}B & \cdots & a_{kl}B \end{pmatrix} \tag{3}$$

Assuming the sizes are such that one can form the matrix products AC and BD, an interesting property, known as the mixed-product property, is:

$$(A \otimes B)(C \otimes D) = (AC) \otimes (BD) \tag{4}$$

The Kronecker product is associative, but not commutative. However, there exist permutation matrices (the shuffle matrices defined in the previous subsection) such that, if A is an $a \times a$ square matrix and B a $b \times b$ square matrix, then [9]:

$$(A \otimes B)P_{a,b} = (B \otimes A)P_{b,a} \tag{5}$$

2.3 Singular value decomposition, image, and kernel

The Singular Value Decomposition (SVD) of an $m \times n$ matrix A is [7]:

$$A = USV^* \tag{6}$$

where U and V are unitary matrices, and S is diagonal. The diagonal of S contains the Singular Values, which are real nonnegative numbers, ranked by decreasing order. The sizes of the matrices are $U(m \times m)$, $S(m \times n)$ and $V(n \times n)$. The SVD is a very useful linear algebra tool because it reveals a great deal about the structure of a matrix.

The image and the kernel of A are defined by:

$$Im(A) = \{y \in \mathbb{C}^m : y = Ax \text{ for some} x \in \mathbb{C}^n\} \tag{7}$$

$$\ker(A) = \{x \in \mathbb{C}^n : Ax = 0\} \tag{8}$$

When used in an algorithm, the notation *null* will also be used for a procedure that computes a matrix whose columns are an orthonormal basis of the kernel of A.

The complement of a subspace $\mathcal{A}$ within a vector space $\mathcal{H}$ is defined by:

$$(\mathcal{A})^c = \{y \in \mathcal{H} : x^* y = 0 \,\text{for all}\, x \in \mathcal{A}\} \tag{9}$$

In an algorithm, if the columns of A are an orthonormal basis of $\mathcal{A}$ then the columns of $B = null(A^*)$ provide an orthonormal basis of $(\mathcal{A})^c$.

The rank of A is its number of nonzero singular values. When programmed on a computer determination of the rank must take into account finite precision arithmetic, which means that "zero" is replaced by "extremely small" (less than a given tolerance value). Let us note $r = rank(A)$. We have

$$\dim\,(Im(A)) = r \tag{10}$$

$$\dim\,(\ker(A)) = n - r \tag{11}$$

An orthonormal basis of $\ker(A)$ is obtained by taking the last $n - r$ columns of the matrix V.

2.4 Joint eigenspaces and joint eigenvalue decomposition (JEVD)

The eigenvalue decomposition of a unitary matrix A is:

$$A = VDV^* \tag{12}$$

where D is a diagonal matrix, the diagonal of which contains the eigenvalues, and V is a unitary matrix whose columns are the eigenvectors.

Let us note E_λ^A the eigenspace of an operator A associated with an eigenvalue λ. The joint eigenspace $E_{\lambda,\mu}^{A,B}$ is:

$$E_{\lambda,\mu}^{A,B} = E_\lambda^A \cap E_\mu^B \tag{13}$$

A property of great interest in quantum information processing is that within $E_{\lambda,\mu}^{A,B}$ (and even within any union of joint eigenspaces) the operators A and B commute.

Determination of the joint eigenspace on a computer may be determined through the complement, because:

$$E_\lambda^A \cap E_\mu^B = \left(\left(E_\lambda^A\right)^c \cup \left(E_\mu^B\right)^c \right)^c \tag{14}$$

Using Matrix-oriented programming languages, such as Matlab or Octave, this requires only a few lines. Let us note A_λ and B_μ matrices whose columns are orthonormal bases of E_λ^A and E_μ^B and $[.]$ the horizontal concatenation of matrices. The following computation procedure provides a matrix C whose columns are an orthonormal basis of $E_{\lambda,\mu}^{A,B}$:

$$C = null\ \left(\left[null(A_\lambda) \quad null\left(B_\mu\right) \right] \right) \tag{15}$$

However, it is not efficient in terms of complexity and in the next sections we will propose faster computational procedures, adapted to each context.

A lower bound on the dimension of a joint eigenspace can be obtained as follows. Let us note n the dimension of the full space. We have, obviously:

$$\dim E_\lambda^A \cup E_\mu^B \leq n \tag{16}$$

and we know that:

$$\dim E_\lambda^A \cup E_\mu^B = \dim E_\lambda^A + \dim E_\mu^B - \dim E_\lambda^A \cap E_\mu^B \tag{17}$$

Combining both equations, we obtain:

$$\dim E_{\lambda,\mu}^{A,B} \geq \dim E_\lambda^A + \dim E_\mu^B - n \tag{18}$$

3. Quantum information principles

A quantum system is described by a state vector $|\psi\rangle \in \mathbb{C}^N$, where N is the dimension of the system. Since in the quantum formalism states $|\psi\rangle$ and $\gamma|\psi\rangle$ are equivalent, for any nonzero complex number γ, the state is usually represented by a normed vector and the global phase is considered irrelevant.

As long as it remains isolated, the evolution of a quantum system is driven by the Schrödinger equation. The latter is a first-order differential equation operating on the quantum state. Its integration shows that the quantum states at times t_1 and t_2 are linked by a unitary matrix U such that $|\psi_2\rangle = U|\psi_1\rangle$. The norm is preserved because U is unitary.

The second kind of evolution, called "measurement," may occur if the system interacts with its environment. A measurement consists of the projection of the state onto a subspace of $\mathbb{C}^N$. When the measurement is controlled, it consists in defining *a priori* a decomposition of the state space into a direct sum of orthogonal subspaces $\bigoplus_i \mathcal{H}_i$. The measurement randomly selects one subspace. The result of the measurement is an identifier of the selected subspace (for instance, its index i). After measurement, the state is projected onto $\mathcal{H}_i$. If P_i is the projection matrix onto $\mathcal{H}_i$, then the state becomes $P_i|\psi\rangle$ (which is then renormalized because the projection does not preserve the norm). The probability of $\mathcal{H}_i$ being selected is the square norm of $P_i|\psi\rangle$.

It is worth noting that a measurement may destroy a part of quantum information (because usually, a projection is an irreversible process), while the unitary evolution is reversible, and as such, preserves quantum information. Consequently, measurements must be used with extreme caution—how to design the system and the measurement device to measure only what is strictly required and not more is one of the difficult problems encountered in quantum information processing.

Quantum systems of special interest for quantum information processing are qubits (quantum bits) and qubit registers. A qubit belongs to a 2D quantum system with state a normed vector of $\mathbb{C}^2$. To highlight links with classical digital computation, it is convenient to note $|0\rangle$ and $|1\rangle$ for the orthonormal basis of $\mathbb{C}^2$. Physically any 2D quantum system can carry a quantum bit. For instance, the spin of an electron is a 2D quantum system, and the spins up and down can be associated with the basic states $|0\rangle$ and $|1\rangle$. A general qubit has an expression:

$$|\psi\rangle = \alpha_0|0\rangle + \alpha_1|1\rangle \tag{19}$$

where α_0 and α_1 are complex numbers subject to $|\alpha_0|^2 + |\alpha_1|^2 = 1$.

A qubit register is a 2^n-D quantum system which, for convenience, is usually referred to as a standard orthonormal basis noted $\{|0...00\rangle, |0...01\rangle, |0...10\rangle, \dots, |1...11\rangle\}$ and then, by analogy with classical digital processing, n is the number of qubits. For instance, for $n = 2$ the basis is $\{|00\rangle, |01\rangle, |10\rangle, |11\rangle\}$, where $|ab\rangle = |a\rangle \otimes |b\rangle$, and the quantum state of the register is:

$$|\psi\rangle = \sum_{(a,b)\in\{0,1\}^2} \gamma_{ab}|ab\rangle \tag{20}$$

Note that, contrary to classical digital registers, the qubits are usually not separable, hence the register must be considered as a whole. We say that the qubits are entangled. However in the special case where the coefficients γ_{ab} can be decomposed in the form $\gamma_{ab} = \alpha_a\beta_b$ the state can be written as a tensor product of the states of two qubits, which can be considered separately. Then, we have:

$$|\psi\rangle = (\alpha_0|0\rangle + \alpha_1|1\rangle) \otimes (\beta_0|0\rangle + \beta_1|1\rangle) \tag{21}$$

4. Application of JEVD to quantum coding

4.1 Principle of quantum coding

The objective of quantum coding is to protect quantum information [10]. In the classical domain, the information can be protected using redundancy—for instance, if we want to transmit bit 0 on a noisy communication channel, we can instead transmit 000 (and, similarly, transmit 111 instead of 1). On the receiver side, if one error has occurred on the channel, for instance, if the second bit is false, we receive 010 instead of 000, from which we can still guess that the most probable hypothesis is that the transmitted word was 000. Of course, if there were two errors the transmitted word could have been 111, but it is assumed that the probability of error is low, hence two errors are less likely than one error. More elaborated channel codes have been proposed, but fundamentally they are all based on the idea of adding redundancy and assuming that the probability of channel error is low.

In the quantum domain, it is impossible to use redundancy because it is impossible to copy a quantum state (this is due to the "no-cloning theorem" [11]). However, we can use entanglement to produce the quantum equivalent of classical redundancy. The principle of quantum coding is shown in **Figure 1**. Assume we want to protect the quantum state $|\psi\rangle$ of a k-qubit register. We add r ancillary qubits initialized to $|0\rangle$ to form an n-qubit register ($n = k + r$). The encoder is represented by a unitary $2^n \times 2^n$

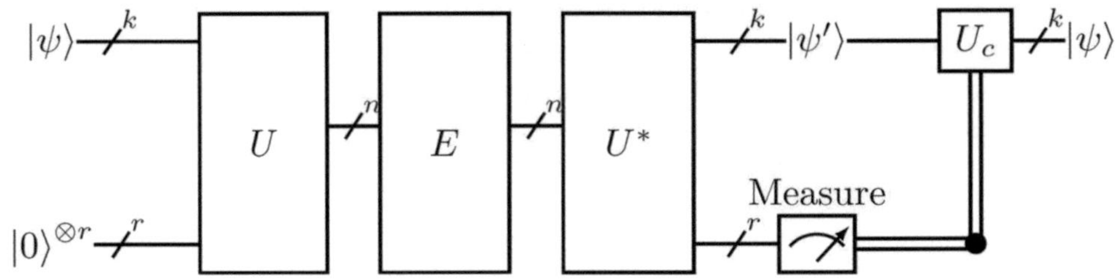

Figure 1.
Principle of quantum coding.

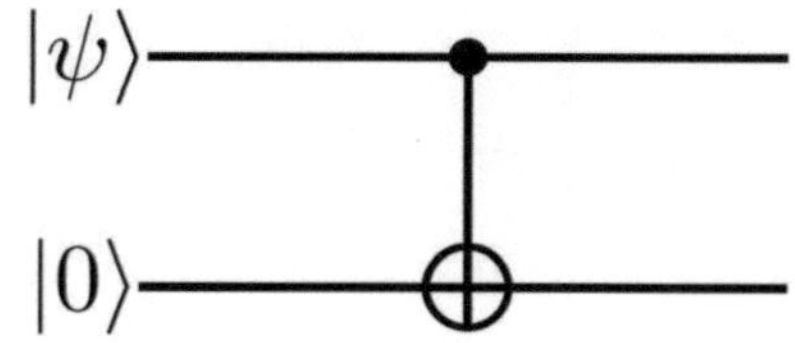

Figure 2.
CNOT quantum gate.

matrix U. Then, errors may occur on the encoded state: they are represented by a unitary matrix E. The decoder is represented by another unitary matrix U^* which is the transpose conjugate of the encoding matrix. Finally, we measure the last r qubits of the decoded state, and, depending on the result of the measurement, we apply the appropriate restoration matrix U_c (which is a unitary matrix of size $2^k \times 2^k$) to the k-qubit register composed of the first k qubits of the decoded state.

As an illustration, let us consider $n = 2$, $k = 1$ and the very simple quantum encoder shown in **Figure 2**. It is a basic quantum circuit known as the CNOT quantum gate, and it is represented by the unitary matrix below:

$$U = \begin{pmatrix} 1 & 0 & 0 & 0 \\ 0 & 1 & 0 & 0 \\ 0 & 0 & 0 & 1 \\ 0 & 0 & 1 & 0 \end{pmatrix} \tag{22}$$

A quantum error on a qubit is described by a 2×2 unitary matrix. It is convenient to decompose the error as a linear sum of the identity and the Pauli matrices below [12]:

$$Z = \begin{pmatrix} 1 & 0 \\ 0 & -1 \end{pmatrix} \quad X = \begin{pmatrix} 0 & 1 \\ 1 & 0 \end{pmatrix} \quad Y = \begin{pmatrix} 0 & i \\ -i & 0 \end{pmatrix} \tag{23}$$

Let us consider that an error may appear on the first encoded qubit and that this error, if present, is represented by the unitary Pauli matrix X. Then, the error matrix which acts on the encoded state is:

$$E = X \otimes I = \begin{pmatrix} 0 & 1 & 0 & 0 \\ 1 & 0 & 0 & 0 \\ 0 & 0 & 0 & 1 \\ 0 & 0 & 1 & 0 \end{pmatrix} \tag{24}$$

It is easy to check that:

$$F = U^* E U = \begin{pmatrix} 0 & 0 & 0 & 1 \\ 0 & 0 & 1 & 0 \\ 0 & 1 & 0 & 0 \\ 1 & 0 & 0 & 0 \end{pmatrix} = X \otimes X \tag{25}$$

The state at the input of the encoder is $[\alpha_0\, \alpha_1]^{\mathrm{T}} \otimes [1 0]^{\mathrm{T}} = [\alpha_0\ 0\ \alpha_1\ 0]^{\mathrm{T}}$. The state at the output of the decoder is, therefore, $[0\ \alpha_1\ 0\ \alpha_0]^{\mathrm{T}}$.

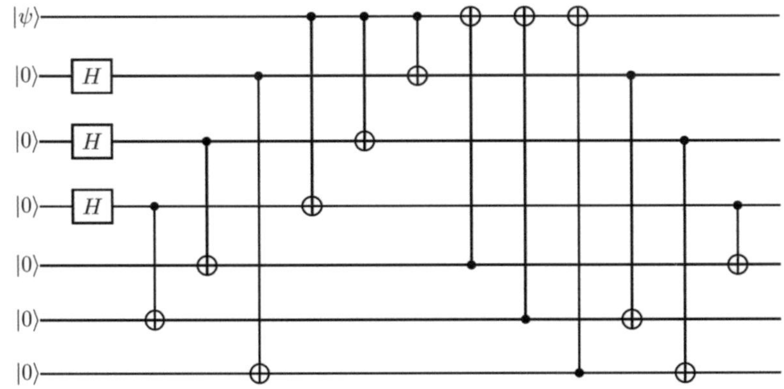

Figure 3.
Steane encoder.

Measuring the second qubit on the output of the decoder consists in decomposing the state space into a direct sum $\mathcal{H}_0 \oplus \mathcal{H}_1$ of two subspaces spanned by $\{|00\rangle, |10\rangle\}$ and $\{|01\rangle, |11\rangle\}$. The result of the measurement will be either 0 or 1 (index of the selected subspace), and by analogy with classical decoding, this result will be called the "syndrome." The projections on these subspaces are $[0\,0]^{\mathrm{T}}$ and $[\alpha_1 \alpha_0]^{\mathrm{T}}$. Then the probability to obtain syndrome 1 is 1.

The measurement then projects the state onto $\mathcal{H}_1$. Note that in this particular case, the information is preserved by the projection. Then, applying the operator $U_c = X$ to the projected state restores the initial state.

Similarly, if there is no error, we can see that F is the identity matrix, then the projections on the subspaces are $[\alpha_0 \alpha_1]^{\mathrm{T}}$ and $[0\,0]^{\mathrm{T}}$. In that case, the syndrome is 0 and the state is projected onto $\mathcal{H}_0$. Correction is done by applying the operator I to the projected state, which is equivalent to doing nothing.

The very simple code used above, as an illustration, cannot correct more complex errors (for instance, an error Z on the first qubit). However, there exist efficient quantum codes, such as the Steane code [13], and the Shor code [14]. A remarkable result of quantum coding theory is that a linear combination of correctable errors is correctable [15].

Figure 3 shows the Steane Encoder, which is a $(n = 7, k = 1, t = 1)$ quantum encoder. This means that it encodes $k = 1$ qubits on $n = 7$ qubits and it is able to correct any error occurring on $t = 1$ encoded qubits. It is built with Hadamard (Eq. (1)) and CNOT (Eq. (22)) quantum gates. From this circuit description, it is possible to obtain the coding matrix U.

4.2 Determination of encoder matrix using JEVD

The problem we address can be stated as follows (see **Figure 1** for the notations)—given a list of n independent Pauli errors E_i with corresponding diagonal outer errors F_i, determine the unitary operator U (quantum encoder) such that:

$$U^* E_i U = F_i \;\; \forall i \in \{1, \ldots, n\} \tag{26}$$

This equation shows that the columns of U are the eigenvectors of E_i. Specification of the code by a small set of Pauli errors is very convenient and the interest of

E_i	F_i
$Z \otimes Z \otimes Z \otimes Z \otimes Z \otimes Z \otimes Z$	$Z \otimes I \otimes I \otimes I \otimes I \otimes I \otimes I$
$X \otimes X \otimes I \otimes I \otimes I \otimes X \otimes X$	$I \otimes Z \otimes I \otimes I \otimes I \otimes I \otimes I$
$X \otimes I \otimes X \otimes I \otimes X \otimes I \otimes X$	$I \otimes I \otimes Z \otimes I \otimes I \otimes I \otimes I$
$X \otimes I \otimes I \otimes X \otimes X \otimes X \otimes I$	$I \otimes I \otimes I \otimes Z \otimes I \otimes I \otimes I$
$Z \otimes Z \otimes I \otimes I \otimes I \otimes Z \otimes Z$	$I \otimes I \otimes I \otimes I \otimes Z \otimes I \otimes I$
$Z \otimes I \otimes Z \otimes I \otimes Z \otimes I \otimes Z$	$I \otimes I \otimes I \otimes I \otimes I \otimes Z \otimes I$
$Z \otimes I \otimes I \otimes Z \otimes Z \otimes Z \otimes I$	$I \otimes I \otimes I \otimes I \otimes I \otimes I \otimes Z$

Table 1.
Collection of Pauli errors.

automatic determination of matrix U is to allow further simulations of the behavior of the quantum code in various configurations.

To illustrate and validate the approach that will be developed below, let us consider the collection of $n = 7$ Pauli errors shown in **Table 1**. Here, to be able to check the results, this collection has been chosen to correspond to the Steane encoder (**Figure 3**), while in a standard application of the method, it would be given *a priori*. The interest is that here we can compute the encoder matrix from the circuit and this will allow us to check that our method produces the correct encoder matrix.

We use n independent equations in which each F_i is a tensor product of I and Z only (including only one Z). Therefore, matrices F_i are diagonal, and their diagonal elements are $+1$ and -1 in equal numbers.

Figure 4 shows the diagonals of the matrices F_i (each row corresponding to one diagonal). Values -1 and $+1$ are represented, respectively, by black and white dots.

Since matrix U does not depend on i in Eq. (26) its columns are joint eigenvectors of the E_i. For instance, in the example above, the 20th column of U is a joint eigenvector of $E_1, E_2, \ldots, E_7$ associated to eigenvalues $+1,+1,-1,+1,+1,-1,-1$ (see **Figure 4**). In the general case, the set of n eigenvalues corresponding to the column c of U is easily obtained by taking the binary representation of $c - 1$ with the mapping $0 \rightarrow +1$ and $1 \rightarrow -1$.

Now, let us consider the determination of column c of U. We know that it is a vector spanning a joint eigenspace of the E_i corresponding to a given set of eigenvalues $\{\lambda_i, i = 1, \ldots, n\}$. For each E_i let A_i denote the $2^n \times 2^{n-1}$ matrix whose orthonormal columns span the eigenspace associated to λ_i and B_i the $2^n \times 2^{n-1}$ matrix whose orthonormal columns span the kernel of A_i (which corresponds to the eigenspace associated to $-\lambda_i$).

Let $\mathcal{Y}_k$ denote the joint eigenspace corresponding to eigenvalues $\{\lambda_j, j = 1, \ldots, k\}$ with $k \in \{1, \ldots, n\}$. We propose Algorithm 1 to efficiently compute the column of U. It computes a series of matrices Y_k whose columns are an orthonormal basis of $\mathcal{Y}_k$. Obviously, the searched column of U is Y_n. For the moment, let us consider that $K(c)$=1 (the optimal value will be discussed later).

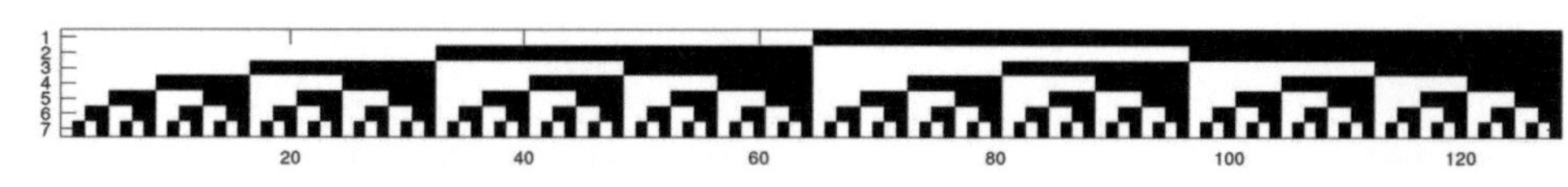

Figure 4.
Diagonals of matrices F_i.

if $K(c) = 1$ **then**
| $Y_1 = A_1$
end
for $k = K(c) + 1$ **to** n **do**
| $C_k = B_k^* Y_{k-1}$
| $Z_k = null(C_k)$
| $Y_k = Y_{k-1} Z_k$

Algorithm 1: Algorithm for determination of a joint eigenspace.

The sizes of the matrices are decreasing with k:
C_k: $2^{n-1} \times 2^{n-k+1}$ Z_k: $2^{n-k+1} \times 2^{n-k}$ Y_k: $2^n \times 2^{n-k}$.
The intuitive ideas under the algorithm are the following:

- $C_k = B_k^* Y_{k-1}$: The orthonormal basis of $\mathcal{Y}_{k-1}$ is projected on the kernel of A_k. The components of the projected vectors are expressed in the orthonormal basis B_k of that kernel. Consequently, $Im(C_k)$ is the projection of $\mathcal{Y}_{k-1}$ on the kernel of A_k, expressed in that kernel.
- A matrix Z_k whose columns are an orthonormal basis of the complement of this projection is determined.
- Finally, the components of the basis vectors are restored to the original space by $Y_k = Y_{k-1} Z_k$

Let us prove that the matrices Y_k have orthonormal columns. This is obviously the case for $k = 1$. Then, by recursion, we have:

$$Y_k^* Y_k = Z_k^* Y_{k-1}^* Y_{k-1} Z_k = I \tag{27}$$

Now, let us prove, by recurrence, that $Im(Y_k) = \mathcal{Y}_k$.
Obviously, this is the case for $k = 1$. Assume this is the case for $k - 1$. We have:

$$Im(Y_k) \subset Im(Y_{k-1}) = \mathcal{Y}_{k-1} \tag{28}$$

We have also:

$$B_k^* Y_k = B_k^* (Y_{k-1} Z_k) = (B_k^* Y_{k-1}) Z_k = C_k Z_k = 0 \tag{29}$$

Then

$$Im(Y_k) \subset \ker(B_k^*) = Im(A_k) \tag{30}$$

From $Im(Y_k) \subset \mathcal{Y}_{k-1}$ and $Im(Y_k) \subset Im(A_k)$ we deduce $Im(Y_k) \subset \mathcal{Y}_k$.
Conversely, assume that a vector x belongs to $\mathcal{Y}_k$. Because $\mathcal{Y}_k \subset \mathcal{Y}_{k-1}$ there exists a vector b such that $x = Y_{k-1} b$ and because $x \in Im(A_k)$ we have also $B_k^* x = 0$

E_i	F_i
$X\otimes X\otimes X\otimes X\otimes X\otimes X\otimes X$	$X\otimes I\otimes I\otimes I\otimes I\otimes I\otimes I$
$Z\otimes I\otimes Z\otimes Z\otimes Z\otimes Z\otimes Z$	$I\otimes X\otimes I\otimes I\otimes I\otimes I\otimes I$
$Z\otimes Z\otimes I\otimes Z\otimes Z\otimes Z\otimes Z$	$I\otimes I\otimes X\otimes I\otimes I\otimes I\otimes I$
$Z\otimes Z\otimes Z\otimes I\otimes Z\otimes Z\otimes Z$	$I\otimes I\otimes I\otimes X\otimes I\otimes I\otimes I$
$X\otimes I\otimes I\otimes I\otimes X\otimes I\otimes I$	$I\otimes I\otimes I\otimes I\otimes X\otimes I\otimes I$
$X\otimes I\otimes I\otimes I\otimes I\otimes X\otimes I$	$I\otimes I\otimes I\otimes I\otimes I\otimes X\otimes I$
$X\otimes I\otimes I\otimes I\otimes I\otimes I\otimes X$	$I\otimes I\otimes I\otimes I\otimes I\otimes I\otimes X$

Table 2.
Additional Collection of Pauli errors.

Then

$$B_k^* Y_{k-1} b = 0 \Rightarrow C_k b = 0 \Rightarrow \exists a : b = Z_k a$$

Therefore $x = Y_{k-1}b = Y_{k-1}Z_k a = Y_k a \Rightarrow x \in Im(Y_k) \Rightarrow \mathcal{Y}_k \subset Im(Y_k)$.

After execution of the algorithm to determine each column of U, there remains an indetermination because the joint eigenvectors (i.e., the columns of U) are determined up to a phase factor. This has no consequence on the performance of the quantum code. However, if we want to fix this residual indetermination, we proposed a fast and simple procedure in ref. [16]. The procedure requires an additional set of n Pauli errors in which each additional F_i is a tensor product of I and X only. As an example, for the Steane code, we use **Table 2**.

After these remaining differences have been removed, we obtain an estimated matrix U that is equal to the true matrix, up to a global phase (**Figure 5**). However, this remaining indetermination does not matter because, as said before, the global phase has no significance in quantum physics. Here we have chosen the global phase so that the encoder matrix is real.

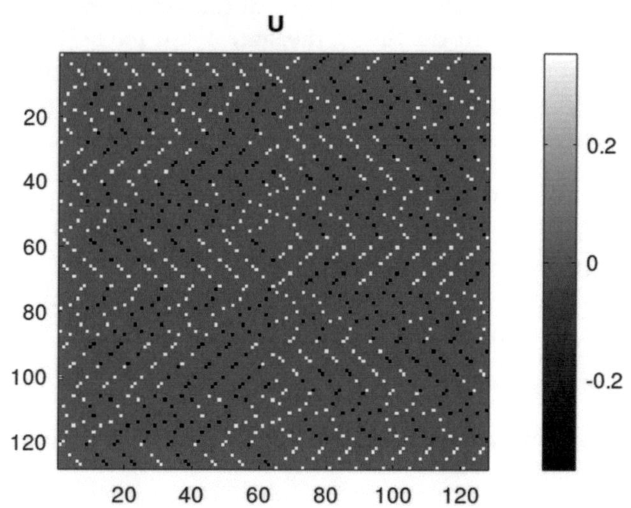

Figure 5.
Estimated Matrix U for the Steane encoder.

Figure 5 shows the matrix computed by our method. We have checked that it is equal to the matrix directly computed from the circuit description.

The programmer may speed up the computation by taking into account the fact that when computing columns c of U, some matrices Y_k have already been computed for other columns and can be reused. For instance, in **Figure 4**, we see that the joint eigenvalues corresponding to columns 19 and 20 are the same, except the last one. Then, when computing column 20, we can set $K(20) = n$ in Algorithm 1 instead of the default value $K(20) = 1$, because the Y_{n-1} for column 20 is the same as for column 19. More generally, Algorithm 2 written in pseudo-Octave code computes the optimal values of the $K(c)$.

$K = [1 \quad 1]$
for $k = 2$ *to* n **do**
 | $K = \text{reshape}(\ [K; k * \text{ones}(1, 2^{k-1})]\ [1\ 2^k])$
end

Algorithm 2: Algorithm for computation of the optimal values $K(c)$.

For instance, for $n = 3$ the algorithm produces $K = [1 \quad 3 \quad 2 \quad 3 \quad 1 \quad 3 \quad 2 \quad 3]$.

5. Application of JEVD to quantum walk search

5.1 Principle of quantum walk search

Let us consider a particle that can move on a graph. In the classical world, at the time t this particle is localized at a node of the graph. It can then randomly choose one of the edges linked to this node to reach one of the adjacent nodes at a time $t + 1$. The repeated iteration of this process is the concept of classical random walk.

A quantum walk [17] relies also on a graph, but contrary to the classical walk, here the particle may be located at many nodes at the same time and can choose many edges simultaneously. At the time t, the state of the particle is then described by a state vector $|\psi_t\rangle$ and the evolution between times t and $t + 1$ is given by a unitary matrix $U = SC$ such that $|\psi_{t+1}\rangle = U|\psi_t\rangle$. The unitary matrices C and S represent, respectively, the choice of the edges and the movement to the adjacent nodes.

In the following, we will consider graphs associated with hypercubes [18]. We will note n the dimension of the hypercube and $N = 2^n$ the number of nodes. **Figure 6** shows the graph corresponding to a hypercube of dimension $n = 3$. It is convenient to label the nodes by binary words. In quantum language, these binary words κ are also used to label the basis vectors of the so-called position space $\mathcal{H}^S$.

The quantum state lives in a Hilbert space built by the tensor product of the position space $\mathcal{H}^S$ (corresponding to the nodes) and the coin space $\mathcal{H}^C$ (corresponding to the possible movements along the edges) $\mathcal{H} = \mathcal{H}^S \otimes \mathcal{H}^C$. The dimensions of these state spaces are $N_e = nN$, N and n.

It is usual to define C as [19]:

$$C = I_N \otimes G_n \tag{31}$$

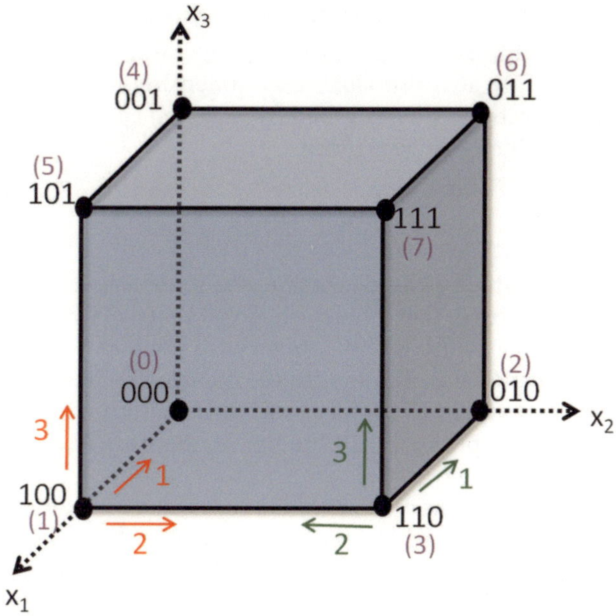

Figure 6.
Hypercube for $n = 3$.

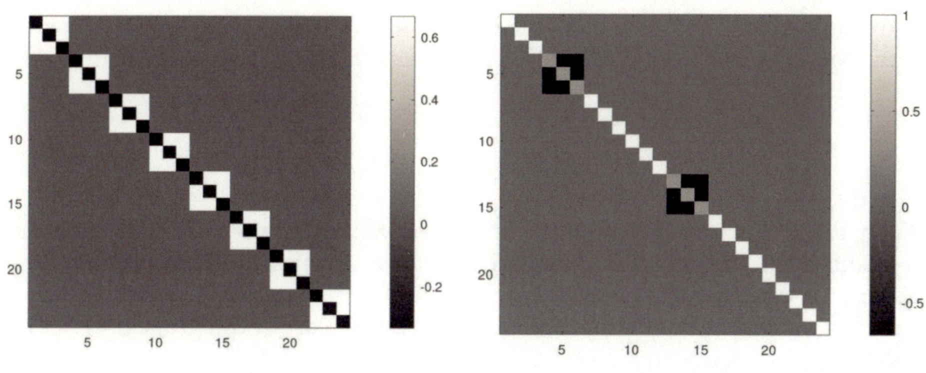

Figure 7.
Matrices C (left) and O (right) for $n = 3$.

where G_n is the $n \times n$ Grover diffusion matrix defined in Section 2. Matrix C obtained for $n = 3$ is shown on **Figure 7**.

The structure of S is more complex. It is convenient to first define it in $\mathcal{H}^C \otimes \mathcal{H}^S$ and then to transpose it to $\mathcal{H}$ using the shuffle matrix $P = P_{n,N}$ (defined in subsection 2.2). Then:

$$S = P\hat{S}P^T \tag{32}$$

where

$$\hat{S} = diag\left(\hat{S}_1, \dots, \hat{S}_n\right) \qquad and \qquad \hat{S}_d = I^{\otimes(n-d)} \otimes X \otimes I^{\otimes(d-1)} \tag{33}$$

The last equation just means that, because a movement along direction d corresponds to an inversion of the dth bit in κ, the shift operator permutes the values associated to nodes that are adjacent along that dimension.

A quantum walk search can be described by repeated application of a unitary evolution operator Q, which can be written:

$$Q = UO \tag{34}$$

Here O is the oracle, which aims at marking the solutions. An example of oracle structure is shown in **Figure 7**. It is a block-diagonal matrix, whose blocks are $-G_n$ when they correspond to a solution and I_n otherwise. Denote M the number of solutions and assume that $M \ll N$ (otherwise the quantum walk search would serve no purpose because the probability of rapidly finding a solution with a classical search would be high). In the example shown in the figure, there are $M = 2$ solutions (located at positions 1 and 4 on **Figure 6**).

Let t denote the number of iterations until a measurement is performed. Starting from an initial state $|\psi_0\rangle$, repeated iterations lead to the state $|\psi_t\rangle = Q^t|\psi_0\rangle$ which is then measured. The theory of quantum walk search [19] shows that the probability of success (that is the probability of obtaining a solution by measurement) oscillates as a function of t. This means that theoretical tools which help to understand and simulate quantum walk search lead to the development of methods to determine the optimal time of measurement.

In the sequel, we will show that JEVD is a fruitful tool in this context. Indeed, set E to be the union of the joint eigenspaces of U and O, and $\overline{E}$ its complement. Inside E, the operators commute. So, if we note with index E the restrictions of the operators to E, we have:

$$Q_E^2 = (U_E O_E)(U_E O_E) = U_E O_E^2 U_E = U_E^2 \tag{35}$$

Then, inside E, there is no significant difference between the effective quantum walk Q and the uniform quantum walk U, because, after each pair of successive iterations, the evolution is identical. Since the uniform quantum walk has no reason to converge to a solution, we deduce that the interesting part of the process lives in the complement of E, that is in $\overline{E}$.

After establishing results about the dimensions of the eigenspaces of U and O, we will show that the concept of joint eigenspaces allows us to establish an upper bound on the dimension of the complement, with the remarkable result that this dimension grows only linearly with n. Then, we propose an algorithm for efficient computation of the joint eigenspaces and, finally, use it to check our theoretical upper bound.

5.2 Eigenspaces of U and O

Set

$$F = H_N \otimes I_n \tag{36}$$

Then matrix F diagonalizes S:

$$FSF = (H_N \otimes I_n) P_{N,n} \hat{S} P_{n,N} (H_N \otimes I_n) \quad (37)$$

$$= P_{n,N} (I_n \otimes H_N) \hat{S} (I_n \otimes H_N) P_{N,n} \quad (38)$$

$$= P^T diag\left(\dots, H_N \hat{S}_d H_N, \dots \right) P \quad (39)$$

The latter term is diagonal because the mixed product property, $H^2 = I$ and $HXH = Z$, shows that:

$$H_N \hat{S}_d H_N = I^{\otimes (n-d)} \otimes Z \otimes I^{\otimes (d-1)} \quad (40)$$

Once more, using the mixed product property, we can also prove that F keeps C unchanged, that is:

$$FCF = C \quad (41)$$

The diagonal of FSF is the concatenation of the binary representation of the numbers 0 to $N-1$ with the mapping $0 \rightarrow (+1)$ and $1 \rightarrow (-1)$. That is:

$$FSF = diag(S_0, \dots, S_\kappa, \dots, S_{N-1}) \quad (42)$$

Note that the diagonal of S_κ contains k times -1 and $n-k$ times $+1$ (where k is the Hamming weight of κ).

Then, because $F^2 = I$, FUF is a block diagonal matrix:

$$FUF = (FSF)(FCF) \quad (43)$$

$$= diag(\dots, S_\kappa, \dots) C \quad (44)$$

Block κ is then

$$U_\kappa = S_\kappa G_n \quad (45)$$

We have:

$$\dim E_-^{U_\kappa} \geq \dim E_{+,-}^{S_\kappa, G_n} \quad (46)$$

$$\geq \dim E_+^{S_\kappa} + \dim E_-^{G_n} - n \quad (47)$$

$$\geq (n-k) + (n-1) - n \quad (48)$$

$$\geq n - k - 1 \quad (49)$$

and

$$\dim E_+^{U_\kappa} \geq \dim E_{-,-}^{S_\kappa, G_n} \quad (50)$$

$$\geq \dim E_-^{S_\kappa} + \dim E_-^{G_n} - n \quad (51)$$

$$\geq k + (n-1) - n \quad (52)$$

$$\geq k - 1 \quad (53)$$

Then, there is only room left for at most 2 eigenvalues, specifically, at most a pair of conjugate ones.

Assume that this pair of eigenvalues exists. Since the diagonal of G_n contains $-1+\frac{2}{n}$, the trace of U_κ is:

$$trace(U_\kappa) = ((n-k)(+1) + k(-1))\left(-1+\frac{2}{n}\right) \tag{54}$$

$$= -n + 2k + 2\left(1 - 2\frac{k}{n}\right) \tag{55}$$

The sum of the eigenvalues is equal to the trace and we already have eigenvalue -1 with multiplicity $n-k-1$ and eigenvalue $+1$ with multiplicity $k-1$. The sum of these $n-2$ eigenvalues is $-n+2k$. Then the sum of the two missing eigenvalues must be $2(1-2\frac{k}{n})$. Let us denote them by λ_k and λ_k^*. We must have $Re(\lambda_k) = 1 - 2\frac{k}{n}$. Then, since $|\lambda_k| = 1$ we have

$$\lambda_k = \left(1 - 2\frac{k}{n}\right) + i\frac{2}{n}\sqrt{k(n-k)} \tag{56}$$

Considering the eigenvalues of $-G_n$ and I_n it is trivial to show that the dimensions of the eigenspaces of the oracle are:

$$\dim E_-^O = M \qquad \text{and} \qquad \dim E_+^O = N_e - M \tag{57}$$

5.3 Upper bound on the dimension of the complement

The eigenvalues of U belong to $\{-1, +1, \lambda_k, \lambda_k^*\}$ where $k \in [1, n-1]$. Then, there are $2 + 2(n-1) = 2n$ eigenspaces of U.

For $j \in [1, 2n]$ let α_j be the dimensions of these eigenspaces and β_j the dimensions of their intersections with E_+^O. An eigenvector of U is in an intersection if and only if it is orthogonal to E_-^O. Then, because the dimension of E_-^O is M, we have $\beta_j \geq \alpha_j - M$. Consequently

$$\sum_{j=1}^{2n} \beta_j \geq \sum_{j=1}^{2n} \alpha_j - 2nM \tag{58}$$

Obviously, we have $\sum_{j=1}^{2n} \alpha_j = N_e$, so that

$$\sum_{j=1}^{2n} \beta_j \geq N_e - 2nM \tag{59}$$

It follows that the dimension of the complement has an upper bound:

$$\dim E_c \leq 2nM \tag{60}$$

This is a remarkable result—despite the fact that the dimension of the Hilbert space grows exponentially ($N_e = n2^n$), the dimension of the complement grows only linearly with n.

5.4 Fast computation of the joint eigenspaces

5.4.1 Introduction

To check our theoretical upper bound, we propose an efficient algorithm for fast computation of the joint eigenspaces.

We have to compute orthonormal bases of joint eigenspaces of U and O. The dimension of E_{-}^{O} is small, hence, it makes sense to define it by an orthonormal basis generating the eigenspace. However, the dimension of E_{+}^{O} is large (greater than $N_e/2$). Hence, it is computationally more efficient to define it by an orthonormal basis of its complement (which is E_{-}^{O}). Indeed dim $E_{-}^{O} \ll$ dim E_{+}^{O}. We then have to design an algorithm adapted to each case.

5.4.2 Intersection of two eigenspaces defined by orthonormal bases

Let us consider a matrix A whose columns are an orthonormal basis of an eigenspace of U, and a matrix B whose columns are an orthonormal basis of E_{-}^{O}. Set p and q to be the number of columns of these matrices (their number of rows being N_e). We want to compute an $N_e \times r$ matrix J whose columns are an orthonormal basis of the joint eigenspace (whose dimension we have set to be r). We propose the algorithm below, which is a straightforward adaptation of Theorem 1 in ref. [20].

First, we compute the $p \times q$ matrix C below:

$$C = A^{*}B \tag{61}$$

Then, we compute the SVD of C:

$$C = U_c S_c V_c^{*} \tag{62}$$

Denote by s_k the singular values (the diagonal elements of S_c) and determine r such that $s_k \geq 1 - \varepsilon$ for $k = 1, \ldots, r$, and $s_k < 1 - \varepsilon$ for $k > r$. Here $\varepsilon \ll 1$ is a very small positive value introduced to take into account the presence of small errors due to computer finite precision arithmetic. Finally:

$$J = AU_c(:,1:r) \tag{63}$$

Or, equivalently, $J = BV_c(:,1:r)$.

5.4.3 Intersection of two eigenspaces, one of them being defined by an orthonormal basis of its complement

Let us consider a matrix A whose columns are an orthonormal basis of an eigenspace of U, and a matrix B whose columns are an orthonormal basis of the complement of E_{+}^{O} (that is E_{-}^{O}). First, we compute the $p \times q$ matrix C (Eq. (61)). Then, we compute the $p \times r$ matrix $(r \leq p)$ Z below:

$$Z = null(C^{*}) \tag{64}$$

λ^O	λ^U	$\mathbf{dim}E^{O,U}_{\lambda^O,\lambda^U}$
−1	any λ_k or λ_k^*	0
+1	$\lambda_0 = +1$	321
+1	λ_1 or λ_1^*	4
+1	λ_2 or λ_2^*	18
+1	λ_3 or λ_3^*	32
+1	λ_4 or λ_4^*	32
+1	λ_5 or λ_5^*	18
+1	λ_6 or λ_6^*	4
+1	$\lambda_7 = -1$	321

Table 3.
Joint eigenspaces of O and U for $n = 7$ and $M = 3$ solutions located at nodes $2, 8, 9$.

and we obtain an $N_e \times r$ matrix J whose columns are an orthonormal basis of $E^{U,O}_{\lambda,+}$ from:

$$J = AZ \tag{65}$$

The justification of the algorithm is as follows. The q columns of C are a basis of the projection of $Im(B)$ into $Im(A)$, the components being expressed in the basis of $Im(A)$ The complement of $Im(C)$ in $Im(A)$ is the desired intersection (expressed in $Im(A)$). The columns of Z are an orthonormal basis of this intersection. Finally, Eq. (65) restores the components in the original space.

5.5 Simulation results

Consider a hypercube of dimension $n = 7$ with $M = 3$ solutions located at nodes $2, 8, 9$. The dimension of the state space is then $N_e = n2^n = 896$. From the discussion above, we know that the dimension of the complement is upper bounded by $2nM = 42$.

The algorithm gives us the dimensions of the joint eigenspaces of U and O (**Table 3**). The sum of the dimensions of the joint eigenspaces is then $\sum_{j=1}^{2n} \beta_j = 858$, from which we obtain the dimension of the complement:

$$\dim E_c = N_e - \sum_{j=1}^{2n} \beta_j = 38 \tag{66}$$

We can see that, as expected, this dimension ($\dim E_c = 38$) is much smaller than the dimension of the original state space ($N_e = 896$). We can also check that it is less than the theoretical upper bound ($2nM = 42$), as expected.

6. Conclusions

The recent growth of research on quantum communications and quantum information processing opens new challenges. In this chapter, we have shown that matrix theory concepts, such as JEVD, are powerful tools to propose new theoretical results as well as efficient simulation algorithms.

In the domain of quantum coding, we have shown how to determine the encoding matrix of a quantum code from a collection of Pauli errors. On a more speculative note to be part of future work concerning interception of quantum channels, it might also be useful to identify the quantum coder used by a noncooperative transmitter.

In the domain of quantum walk search, thanks to JEVD we have proved that there exists a small subspace of the whole Hilbert space which captures the essence of the search process, and we have given an algorithm that allows us to check this result by simulation.

Acknowledgements

The authors thank the IBNM (Institut Brestois du Numérique et des Mathématiques), CyberIoT Chair of Excellence, for its support.

Abbreviations

[qubit]	Quantum bit
[JEVD]	Joint EigenValue Decomposition
[SVD]	Singular Value Decomposition

Author details

Gilles Burel[1*], Hugo Pillin[1,2], Paul Baird[2], El-Houssaïn Baghious[1] and Roland Gautier[1]

1 Lab-STICC, University of Brest, Brest, France

2 LMBA, University of Brest, Brest, France

*Address all correspondence to: gilles.burel@univ-brest.fr

References

[1] Humble TS. Quantum security for the physical layer. IEEE Communications Magazine. 2013;**51**(8):56-62

[2] Liao SK et al. Satellite-to-ground quantum key distribution. Nature. 2017; **549**:43-47

[3] Ren JG et al. Ground-to-satellite quantum teleportation. Nature. 2017; **549**:70-73

[4] Valivarthi R et al. Quantum teleportation across a metropolitan fibre network. Nature Photonics. 2016;**10**: 676-680

[5] Yin J et al. Quantum teleportation and entanglement distribution over 100-kilometre freespace channels. Nature. 2012;**488**:185-188

[6] Microsoft. The Q# programming language user guide [Internet] 2022 Available from: https://docs.microsoft.com/en-us/azure/quantum/user-guide/?view=qsharp-preview [Accessed: January 11, 2022]

[7] Golub GH, Van Loan CF. Matrix Computations. 3rd ed. Baltimore and London: The John Hopkins University Press; 1996

[8] Grover LK. Quantum mechanics helps in searching for a needle in a haystack. Physical Review Letters. 1997; **79**(2):325

[9] D'Angeli D, Donno A. Shuffling Matrices, Kronecker Product and Discrete Fourier Transform. Discrete Applied Mathematics. 2017;**233**:1-18

[10] Raussendorf R. Key ideas in quantum error correction. Philosophical Transactions of the Royal Society A. 2012;**370**:4541-4565

[11] Wootters W, Zurek W. A single quantum cannot be cloned. Nature. 1982; **299**(5886):802-803. DOI: 10.1038/299802a0

[12] Nielsen MA, Chuang IL. Quantum Computation and Quantum Information. Cambridge, UK: Cambridge University Press; 2010

[13] Steane A. Multiple-particle interference and quantum error correction. Proceeding of the Royal Society of London. 1996;**452**(1954): 2551-2577. DOI: 10.1098/rspa.1996.0136

[14] Shor PW. Scheme for reducing decoherence in quantum computer memory. Physical Review A. 1995;**52**(4): R2493-R2496. DOI: 10.1103/PhysRevA.52.R2493

[15] Calderbank AR, Shor PW. Good quantum error-correcting codes exist. Physical Review A. 1996;**54**(2): 1098-1105. DOI: 10.1103/physreva.54.1098

[16] Burel G, Pillin H, Baghious EH, Baird P, Gautier R. Identification Of Quantum Encoder Matrix From A Collection Of Pauli Errors. Ho Chi Minh city, Vietnam: Asia-Pacific Conference on Communications; 2019

[17] Kempe J. Quantum random walks – An introductory overview. Contemporary Physics. 2003;**44**(4):307-327. DOI: 10.1080/00107151031000110776

[18] Moore C, Russell A. Quantum Walks on the Hypercube. Lecture Notes in Computer Science. Vol. 2483. New York: Springer. DOI: 10.1007/3-540-45726-7_14

[19] Shenvi N, Kempe J, Whaley KB. Quantum random-walk search

algorithm. Physical Review A. 2003;**67**: 052307

[20] Bjorck A, GolubG. Numerical methods for computing angles between linear subspaces. Mathematics of Computation. 1973;**27**:123. DOI: 10.2307/2005662

Chapter 9

Transformation Groups of the Doubly-Fed Induction Machine

Giovanni F. Crosta and Goong Chen

Abstract

Three-phase, doubly-fed induction (*DFI*) machines are key constituents in energy conversion processes. An ideal *DFI* machine is modeled by inductance matrices that relate electric and magnetic quantities. This work focuses on the algebraic properties of the mutual (rotor-to-stator) inductance matrix $\mathbf{L}_{sr}$: its kernel, range, and left zero divisors are determined. A formula for the differentiation of $\mathbf{L}_{sr}$ with respect to the rotor angle θ_r is obtained. Under suitable hypotheses $\mathbf{L}_{sr}$ and its derivative are shown to admit an exponential representation. A recurrent formula for the powers of the corresponding infinitesimal generator $\mathbf{A}_0$ is provided. Historically, magnetic decoupling and other requirements led to the Blondel-Park transformation which, by mapping electric quantities to a suitable reference frame, simplifies the *DGI* machine equations. Herewith the transformation in exponential form is axiomatically derived and the infinitesimal generator is related to $\mathbf{A}_0$. Accordingly, a formula for the product of matrices is derived which simplifies the proof of the Electric Torque Theorem. The latter is framed in a Legendre transform context. Finally, a simple, "realistic" machine model is outlined, where the three-fold rotor symmetry is broken: a few properties of the resulting mutual inductance matrix are derived.

Keywords: mutual inductance matrix, Blondel-Park transformation, exponential representation, infinitesimal generator, zero divisors, circulants, broken symmetry

1. Introduction

Three-phase, doubly-fed induction (*DFI*) machines have a long history [1–4] and continue to be key constituents in energy conversion processes [5, 6]. Motivation for modeling a *DFI* generator comes from the need to deal with intermittency in the primary energy supply (e.g., the wind field) and with uncertainty in the load (i.e., the grid). Similarly, the modeling and control of a *DFI* motor can improve the efficiency and reliability of electric-to-mechanical work conversion. The equations modeling the ideal *DFI* machine have been studied for more than a century. Results in modeling and control [7–13], including those which derive from numerical simulation, demonstrate how attention to the *DFI* machine is being continuously paid. In essence, the ideal three-phase machine model centers on two matrices, on which this work focuses: the rotor-to-stator mutual inductance matrix, $\mathbf{L}_{sr}[.]$, which depends on the "rotor angle" θ_r and characterizes the machine itself, and the Blondel-Park transformation matrix, $\mathbf{K}[.]$, which depends on another angle and describes a change of variables, from the $\{abc\}$ reference frame (Section 2) to the $\{dq0\}$ reference frame (Section 4). Both matrices, $\mathbf{L}_{sr}[.]$ and $\mathbf{K}[.]$, appear in the Electric Torque Theorem (*ETT*) which relates mechanical to electrical variables and as such represents the *raison d'être* of the *DFI* machine. Stated in the $\{abc\}$ frame, the *ETT* is a straightforward application of energy balance, once a Legendre transformation (Section 5.2) has been introduced and co-energy accordingly defined. Instead, the proof, in fact the translation of the *ETT* in the $\{dq0\}$ frame (Section 5.3), requires all relevant properties of $\mathbf{L}_{sr}[.]$ and $\mathbf{K}[.]$ to be known. For this reason, in Section 3 the kernel, the range (Proposition 1), the classical adjoint and the left zero divisors (Proposition 3) of $\mathbf{L}_{sr}[.]$ are determined. Derivation benefits from $\mathbf{L}_{sr}[.]$ being a circulant matrix [14] (Lemma 1) and from its eigenvalues representing the discrete Fourier transform of a 3-sequence [15] (Proposition 2). Special attention need two constant matrices, $\mathbf{A}_0$ and its square (Lemma 2), because they relate differentiation of $\mathbf{L}_{sr}$ with respect to θ_r to multiplication (Theorem 1). In a suitable subspace of $\mathbb{R}^3$, $\mathbf{L}_{sr}$ admits an exponential representation (Theorem 2) with $\mathbf{A}_0$ as infinitesimal generator. Section 4 is devoted to $\mathbf{K}[.]$: its structure, as well as its exponential representation with generator $-\mathbf{A}_0$, is inferred by satisfying, in sequence, a list of requirements (Proposition 4 and Theorem 4). The key formula for the product of matrices (Theorem 5) is then applied to prove the *ETT* in the $\{dq0\}$ frame in one step (Theorem 6). An attempt is finally made in Section 6 to deal with a "realistic" machine model, where the three-fold rotor symmetry is broken: the b and c rotor axes are misaligned by angles ϵ_b and ϵ_c. To second-order in ϵ_b and ϵ_c there exists a constant $\mathbf{B}$ which relates differentiation to multiplication of the approximate inductance matrix (Proposition 7).

2. The ideal doubly-fed induction machine

Definition 1. (*Three-phase, ideal DFI machine*) [3, 4]. A three-phase *DFI* machine is said ideal whenever its stator and rotor windings exhibit three-fold symmetry. Moreover, magnetomotive forces and flux waves created by the windings are sinusoidally distributed and windings give rise to a linear electric network.

Remark 1. (*Neglected phenomena*). Higher harmonics, hysteresis, and eddy currents are excluded by the model. Deviations from three-fold symmetry will be addressed in Section 6.

Notation. (*$\{abc\}$ frames*). The most natural frames where three-phase stator and rotor voltages and currents can be represented are the $\{abc\}$ frames. For example, the stator currents are an ordered triple which one agrees to represent as a vector

$$\vec{j}_{\{abc\}s} = \begin{bmatrix} j_{as} & j_{bs} & j_{cs} \end{bmatrix}^{\mathrm{Trs}} \in \mathbb{R}^3, \tag{1}$$

whose components are functions of time $t \in \mathfrak{T}$. A similar notation will hold for other electric quantities. Unless otherwise specified, all vectors are understood in $\mathbb{R}^3$.

Hp. (*Function class*). Dependence of all quantities of interest on time is assumed as smooth as required.

Definition 2. (*Balanced triple*). An $\{abc\}$ current triple is balanced or is a trivial zero sequence, whenever

$$j_a + j_b + j_c = 0, \quad \forall t \in \mathfrak{T}. \tag{2}$$

Such sequences define the subspace $\mathfrak{B} \subset \mathbb{R}^3$, a plane through the origin; the corresponding notation is $\vec{j}_{\mathfrak{B}} \in \mathfrak{B}$.

Notation.

$\theta_r[t] \in [0, 2\pi)$ is the electric rotor angle at a time t, formed by the rotor *ar* axis with respect to the stator *as* axis.

$\beta_r[t] \in [0, 2\pi)$ is the electric rotor angle at a time t, formed by d axis with respect to the rotor *ar* axis (Section 4).

$\beta_s[t] \in [0, 2\pi)$ is the electric rotor angle at a time t, formed by d axis with respect to the stator *as* axis (Section 4).

$\vec{j}'_{\{abc\}r} := \frac{N_r}{N_s} \vec{j}_{\{abc\}r}$ is the stator-referred (*ständer-bezogen*) vector of rotor currents, where N_r and N_s are the rotor and stator turns.

A boldface, roman capital denotes a matrix in $\mathcal{M}[N_{\text{row}}, N_{\text{col}}]$ of $N_{\text{row}}\ (\geq 2)$ rows $\times N_{\text{col}}\ (\geq 2)$ columns.

$a^*_{m,n}$ is the cofactor of entry $\{m, n\}$ in $\mathbf{A} \in \mathcal{M}[N, N]$, where $N_{\text{row}} = N_{\text{col}} = N\ (\geq 2)$.

$[a^*_{m,n}]$ is the corresponding matrix.

adj$[\mathbf{A}]$ is the matrix adjoint to $\mathbf{A} \in \mathcal{M}[N, N]$, obtained by transposing $[a^*_{m,n}]$.

$\vec{u})(\vec{v}$ is the dyadic product of the column vector $\vec{u}$ by the row vector $\vec{v}$, both $\in \mathbb{R}^3$.

$\mathbf{1}_3$ is the 3 × 3 identity matrix.

The end of a *Proof* is marked by ⊳⊲, that of a general statement or of a Remark by ◊.

The electrical network equations which describe the dynamics of the three-phase, ideal *DFI* machine are well-known [3, 4], as a consequence they are omitted.

3. Group properties of the mutual inductance matrix

By letting $\varphi \equiv \frac{2}{3}\pi$, the rotor-referred (*läufer-bezogene*) form of the mutual inductance matrix is [3, 4]

$$\mathbf{L}_{sr}[\theta_r] = \begin{bmatrix} \cos[\theta_r] & \cos[\theta_r + \varphi] & \cos[\theta_r - \varphi] \\ \cos[\theta_r - \varphi] & \cos[\theta_r] & \cos[\theta_r + \varphi] \\ \cos[\theta_r + \varphi] & \cos[\theta_r - \varphi] & \cos[\theta_r] \end{bmatrix} \tag{3}$$

Given $\mathbf{L}_{sr}[\theta_r]$ one defines the stator-referred (*ständer-bezogen*) rotor-to-stator mutual inductance matrix

$$\mathbf{L}'_{sr}[\theta_r] := \frac{N_r}{N_s} L_{ms} \mathbf{L}_{sr}[\theta_r]\ , \tag{4}$$

where $L_{ms}(> 0)$ is a constant parameter. Hereinafter, the dependence of the involved matrices and of related quantities on θ_r will be shown only if mandatory. The following properties hold because rows two and three of $\mathbf{L}_{sr}$ are right shift-circular permutations of the first row and because all row-wise (and column-wise) sums of $\mathbf{L}_{rs}$ vanish.

Proposition 1. (*Eigenvalues of* $\mathbf{L}_{sr}$ *and their implications*).

- The eigenvalues of $\mathbf{L}_{sr}$ are $\mu_0 = 0$, $\mu_\pm = \frac{3}{2} e^{\pm i\theta_r}$.

- $\det[\mathbf{L}_{sr}] = 0, \forall\theta_r$,
- $\dim[\mathrm{Ker}[\mathbf{L}_{sr}]] = 1$ and $\mathrm{Ker}[\mathbf{L}_{sr}] = \left\{\vec{j} \in \mathbb{R}^3 |\ j_1 = j_2 = j_3\right\} := \mathfrak{K}_{sr}, \forall\theta_r$.
- $\dim[\mathrm{range}[\mathbf{L}_{sr}]] = 2$ and $\mathrm{range}[\mathbf{L}_{sr}] = \mathfrak{B}, \forall\theta_r$,

Remark 2. (*Orthogonal decomposition of* $\mathbb{R}^3$). The last two properties in the list translate the orthogonal decomposition

$$\begin{array}{rcccc} \mathbb{R}^3 & = & \mathrm{Ker}[\mathbf{L}_{sr}] & \oplus & \mathrm{range}[\mathbf{L}_{sr}] \\ & = & \mathfrak{K}_{sr} & \oplus & \mathfrak{B} \\ \vec{j} & = & \vec{j}_{\mathfrak{K}_{sr}} & + & \vec{j}_{\mathfrak{B}} \end{array} , \tag{5}$$

where the straight line $\mathfrak{K}_{sr} : \{j_1 = j_2 = j_3\}$ is the normal to the plane $\mathfrak{B} : \{j_1 + j_2 + j_3 = 0\}$ of Eq. (2).

Lemma 1 (*Eigenvalues of a permutation matrix*, pp. 65–66 of M. Marcus and H. Minc's textbook [14]). For a general $N(\geq 2)$ and for an $N \times N$ matrix $\mathbf{P}$, a.k.a. "circulant", which results from the right shift-circular permutation of the first row $[c_0\ c_{N-1}\ c_{N-2}\ \dots\ c_1]$, one denotes $\epsilon := \mathrm{e}^{\mathrm{i}2\pi/N}$ and introduces the polynomial $\psi[.]$ of degree $N - 1$ in the complex variable ζ

$$\psi[\zeta] := \sum_{n=0}^{N-1} c_n \zeta^n \ . \tag{6}$$

The possibly multiple eigenvalues $\{\mu_k | 0 \leq k \leq N-1\}$ of $\mathbf{P}$ are obtained by letting $\zeta = \epsilon^m$ and evaluating $\psi[\epsilon^m]$ for $m = 1, 2, \dots, N$. Since $\epsilon^m := \mathrm{e}^{\mathrm{i}2\pi m/N}$, there exists only one value of m, denoted by ℓ, at which all powers of ϵ appearing in $\psi[.]$ are equal: $\epsilon^\ell = \epsilon^{2\ell} = \epsilon^{3\ell} = \dots = \epsilon^{(N-1)\ell} = 1$. Such value is $\ell = N$. Therefore

$$\psi\left[\epsilon^N\right] = \sum_{n=1}^{N-1} c_n = \psi\left[\epsilon^0\right] \ \text{when} \ \ell = N \ . \tag{7}$$

If the additional property

$$\sum_{n=1}^{N-1} c_n = 0 \tag{8}$$

is exhibited by the rows of $\mathbf{P}$, then

$$\psi\left[\epsilon^N\right] = 0 = \mu_0 \ . \tag{9}$$

Such eigenvalue is algebraically (and geometrically) simple.

Remark 3. (*Features of* $\psi[.]$, *k and n*). The polynomial $\psi[.]$ shall not be confused with any of the polynomials annihilated by $\mathbf{P}$. There is no correspondence between the eigenvalue label k and the ordering of powers induced by m.

Proof of Proposition 1. Since $\mathbf{L}_{sr}[\theta_r]$ is a circulant matrix of the sequence $\{c_0\ c_2\ c_1\}$

$$\mathbf{L}_{sr}[\theta] = \begin{bmatrix} c_0 & c_2 & c_1 \\ c_1 & c_0 & c_2 \\ c_2 & c_1 & c_0 \end{bmatrix} \tag{10}$$

and

$$c_0 + c_1 + c_2 = 0 \ , \ \forall \theta_r \in [0, 2\pi] \ , \tag{11}$$

then Lemma 1 applies with $N = 3$. Hence $\ell = 3$. In the first place, μ_0 is independent of θ_r. Next, one verifies the other two eigenvalues, $\mu_\pm = \frac{3}{2} e^{\pm i\theta_r}$, are respectively returned by $\psi[\epsilon]$ and by $\psi[\epsilon^2]$ and do instead depend on θ_r. The property $\dim[\mathrm{Ker}[\mathbf{L}_{sr}]] = 1$ derives from the algebraic simplicity of $\mu_0 = 0$. From Eq. (11) one deduces $\mathrm{Ker}[\mathbf{L}_{sr}] = \mathfrak{K}_{sr}$. The properties of range$[\mathbf{L}_{sr}]$ are not independent of those of $\mathrm{Ker}[\mathbf{L}_{sr}]$: namely, they follow from orthogonality, as highlighted by Eq. (5). ⊳⊲

Proposition 2. (*Eigenvalues of a circulant as the discrete* Fourier transform of a 3-sequence [15]). Let ϵ be as in Lemma 1. The discrete Fourier transform $\mathbf{b}^{(3)} := \{b_0 \ b_1 \ b_2\}$ of a 3-sequence $\mathbf{c}^{(3)} := \{c_0 \ c_1 \ c_2\}$ is obtained, in terms of row vectors, by

$$(b_0 \ b_1 \ b_2) = (c_0 \ c_1 \ c_2) \cdot \begin{bmatrix} 1 & 1 & 1 \\ 1 & \epsilon & \epsilon^2 \\ 1 & \epsilon^2 & \epsilon \end{bmatrix} := (c_0 \ c_1 \ c_2) \cdot \mathbf{T} \ . \tag{12}$$

The circulant $\mathbf{P}\left[\mathbf{c}^{(3)}\right]$ assembled from $\mathbf{c}^{(3)}$ is diagonalized by $\mathbf{T}$ according to

$$\mathbf{P}\left[\mathbf{c}^{(3)}\right] = \mathbf{T}^{-1} \cdot \begin{bmatrix} b_0 & 0 & 0 \\ 0 & b_1 & 0 \\ 0 & 0 & b_2 \end{bmatrix} \cdot \mathbf{T} \ . \tag{13}$$

Application to $\mathbf{L}_{sr}$ of Eq. (10) requires a permutation of the first row:

$$(\mu_0 \ \mu_1 \ \mu_2) = (c_0 \ c_2 \ c_1) \cdot \begin{bmatrix} 1 & 0 & 0 \\ 0 & 0 & 1 \\ 0 & 1 & 0 \end{bmatrix} \cdot \mathbf{T} \ . \tag{14}$$

Lemma 2 (*The matrix* $\mathbf{A}_0$ *and its properties*). Let the matrix $\mathbf{A}_0$ be defined by

$$\mathbf{A}_0 := \frac{1}{\sqrt{3}} \begin{bmatrix} 0 & -1 & 1 \\ 1 & 0 & -1 \\ -1 & 1 & 0 \end{bmatrix} \ . \tag{15}$$

Its properties are the following.

- (*Determinant, rank, eigenvalues, eigenspaces*).

$$\det \mathbf{A}_0 = 0 \ ; \ \mathrm{rank}[\mathbf{A}_0] = 2 \ , \tag{16}$$

$$\lambda_0 = 0 \ , \lambda_1 = -i \ , \ \lambda_2 = +i \ . \tag{17}$$

There is an eigenspace of dimension one, $\mathcal{X}_0[\mathbf{A}_0]$, which corresponds to λ_0:

$$\mathcal{X}_0[\mathbf{A}_0] = \text{Ker}[\mathbf{A_0}] = \text{Ker}[\mathbf{L}_{sr}] = \mathfrak{K}_{sr}. \tag{18}$$

- (*Left zero divisors*). The constant, nontrivial left zero divisors of $\mathbf{A}_0$ are given by dyads

$$\mathbf{Z}\left[\vec{c}\right] = \vec{c})(\vec{1} \tag{19}$$

where c_k, $k = 1, 2, 3$, are real constants with $c_k \neq 0$ for at least one k.

- (*Dyadic representation*). The matrix $\mathbf{A}_0$ admits **no** "algebraic" dyadic representation of the form

$$\mathbf{A}_0 = \vec{a})(\vec{b} \text{ (false)} \tag{20}$$

with constant $a_k, b_k \in \mathbb{R}$, $k = 1, 2, 3$.

- (*Sign reversal*). There exists no left zero divisors of $\mathbf{A}_0$ which, added to $\mathbf{A}_0$, reverses its sign.

- (*Recurrent formula for powers of* $\mathbf{A}_0$). Given $\mathbf{A}_0$ and

$$\mathbf{A}_0^2 = -\mathbf{1}_3 + \frac{1}{3}\vec{1})(\vec{1}, \tag{21}$$

the powers of $\mathbf{A}_0$ are obtained from

$$\mathbf{A}_0^n = (-1)^{1+((n-3)\%4)/2}\mathbf{A}_0^{2-(n-2)\%2}, \quad n \geq 3, \tag{22}$$

where $(n-3)\%4$ stands for the remainder from integer division of $(n-3)$ by 4 and the "/" (slash) denotes division between integers; similarly, $(n-2)\%2$ stands for the remainder from integer division of $(n-2)$ by 2.

Proof of Lemma 2. The properties described by Eqs. (16)-(19) are immediately verified, as well as the nonexistence of an algebraic dyadic representation. The statement about sign reversal is proved by contradiction. To obtain the recurrent formula for powers of $\mathbf{A}_0$ one computes $\mathbf{A}_0^3$ $(= -\mathbf{A}_0)$ and $\mathbf{A}_0^4$ $(=-\mathbf{A}_0^2)$, then one examines the higher powers $\mathbf{A}_0$ and the sequence formed by their signs. Since the sequence has period 4 and reads $- - + + - - + + \cdots$, then the exponents of both $\mathbf{A}_0$ and (-1) in Eq. (22) can be determined. ⊳⊲

Remark 4, to Lemma 2.

- Let $\vec{a})$ in Eq. (20) be replaced by $\nabla)$, then there exists a C^1, divergence-free vector field $\vec{f}$ giving rise to the "differential" dyadic representation of $\mathbf{A}_0$

$$\mathbf{A}_0 = \frac{1}{\sqrt{3}}\nabla)(\vec{f}. \tag{23}$$

The system of first-order linear partial differential equations to which $\vec{f}$ is the solution is obtained by comparing like terms in the arrays.

- As already noticed in the proof, $\mathbf{A}_0$ cannot be a left zero divisor of itself. (In fact, $\mathbf{A}_0^2$ is given by Eq. (21)).

- Obviously, there is no way of including $n = 0$ in any recurrent formula for the powers of $\mathbf{A}_0$, of which Eq. (22) is an example because $\det[\mathbf{A}_0] = 0$.

- As one can easily verify, the eigenvalues of $\mathbf{A}_0^2$ are $\lambda_0 = 0$, $\lambda_1 = -1$. The latter has algebraic multiplicity $\alpha_1 = 2$ and geometric multiplicity $\gamma_1 = 1$. As a consequence, its eigenspace, $\mathcal{X}_1\left[\mathbf{A}_0^2\right]$, not only has a dimension $\alpha_1 - \gamma_1 + 1 = 2$ but complies with

$$\mathcal{X}_1\left[\mathbf{A}_0^2\right] = \mathfrak{B} \tag{24}$$

as well. In other words, by recalling Eqs. (5), (18), and (21),

$$\mathbf{A}_0^2 = -\mathbf{1}_3 \restriction_{\mathfrak{B}} \text{ or } \mathbf{A}_0^2 \cdot \vec{\psi} = -\vec{\psi}_{\mathfrak{B}}, \ \forall \vec{\psi} \in \mathbb{R}^3 \tag{25}$$

i.e., $\mathbf{A}_0^2$ coincides with $-\mathbf{1}_3$ restricted to the subspace $\mathfrak{B}$. ◊

Proposition 3. (*The matrix* $\mathbf{L}_{sr}[.]$: *trigonometric decomposition and classical adjoint; the left zero divisors of* $\mathbf{L}_{sr}[.]$, *their kernel and range*).

- (*The matrices C and S*). $\mathbf{L}_{sr}[\theta_r]$ is a linear combination of trigonometric functions according to

$$\mathbf{L}_{sr}[\theta_r] = \mathbf{C}\cos[\theta_r] + \mathbf{S}\sin[\theta_r], \tag{26}$$

where $\mathbf{C}$ and $\mathbf{S}$ are the constant, 3×3 matrices

$$\mathbf{C} = \frac{3}{2}\mathbf{1}_3 - \frac{1}{2}\vec{1})(\vec{1} \quad ; \quad \mathbf{S} = \begin{bmatrix} 0 & -1 & 1 \\ 1 & 0 & -1 \\ -1 & 1 & 0 \end{bmatrix} \frac{\sqrt{3}}{2}. \tag{27}$$

- (*Relations between* $\mathbf{A}_0$, $\mathbf{C}$ *and* $\mathbf{S}$).

$$\mathbf{C} = -\frac{3}{2}\mathbf{A}_0^2, \quad \mathbf{S} = \frac{3}{2}\mathbf{A}_0. \tag{28}$$

- (*The classical adjoint matrix*). The classical adjoint to $\mathbf{L}_{sr}[\theta_r]$ ($\equiv$ transpose of the cofactor matrix) is

$$\operatorname{adj}[\mathbf{L}_{sr}[\theta_r]] = \frac{3}{4}\vec{1})(\vec{1} \ , \ \forall \theta \in [0, 2\pi] \ . \tag{29}$$

- (*Left divisors as dyads*). If $f_k[.]$, $k = 1, 2, 3$, denote real-valued functions of class $\mathcal{C}^M([0, 2\pi])$ (for some M), one of which, at least, does not vanish identically, then a left zero divisor $\mathbf{Z}$ (*linker Nullteiler*) of $\mathbf{L}_{sr}$ is a rank one dyad

$$\mathbf{Z}\left[\vec{f}[\theta]\right] = \vec{f}[\theta])(\vec{1} \tag{30}$$

forming an algebra $\{\mathbf{Z}\} \equiv \mathfrak{Z}$.

- (*Kernel of the* $\mathbf{Z}$*'s*).

$$\mathrm{Ker}[\mathbf{Z}] = \mathfrak{B} \ , \ \forall \mathbf{Z} \in \mathfrak{Z}. \tag{31}$$

Proof of Proposition 3. The identification of **C** and **S** follows from expanding the $\cos\left[.\pm\frac{2}{3}\pi\right]$ entries in $\mathbf{L}_{sr}$. In order to determine $\mathrm{adj}[\mathbf{L}_{sr}]$ one starts from a relation which is one of the many formulas due to Laplace [16]

$$\mathrm{adj}[\mathbf{A}]\cdot\mathbf{A} = \det[\mathbf{A}]\mathbf{1}_N = \mathbf{A}\cdot\mathrm{adj}[\mathbf{A}] \tag{32}$$

and holds for a general $\mathbf{A}\in\mathcal{M}[N\times N]$; then one recalls $\det[\mathbf{L}_{sr}] = 0$ and the θ_r-invariance of μ_0 of Eq. (9): one can thus compute the classical adjoint to either constant matrix, **C** or **S**, whichever is simpler to deal with; the result is the dyad on the right side of Eq. (29), a result which holds $\forall\theta_r$. The search for left zero divisors of $\mathbf{L}_{sr}$ as dyads like that of Eq. (30) is suggested by Eqs. (29) and (32) because $\mathrm{adj}[\mathbf{L}_{sr}]$ must be a zero divisor of $\mathbf{L}_{sr}$. The most general form of a left zero divisor **Z** is inferred from Eq. (11): since all column-wise sums of $\mathbf{L}_{sr}$ vanish, the columns of Z must be equal. Therefore such divisor, if non-trivial, has rank one and is obtained from the dyadic product of Eq. (30). Finally, Eq. (32) and the orthogonal decomposition Eq. (5) imply

$$\mathrm{Ker}[\mathrm{adj}[\mathbf{L}_{sr}]] = \mathrm{Ker}[\mathbf{Z}] = \mathrm{range}[\mathbf{L}_{sr}] \ \ (=\mathfrak{B}) \ . \tag{33}$$

$$\mathrm{Ker}[\mathbf{L}_{sr}] = \mathrm{range}[\mathrm{adj}[\mathbf{L}_{sr}]] = \mathrm{range}[\mathbf{Z}] \ \ (=\mathfrak{K}_{sr}) \ . \tag{34}$$

⊳⊲

Remark 5. *(Duality; divisors; other properties of* $\mathbf{L}_{sr}$*).*

- From Eqs. (32–34) one says $\mathbf{L}_{sr}$ and $\mathrm{adj}[\mathbf{L}_{sr}]$ are dual to each other.
- The constant, nontrivial left zero divisors of $\mathbf{L}_{sr}$ are those of Eq. (19). By consistency, the matrices **C** and **S** of Eq. (26) not only have the same classical adjoint as $\mathbf{L}_{sr}[\theta_r]$ has but have all and the same zero divisors, because Eq. (26) holds $\forall\theta_r$.
- Right zero divisors are obtained by transposing the left ones.
- Nonexistence of the representation of Eq. (20) prevents $\mathbf{A}_0$ it from being a left zero divisor of $\mathbf{L}_{sr}[.]$. Nor can $\mathbf{A}_0$ be, as Eqs. (28) and (22) show, a divisor of either **C** or **S** taken separately.
- If $\mathfrak{L}_{sr}$ stands for the subspace of functions $\overrightarrow{f}[.]\in\left(C^0([0,2\pi])\right)^3$ complying with $\oint\mathbf{L}_{sr}[\theta_r]\cdot\overrightarrow{f}[\theta_r]\mathrm{d}\theta_r = \vec{0}$, and if $a_m^{(k)}, m = 0,1,2,\cdots, b_m^{(k)}, m = 1,2,\cdots$, are the cosine and, respectively, the sine Fourier coefficients of the (real-valued) components $f_k[.], k = 1,2,3$ of $\overrightarrow{f}[.]$, then

$$\overrightarrow{f}\in\mathfrak{L}_{sr}\Leftrightarrow\left\{\left\{a_1^{(1)} = a_1^{(2)} = a_1^{(3)}\right\} \ \text{and} \ \left\{b_1^{(1)} = b_1^{(2)} = b_1^{(3)}\right\}\right\}. \tag{35}$$

◊

Remark 6. (*The physical meaning of* $\mathbf{L}_{sr}$, **C** *and* **S**). As Eq. (5) suggests, given any instantaneous current vector $\overrightarrow{j}[t]\in\mathbb{R}^3$, left multiplication by $\mathbf{L}_{sr}[\theta_r[t]]$ returns a balanced current triple $\overrightarrow{j}_{\mathfrak{B}}[t]\in\mathfrak{B}$. This follows from the three-fold symmetry of the ideal

DFI machine, mirrored by the structure of $\mathbf{L}_{sr}[.]$. Moreover, the $\mathbf{C}$ term of Eq. (26) represents the opposite of reactive torque, whereas the $\mathbf{S}$ term represents active torque.◊

Theorem 1. (*Differentiation of* $\mathbf{L}_{sr}$ *with respect to* θ_r). Let $\mathbf{A}_0$ and $\mathbf{Z}\left[\vec{c}\right]$ be respectively given by Eqs. (15) and (19). Then the derivative of $\mathbf{L}_{sr}$ with respect to θ_r is the set-valued map

$$\left(\mathbf{A}_0 + \mathbf{Z}\left[\vec{c}\right]\right) \cdot \mathbf{L}_{sr}[\theta_r] \in \left(\frac{\partial \mathbf{L}_{sr}}{\partial \theta_r}\right)[\theta_r]. \tag{36}$$

Notation: a convenient notation is $\left(\frac{\partial \mathbf{L}_{sr}}{\partial \theta_r}\right)[\theta_r] \equiv \mathbf{M}[\theta_r]$.

Proof of Theorem 1. Differentiation of $\mathbf{L}_{sr}[\theta_r]$ as represented by Eq. (26), and the use of Eq. (28) yield

$$\left(\frac{\partial \mathbf{L}_{sr}}{\partial \theta_r}\right)[\theta_r] = \frac{3}{2}\mathbf{A}_0 \cos[\theta_r] + \frac{3}{2}\mathbf{A}_0^2 \sin[\theta_r]. \tag{37}$$

One seeks for a constant matrix $\mathbf{B}_1$ which complies with

$$\frac{3}{2}\mathbf{A}_0 \cos[\theta_r] + \frac{3}{2}\mathbf{A}_0^2 \sin[\theta_r] = \mathbf{B}_1 \cdot \mathbf{C}\cos[\theta_r] + \mathbf{B}_1 \cdot \mathbf{S}\sin[\theta_r]. \tag{38}$$

The application of Eq. (22) leads to

$$\mathbf{B}_1 = \mathbf{A}_0. \tag{39}$$

Then, the whole set of constant matrices $\mathbf{B}$ complying with $\frac{\partial \mathbf{L}_{sr}}{\partial \theta_r}\mathbf{B} \cdot \mathbf{L}_{sr}$ is obtained by adding to $\mathbf{B}_1$ a left zero divisor $\mathbf{Z}\left[\vec{c}\right]$ as of Eq. (19)

$$\mathbf{B} = \mathbf{A}_0 + \mathbf{Z}\left[\vec{c}\right]. \tag{40}$$

The result justifies the notation for $\frac{\partial \mathbf{L}_{sr}}{\partial \theta_r}$ of Eq. (36) as a set-valued map. ⊳⊲

The above Eq. (36) means $\frac{\partial \mathbf{L}_{sr}[\theta_r]}{\partial \theta_r} \cdot \vec{j} = \left(\mathbf{A}_0 + \mathbf{Z}\left[\vec{c}\right]\right) \cdot \mathbf{L}_{sr}[\theta_r] \cdot \vec{j}, \forall \vec{j} \in \mathbb{R}^3$. In spite of this last relation, the search for an exponential representation of $\mathbf{L}_{sr}[.]$ and for a one-parameter (θ_r) group acting on the whole of $\mathbb{R}^3$ is ill-posed. Namely, $\mathbf{L}_{sr}[0]$ is not invertible, hence one cannot normalize $\mathbf{L}_{sr}$ by $\mathbf{L}_{sr}[0]$ and no unit element of the group can be defined. To a greater extent, the search for a generator for the group would make no sense. Nonetheless, an exponential representation is obtained in the subspace $\mathfrak{B}$.

Theorem 2. (*Exponential representations on* $\mathfrak{B}$). If $\vec{j} \in \mathfrak{B}$ then the following hold.

$$\mathbf{L}_{sr}[\theta_r] \cdot \vec{j} = \frac{3}{2}e^{\theta_r \mathbf{A}_0} \cdot \vec{j} \quad , \quad \forall \vec{j} \in \mathfrak{B} \tag{41}$$

and

$$\mathbf{M}[\theta_r] \cdot \vec{j} = \mathbf{M}[0] \cdot e^{\theta_r \mathbf{A}_0} \cdot \vec{j} \quad , \quad \forall \vec{j} \in \mathfrak{B} \tag{42}$$

with

$$\mathbf{M}[0] = \frac{3}{2}\mathbf{A}_0. \tag{43}$$

Proof of Theorem 2. A matrix $\mathbf{J}[\theta_r]$ is sought for, which, like $\mathbf{L}_{sr}[\theta_r]$, splits into a $\cos[.]$ and a $\sin[.]$ term as in Eq. (26) and, unlike $\mathbf{L}_{sr}[.]$, satisfies $\mathbf{J}[0] = \mathbf{1}_3$. As Eqs. (28) suggest, one solution is

$$\mathbf{J}[\theta_r] = \mathbf{1}_3 \cos\theta_r + \frac{2}{3}\mathbf{S}\sin\theta_r = \frac{2}{3}\left(\mathbf{C} + \frac{1}{2}\vec{1}\right)\left(\vec{1}\right)\cos\theta_r + \frac{2}{3}\mathbf{S}\sin\theta_r \ . \tag{44}$$

Next, one requires θ_r-differentiation to coincide with the multiplication of $\mathbf{J}[.]$ by a constant matrix $\mathbf{H}$

$$\left(\frac{\partial \mathbf{J}}{\partial \theta_r}\right)[\theta_r] = \mathbf{H}\cdot\mathbf{J}[\theta_r] \ . \tag{45}$$

By identifying terms like trigonometric functions one obtains the pair

$$\left\{\mathbf{H} = \frac{2}{3}\mathbf{S} \ ; \ \frac{2}{3}\mathbf{H}\cdot\mathbf{S} = -\mathbf{1}_3\right\} \text{ i.e. } \left\{\mathbf{H}^2\cdot\vec{j} = -\mathbf{1}_3\cdot\vec{j} \ , \ \forall\vec{j}\in\mathfrak{B}\right\} \ . \tag{46}$$

One solution, *modulo* left zero divisors, comes from the properties of $\mathbf{A}_0^2$ in Eq. (25):

$$\mathbf{H} = \mathbf{A}_0 \ . \tag{47}$$

Hence $\mathbf{J}[\theta_r] = \mathrm{e}^{\theta_r \mathbf{A}_0}$. The proposed representation of $\mathbf{L}_{sr}[.]$ is

$$\mathbf{L}_{sr}[\theta_r] = \frac{3}{2}\mathrm{e}^{\theta_r \mathbf{A}_0} - \frac{1}{2}\vec{1})(\vec{1}\cos\theta_r \ . \tag{48}$$

Consistency with $\frac{\partial \mathbf{L}_{sr}}{\partial \theta_r}[.]$ implies

$$\left(\frac{\partial \mathbf{L}_{sr}}{\partial \theta_r}\right)[\theta_r] = \frac{3}{2}\mathbf{A}_0\cdot\mathbf{J}[\theta_r] + \frac{1}{2}\vec{1})(\vec{1}\sin\theta_r \ . \tag{49}$$

Since $\vec{j}\in\mathfrak{B}$, then the rightmost dyads in Eqs. (48) and (49) return $\vec{0}$ when right multiplied by $\vec{j}$. Replacing $\mathbf{A}_0$ in Eq. (47) by the $\mathbf{B}$ of Eq. (40) does not change the results (Eqs. (48) and (49)) because

$$\mathrm{e}^{\theta_r \mathbf{Z}[\vec{c}]}\cdot\vec{j} = \vec{0} \ , \ \forall\vec{j}\in\mathfrak{B} \ , \ \forall\mathbf{Z}\left[\vec{c}\right]\in\mathfrak{Z} \ . \tag{50}$$

⊳⊲

Theorem 3. (*Symmetry properties*).

$$\mathbf{L}_{sr}^{\mathrm{Trs}}[\theta_r] = \mathbf{L}_{sr}[-\theta_r] \tag{51}$$

$$\mathbf{M}^{\mathrm{Trs}}[\theta_r] = -\mathbf{M}[-\theta_r] \tag{52}$$

$$[\mathbf{L}_{sr}, \quad \mathbf{A}_0] = \mathbf{0}. \tag{53}$$

Proof of Theorem 3. The first two relations are immediate. The third one follows from the representation of $\mathbf{L}_{sr}$ as linear combination of powers of $\mathbf{A}_0$ according to Eqs. (26) and (28). ▷◁

4. The Blondel-Park transformation and the rotation group

4.1. Axiomatics of the transformation

The Kirchhoff voltage and current laws bring redundancy into the $\{abc\}$ frame representations. In order to remove said redundancy, another frame, called $\{dq0\}$, is introduced, where only two components of a vector shall matter, the direct one, d, and the quadrature component, q.

Definition 3. (*The* $dq0$ *frame*). Let $\{dq0\}$ denote a reference frame for electric quantities of axes d and q, subject to five specifications.

d.1) The new frame shall be suitable to represent both stator-referenced and rotor-referenced quantities.

d.2) The component of a stator-referenced quantity with respect to both the *direct* or d axis and the stator *as* axis shall be represented by the same function of angle, evaluated at arguments which differ by β_s. Similarly for variables pertaining to the rotor *ar*: the phase difference shall be β_r.

d.3) The above angles are related by

$$\beta_s = \beta_r + \theta_r. \tag{54}$$

d.4) The *quadrature* or q axis shall be orthogonal to d in the $L^2([0, 2\pi])$ sense: if $w_d[.]$ and $w_q[.]$ are the d- and q-components of a (generally complex-valued) signal $\vec{w}[.]$ which depends on η, then: $\int_0^{2\pi} w_d[\eta]^* w_q[\eta] \mathrm{d}\eta = 0$.

d.5) The third entry $w_0[.]$ of a vector $\vec{w}[.]$ in the $\{dq0\}$ frame shall be equal to the sum of its $\{abc\}$ components. (For this reason, such a sum is called "zero sequence", or *Nullfolge*, and may be trivial or not).

Problem 1. (*The* $\{abc\}$ *to* $\{dq0\}$ *transformation problem* [3, 4, 6]). Find a transformation $\mathbf{K}[.]$ that maps a vector $\vec{w}_{abc}$ (a physical quantity) from the $\{abc\}s$ and, respectively, the $\{abc\}r$ frames to a vector $\vec{w}_{dq0}$ in the $\{dq0\}$ frame, as specified by Definition 3 and

K.1) is invertible and linear;

K.2) conserves instantaneous electric power;

K.3) has the same functional form for both stator and rotor quantities,

K.4) depends at most on one real parameter, an "electric angle", which may be different for stator or rotor quantities;

K.5) is of class C^1 at least with respect to that parameter;

K.6) magnetically decouples flux linkages [6].

Proposition 4. (*Matrix representation*). A solution to Problem 1 which applies to a three-phase machine exists and is the Blondel-Park [1, 2, 7] transformation $\mathbf{K}[.]$

$$\mathbf{K}[\eta] := \sqrt{\frac{2}{3}} \begin{bmatrix} \cos[\eta] & \cos[\eta-\varphi] & \cos[\eta+\varphi] \\ -\sin[\eta] & -\sin[\eta-\varphi] & -\sin[\eta+\varphi] \\ \frac{1}{\sqrt{2}} & \frac{1}{\sqrt{2}} & \frac{1}{\sqrt{2}} \end{bmatrix}, \tag{55}$$

where η stands for an electrical angle. One has

$$\vec{w}_{dq0}[t] = \mathbf{K}[\eta[t]] \cdot \vec{w}_{abc}[t] \tag{56}$$

and $w_0[t] = (w_a + w_b + w_c)[t], \forall t$ and, if $\vec{w}[t]$ has a period 2π, $\oint w_d[t] w_q[t] \mathrm{d}t = 0$.

Proof of Proposition 4. The structure of $\mathbf{K}[.]$ can be inferred by satisfying, in sequence, requirements $K.1$, $K.2$, $K.6$, $d.4$, $d.5$. The Ansatz

$$\mathbf{K}[\eta] = \mathbf{K}_0 \cdot \mathrm{e}^{\eta \mathbf{F}}, \tag{57}$$

where $\mathbf{K}_0 \equiv \mathbf{K}[0]$ and $\mathbf{F}$ is a constant matrix, is shown to be consistent with all requirements, hence the entries of $\mathbf{K}_0$ and $\mathbf{F}$ can be identified. No further details can be provided for reasons of space. ⊳⊲

Remark 7 to Proposition 4. (*On the exponential representation of* $\mathbf{K}[\eta]$). As a result of work at proving Proposition 4, $\mathbf{K}[\eta]$ defines a one-parameter (η) group of unitary (power preserving) transformations, represented by Eq. (57). Since $\mathbf{K}_0$ is invertible, then

$$\mathbf{R}[\eta] := \mathbf{K}_0^{-1} \cdot \mathbf{K}[\eta] \tag{58}$$

and $\mathbf{R}[0]$ is the unit element. The existence of the composition law is implied by the Ansatz. Obviously, $\det[\mathbf{K}[\eta]] = 1$ implies $\det[\mathbf{R}[\eta]] = 1, \forall \eta$. ◊.

Theorem 4. (*Infinitesimal generator*). The infinitesimal generator $\mathbf{F}$ of $\mathbf{K}[.]$ is $\mathbf{K}_0$-similar to the opposite of the infinitesimal generator $\mathbf{A}_3$ of rotations about the $\hat{\mathbf{x}}_3$ axis of $\mathbb{R}^3$ according to

$$\mathbf{F} = -\mathbf{K}[0]^{-1} \cdot \mathbf{A}_3 \cdot \mathbf{K}[0] \tag{59}$$

and is related to the $\mathbf{A}_0$ of Eq. (41) by

$$\mathbf{F} = -\mathbf{A}_0. \tag{60}$$

Proof of Theorem 4. From Eq. (55)

$$\frac{\mathrm{d}\mathbf{K}[\eta]}{\mathrm{d}\eta} = \sqrt{\frac{2}{3}} \begin{bmatrix} -\sin[\eta] & -\sin\left[\eta - \frac{2\pi}{3}\right] & -\sin\left[\eta + \frac{2\pi}{3}\right] \\ -\cos[\eta] & -\cos\left[\eta - \frac{2\pi}{3}\right] & -\cos\left[\eta + \frac{2\pi}{3}\right] \\ 0 & 0 & 0 \end{bmatrix} \tag{61}$$

and the constant matrix $\mathbf{B}$ satisfying $\frac{\mathrm{d}\mathbf{K}[\eta]}{\mathrm{d}\eta} = \mathbf{B} \cdot \vec{\mathbf{K}}[\eta]$ reads

$$\mathbf{B} = \begin{bmatrix} 0 & 1 & 0 \\ -1 & 0 & 0 \\ 0 & 0 & 0 \end{bmatrix} = -\mathbf{A}_3 \ . \tag{62}$$

Next, the infinitesimal generator of $\mathbf{R}[.]$ Eq. (58) is identified according to

$$\frac{\mathrm{d}\mathbf{R}[\eta]}{\mathrm{d}\eta} = \mathbf{K}_0^{-1} \cdot \mathbf{B} \cdot \mathbf{K}[\eta] = \mathbf{K}_0^{-1} \cdot \mathbf{B} \cdot \mathbf{K}_0 \cdot \mathbf{R}[\eta] := \mathbf{F} \cdot \mathbf{R}[\eta]. \tag{63}$$

In other words, the matrix $\mathbf{F} := \mathbf{K}_0^{-1} \cdot \mathbf{B} \cdot \mathbf{K}_0$ is the sought for infinitesimal generator of the group $\mathbf{R}[.]$. This proves Eq. (59). The relation between $\mathbf{F}$ and $\mathbf{B}$ is to be expected (e.g., § 2.5 of Altmann's textbook [17]). Finally, Eq. (60) follows from direct verification. ⊳⊲

4.2. The product of matrices formula

Theorem 5. (*The formula*). Equations (54), (57) and (42) imply

$$\mathbf{K}[\beta_s] \cdot \left(\frac{\partial \mathbf{L}_{sr}}{\partial \theta_r}\right)[\theta_r] \cdot \mathbf{K}[\beta_r]^{-1} = \mathbf{K}_0 \cdot \mathbf{M}[0] \cdot \mathbf{K}_0^{-1} = \frac{3}{2}\mathbf{A}_3 \quad . \tag{64}$$

Proof of Theorem 5. The proof branches out according to which current triple is being dealt with.

- (*Balanced current triple ≡ trivial zero sequence*). Let $\vec{j}_{\{abc\}} \in \mathfrak{B}$, then, by Eqs. (42), (57), and (60) and applying transposition

$$\begin{aligned}\mathbf{K}[\beta_s] \cdot \left(\frac{\partial \mathbf{L}_{sr}}{\partial \theta_r}\right)[\theta_r] \cdot \mathbf{K}[\beta_r]^{-1} &= \mathbf{K}[\beta_s] \cdot \mathbf{M}[\theta_r] \cdot \mathbf{K}[\beta_r]^{-1} = \mathbf{K}_0 \cdot \mathrm{e}^{-\beta_s \mathbf{A}_0} \cdot \mathbf{M}[\beta_r + \theta_r] \cdot \mathbf{K}_0^{-1} = \\ &= \mathbf{K}_0 \cdot \mathbf{M}[-\beta_s + \beta_r + \theta_r] \cdot \mathbf{K}_0^{-1} = \mathbf{K}_0 \cdot \mathbf{M}[0] \cdot \mathbf{K}_0^{-1} = \frac{3}{2}\mathbf{A}_3.\end{aligned} \tag{65}$$

- (*General current triple*). For general $\vec{j}_{\{abc\}} \in \mathbb{R}^3$ no exponential representation is available. In analogy with Eq. (26) one identifies the constant matrices $\mathbf{P}, \mathbf{Q}$ and $\mathbf{R}$ giving rise to $\mathbf{K}[\eta] = \sqrt{\frac{2}{3}}\mathbf{P}\cos\eta + \sqrt{\frac{2}{3}}\mathbf{Q}\sin\eta + \frac{1}{\sqrt{3}}\mathbf{R}$. The product $\mathbf{M}[\theta_r] \cdot \mathbf{K}^{-1}[\beta_r]$, after simplification, turns out to be an affine function of $\cos[\theta_r + \beta_r]$ and $\sin[\theta_r + \beta_r]$ which in turn depend on angle sums: products of the involved matrices hide an addition formula for angles on which trigonometric functions depend. Taking Eq. (54) into account, left multiplication by $\mathbf{K}[\beta_s]$ leads to a polynomial in $\cos[\beta_s]$ and $\sin[\beta_s]$ with coefficients like $\mathbf{P} \cdot \mathbf{C} \cdot \mathbf{Q}^{\mathrm{Trs}}$, $\mathbf{R} \cdot \mathbf{C} \cdot \mathbf{P}^{\mathrm{Trs}}$ and so forth. All β_s-dependent terms in the polynomial disappear. Eventually, the only non-zero term is $\frac{2}{3}\mathbf{P} \cdot \mathbf{C} \cdot \mathbf{Q}^{\mathrm{Trs}} = \frac{3}{2}\mathbf{A}_3$, a constant. ⊳⊲

Remark 8. (*Prior results*). To the best of the authors' knowledge, the role of the Blondel-Park transformation in realization theory was pointed out by J.L. Willems [8], who derived the exponential representation of $\mathbf{K}[.]$ while obtaining a time-invariant system from time-varying electric machine equations. The group properties of $\mathbf{K}[.]$ have been known for some time (e.g., [9], p. 1060). Instead, the relation of $\mathbf{F}$ to $\mathbf{A}_0$, at least in the form of Eq. (60), the relation between exponential representations of $\mathbf{K}[.]$ and $\mathbf{L}_{sr}[.]$, and the roles played by the left zero divisors of $\mathbf{L}_{sr}[.]$ and by the subspace $\mathfrak{B}$, seem to have been overlooked so far.◊

5. Electric torque

5.1. The electric torque law in the $\{abc\}$ frame

From the principles of analytical mechanics, the following relation can be deduced [3, 4] for the ideal *DFI* machine in generator mode. The relation involves previously defined quantities, namely stator and rotor currents and a machine parameter, the $\mathbf{L}'_{sr}[.]$ of Eq. (4), and a quantity, the electric torque $\mathcal{T}_{el,g}$, which has not yet been mentioned herewith. As a consequence, the relation can be regarded as the physical "law" which defines $\mathcal{T}_{el,g}$.

Definition 4. (*Electric torque in the $\{abc\}$ frame*). Let the ideal *DFI* machine have P poles and be described by current vectors $\vec{j}_{\{abc\}s}$ and $\vec{j}'_{\{abc\}r}$. The electric torque in generator mode is defined by

$$\mathcal{T}_{el,g} = -\frac{P}{2}\ \vec{j}_{\{abc\}s}^{\ \mathrm{Trs}} \cdot \frac{\partial \mathbf{L}'_{sr}[\theta_r]}{\partial \theta_r} \cdot \vec{j}'_{\{abc\}r}\ . \tag{66}$$

The relevance of Eq. (66) sits in the link it establishes between electric quantities and a mechanical one: in generator mode, it is the torque produced by, usually a working fluid, on the *DFI* machine shaft which, through a suitably excited rotor, gives rise to electric currents in the stator coils; in motor mode power flow from machine coils to the shaft is reversed. All machine control laws rely on Eq. (66) in order to be implemented.

5.2. Co-energy and the Legendre transform

Ansatz. (*Internal energy*). If S is entropy and T is temperature, then the first differential of internal energy U of an electric machine with one mechanical degree of freedom, θ_m, at constant volume V and numbers of moles $\vec{N}$, is

$$\mathrm{d}U = T\mathrm{d}S + \vec{j} \cdot \mathrm{d}\vec{\lambda} - \mathcal{T}_{el,m}\mathrm{d}\theta_m, \tag{67}$$

where $\vec{\lambda}$ is the vector of flux linkages, $\mathcal{T}_{el,m}$ is mechanical torque in motor mode (opposite to that in generator mode), and θ_m is the shaft angle.

A consequence of the Ansatz is the following.

Proposition 5. (*Relation between motor torque and internal energy*).

$$\mathcal{T}_{el,m} = -\left(\frac{\partial U}{\partial \theta_m}\right)_{S,V,\vec{N},\vec{\lambda}}\ . \tag{68}$$

Definition 5. (Legendre *transform of energy with respect to flux linkage*). Let $\Lambda \subset IR^3$ be a subset where U is at least twice differentiable and convex with respect to $\vec{\lambda}$ and let $\vec{p}$ denote the variable conjugate to $\vec{\lambda}$. Then the Legendre transform $\mathcal{Y}$ of energy U with respect to $\vec{\lambda}$ is defined by

$$\mathcal{Y}\left[S, V, \vec{N}, \vec{p}, \theta_m\right] := \sup_{\vec{\lambda} \in \Lambda} \left(\vec{p} \cdot \vec{\lambda} - U\left[S, V, \vec{N}, \vec{\lambda}, \theta_m\right]\right). \tag{69}$$

Remark 9. (*Conjugate variables; motor torque; differential geometric setting*).

- $\vec{p}$ coincides with $\vec{j}$ and one has

$$\mathcal{Y} + U = \vec{j} \cdot \vec{\lambda}. \tag{70}$$

- Motor torque can thus be rewritten as

$$\mathcal{T}_{el,m} = \left(\frac{\partial \mathcal{Y}}{\partial \theta_m}\right)_{S,V,\vec{N},\vec{j}}. \tag{71}$$

- The extensive variables on which *energy* depends are S, $\vec{N}$, $\vec{\lambda}$ and θ_m, and as such are coordinates of the dynamical system's manifold $\mathcal{N}$. Instead, the intensive variables T, $\vec{\mu}$ (vector of chemical potentials), $\vec{j}$ and $\mathcal{T}_{el,m}$ belong to the system's co-tangent bundle $T^*\mathcal{N}$ [11, 13]. Upon a multivariate Legendre transformation, as many extensive variables can be replaced by their conjugates, which are intensive variables.◊

Remark 10. ($\mathcal{Y}$ *vs.* W'_{fld}). By identifying U with "the energy W_{fld} stored in the coupling fields" of an electric machine having P poles, the rotor of which forms the mechanical angle θ_m in the stator frame, one has the following relations:

- the electrical angle θ_r is related to the mechanical angle θ_m by $\theta_r = \frac{P}{2}\theta_m$ (multiplier effect of P);

- usually [3, 4] $\mathcal{T}_{el,m}$ is related to "co-energy" $W'_{fld}\left[\vec{j}_{\{abc\}_s}, \theta_r\right]$ by

$$\mathcal{T}_{el,m} = \frac{\partial W'_{fld}\left[\vec{j}_{\{abc\}_s}, \theta_r\right]}{\partial \theta_m} = \frac{P}{2}\frac{\partial W'_{fld}\left[\vec{j}_{\{abc\}_s}, \theta_r\right]}{\partial \theta_r}. \tag{72}$$

In other words,

$$W'_{fld} = \mathcal{Y}\Big|_{S,V,\vec{N},\vec{j}}. \tag{73}$$

◊

Remark 11. (*Models of real machines*). The relation between $\mathcal{Y}$ and torque applies to any machine and can, in principle, deal with any functional dependence between $\vec{\lambda}$ and $\vec{j}$. Nonlinear $\vec{\lambda}\left[\vec{j}\right]$ relations [9, 10, 12] become of interest when saturation of the magnetic circuit has to be modeled. Hysteresis and the related energy losses pose further difficulties. ◊

5.3. The electric torque theorem in the $\{dq0\}$ frame

Translating Eq. (66) into the $\{dq0\}$ relies on relations between $\mathbf{K}[.]$-transformed current vectors which involve all three angles, β_s , β_r , θ_r. Translation is made remarkably simpler by Theorem 5.

Theorem 6. (*Electric torque in the dq0 frame*). For an ideal *DFI* machine, the electric torque in generator mode and in the $\{dq0\}$-frame is the following bilinear form for the matrix $\mathbf{A}_3$:

$$\mathcal{T}_{el,g} = -\frac{P}{2}\,\frac{3}{2}\,\frac{N_r}{N_s}L_{ms}\begin{bmatrix} j_{ds} & j_{qs} & \checkmark \end{bmatrix} \cdot \mathbf{A}_3 \cdot \begin{bmatrix} j'_{dr} \\ j'_{qr} \\ \checkmark \end{bmatrix} \tag{74}$$

which simplifies to

$$\mathcal{T}_{el,g} = +\frac{P}{2}\,\frac{3}{2}\,\frac{N_r}{N_s}L_{ms}\,\left(j_{ds}\;\; j'_{qr} - j_{qs}\;\; j'_{dr}\right). \tag{75}$$

Proof of Theorem 6. It suffices to combine Eqs. (66), (56) and (64). The matrix $\mathbf{A}_3$ makes the 3rd entries of current vectors not relevant (✓). ▷◁

6. A "realistic" machine model

Real machines deviate from the hypotheses which have led to the relatively simple form of the equations discussed so far. A satisfactory model shall account for one or more of the following features:

(*a*) the effects of tooth saliency and slots on the linked fluxes,

(*b*) deviations from three-fold symmetry,

(*c*) the instantaneous dependence of self-and mutual inductances on current, when the magnetic material is not linear,

(*d*) memory effect in a non-linear, hysteretic magnetic material.

Models which, step-wise, account for features (*a*) to (*d*) are "realistic" in the sense of Fitzgerald and Kingsley [3]. Features listed under (*a*) are relatively simple to model if three-fold symmetry is assumed: very briefly, higher harmonics are introduced which, because of linearity, can be dealt with separately. Instead, broken symmetry may be of some interest: the model outlined herewith focuses on feature (*b*) and consists of constructing a "realistic" mutual inductance matrix, then determining its algebraic (determinant, eigenvalues) and analytical (θ_r-derivative) properties.

Definition 6. (*Broken symmetry in the rotor*). At fixed θ_r the rotor *ar* axis forms angles θ_r, $\theta_r - \varphi$ and $\theta_r + \varphi$ with the *as*, *bs* and *cs* axes respectively. With ϵ_b and ϵ_c satisfying

$$0 \leq \left|\frac{3\epsilon_b}{2\pi}\right| \;,\; \left|\frac{3\epsilon_c}{2\pi}\right| < < 1 \tag{76}$$

the rotor *br* axis forms angles $\theta_r + \varphi + \epsilon_b$, $\theta_r + \epsilon_b$ and $\theta_r - \varphi + \epsilon_b$ with the *as*, *bs* and *cs* axes respectively. Similar relations hold for the rotor *cr* axis in terms of ϵ_c.

As a consequence the mutual inductance matrix is

$$\mathbf{L}_{sr}[\theta_r;\epsilon_b,\epsilon_c] = \begin{bmatrix} \cos[\theta_r] & \cos[\theta_r+\varphi+\epsilon_b] & \cos[\theta_r-\varphi+\epsilon_c] \\ \cos[\theta_r-\varphi] & \cos[\theta_r+\epsilon_b] & \cos[\theta_r+\varphi+\epsilon_c] \\ \cos[\theta_r+\varphi] & \cos[\theta_r-\varphi++\epsilon_b] & \cos[\theta_r+\epsilon_c] \end{bmatrix} . \quad (77)$$

Because of broken symmetry, $\mathbf{L}_{sr}[\theta_r;\epsilon_b,\epsilon_c]$ is no longer circulant. However, its column-wise entries add to zero and the following properties hold.

Proposition 6. (*Kernel, adjoint, zero divisors for general ϵ_b and ϵ_c*).

$$\begin{aligned} &\mathrm{Ker}[\mathbf{L}_{sr}[\theta_r;\epsilon_b,\epsilon_c]] = \mathfrak{K}_{sr} \ , \ \mathbf{Z}\left[\overrightarrow{f}[\theta]\right] = \overrightarrow{f}[\theta])(\overrightarrow{1} \ , \\ &\mathrm{adj}[\mathbf{L}_{sr}[\theta_r;\epsilon_b,\epsilon_c]] \in \mathfrak{Z} \ , \ \mathrm{Ker}[\mathrm{adj}[\mathbf{L}_{sr}[\theta_r;\epsilon_b,\epsilon_c]]] = \mathfrak{B} \ . \end{aligned} \quad (78)$$

To second order in ϵ_b and ϵ_c, $\mathbf{L}_{sr}[\theta_r;\epsilon_b,\epsilon_c]$ is approximated by

$$\begin{aligned} &\mathbf{L}_{sr}[\theta_r;\epsilon_b,\epsilon_c] \simeq \mathbf{G}^{(2)}_{\epsilon_b,\epsilon_c} = \\ &= \mathbf{C}^{(2)}_{\epsilon_b,\epsilon_c}\cos[\theta_r] + \mathbf{S}^{(2)}_{\epsilon_b,\epsilon_c}\sin[\theta_r] + \mathbf{C}^{(1)}_{\epsilon_b,\epsilon_c}\cos[\theta_r] + \mathbf{S}^{(1)}_{\epsilon_b,\epsilon_c}\sin[\theta_r] \ , \end{aligned} \quad (79)$$

where the four new matrices have to be defined. $\mathbf{C}^{(2)}_{\epsilon_b,\epsilon_c}$ and $\mathbf{S}^{(2)}_{\epsilon_b,\epsilon_c}$ are obtained from $\mathbf{C}$ and $\mathbf{S}$ of Eq. (27) when their second columns are multiplied by $\left(1-\frac{1}{2}\epsilon_b^2\right)$ and their third columns are multiplied by $\left(1-\frac{1}{2}\epsilon_c^2\right)$. Similarly, $\mathbf{C}^{(1)}_{\epsilon_b,\epsilon_c}$ and $\mathbf{S}^{(1)}_{\epsilon_b,\epsilon_c}$ derive from splitting the $\cos[\theta_r]$ and $\sin[\theta_r]$ terms in the following matrix

$$\begin{bmatrix} 0 & -\epsilon_b\sin[\theta_r+\varphi] & -\epsilon_c\sin[\theta_r-\varphi] \\ 0 & -\epsilon_b\sin[\theta_r] & -\epsilon_c\sin[\theta_r+\varphi] \\ 0 & -\epsilon_b\sin[\theta_r-\varphi] & -\epsilon_c\sin[\theta_r] \end{bmatrix} \mathbf{C}^{(1)}_{\epsilon_b,\epsilon_c}\cos[\theta_r] + \mathbf{S}^{(1)}_{\epsilon_b,\epsilon_c}\sin[\theta_r]. \quad (80)$$

As a consequence, a property can be stated about the derivative of $\mathbf{G}^{(2)}_{\epsilon_b,\epsilon_c}$.

Proposition 7. (*Differentiation as multiplication*). At least to second order in ϵ_b and ϵ_c, there exists a matrix $\mathbf{B}$, independent of θ_r, by which the differentiation of $\mathbf{G}^{(2)}_{\epsilon_b,\epsilon_c}$ is represented as multiplication

$$\left(\frac{\partial \mathbf{G}^{(2)}_{\epsilon_b,\epsilon_c}}{\partial\theta_r}\right)[\theta_r]\cdot\overrightarrow{j} = \mathbf{B}\cdot\mathbf{G}^{(2)}_{\epsilon_b,\epsilon_c}[\theta_r]\cdot\overrightarrow{j} \ , \ \forall\overrightarrow{j}\in\mathbb{R}^3 \ . \quad (81)$$

Such matrix complies with

$$\mathbf{B}^2\cdot\left(\mathbf{C}^{(2)}_{\epsilon_b,\epsilon_c}+\mathbf{C}^{(1)}_{\epsilon_b,\epsilon_c}\right) = -\left(\mathbf{C}^{(2)}_{\epsilon_b,\epsilon_c}+\mathbf{C}^{(1)}_{\epsilon_b,\epsilon_c}\right) \ . \quad (82)$$

In particular, to 1st order in ϵ_b and ϵ_c

$$\mathbf{B}^2\cdot\left(\mathbf{C}+\mathbf{C}^{(1)}_{\epsilon_b,\epsilon_c}\right) = -\left(\mathbf{C}+\mathbf{C}^{(1)}_{\epsilon_b,\epsilon_c}\right) \ . \quad (83)$$

7. Conclusion

In view of the large amounts of power converted from electric to mechanical or *vice-versa*, mathematical methods for electric machinery have to undergo continuous investigation and, possibly, improvement. Model errors, although "small" in relative terms, may translate into large amounts of mishandled power. To date, control methods and the corresponding algorithms are satisfactory in the low frequency (tens of Hz) range: better performance is needed to deal with the higher (thousands of Hz) frequency components of a transient [18]. This work has focused on the basics of the ideal *DFI* machine model, where linearity and three-fold symmetry are the main features. As a result, the electric torque theorem has been stated in the $\{dq0\}$ frame without any restriction on the $\vec{j}$'s. The product of matrices formula has accordingly simplified the proof. Some of the properties derived in the ideal case have been shown to hold even if symmetry is broken.

Author details

Giovanni F. Crosta[1,2*] and Goong Chen[3]

1 Department of Earth- and Environmental Sciences, University of Milan-Bicocca, Milan, Italy

2 Biblioteca Quadrelli-Crosta, Milan, Italy

3 Department of Mathematics, Texas A&M University, College Station, TX, USA

*Address all correspondence to: giovanni_crosta@uml.edu

References

[1] Blondel A. Synchronous Motor and Converters - Part III. New York: McGraw-Hill; 1913

[2] Park RH. Two-reaction theory of synchronous machines - Part I. AIEE Transactions. 1929;**48**(2):716-730

[3] Fitzgerald AE, Kingsley C Jr. Electric Machinery. 2nd ed. New York, NY: McGraw-Hill; 1961. p. 568

[4] Krause PC. Analysis of Electric Machinery. New York, NY: McGraw-Hill; 1986

[5] Hau E. Windkraftanlagen - 4. Auflage. Berlin: Springer; 2008

[6] Elkington K. Modelling and Control of Doubly Fed Induction Generators in Power Systems [Thesis]. KTH School of Engineering: Stockholm, SE; 2009

[7] Blaschke F. The principle of field orientation as applied to the new transvector closed loop control system for rotating field machines. Siemens Review. 1972;**39**:217-220

[8] Willems JL. A system theory approach to unified electric machine analysis. International Journal of Control. 1972;**15**(3):401-418

[9] Liu XZ, Verghese GC, Lang JH, Önder MK. Generalizing the Blondel-Park transformation of electrical machines: Necessary and sufficient conditions. IEEE Transactions on Circuits and Systems. 1989;**36**(8): 1058-1067

[10] Youla DC, Bongiorno JJ Jr. A Floquet theory of the general linear rotating machine. IEEE Transactions on Circuits and Systems. 1980;**27**(1):15-19

[11] Maschke BM, van der AJS, Breedveld PC. An intrinsic Hamiltonian formulation of the dynamics of LC-circuits. IEEE Transactions on Circuits and Systems I: Fundamental Theory and Applications. 1995;**42**(2):73-82

[12] Sullivan CR, Sanders SR. Models for induction machines with magnetic saturation of the main flux path. IEEE Transactions on Industry Applications. 1995;**31**(4):907-917

[13] Eberard D, Maschke B, van der AJ S. Conservative systems with ports on contact manifolds. In: Proceedings of the 16th IFAC World Congress; July 4–8, 2005. IFAC: Prague, CZ; 2005

[14] Marcus M, Minc H. A Survey of Matrix Theory and Matrix Inequalities. New York, NY: Dover; 1992. p. 180

[15] Rueppel AR. Stream ciphers. In: Simmons G, editor. Contemporary Cryptology the Science of Information Integrity. New York, NY: IEEE Press; 1992. pp. 65-134

[16] Schlesinger E. Algebra lineare e geometria. Bologna, IT: Zanichelli; 2011. p. 502

[17] Altmann SL. Rotations, Quaternions and Double Groups. Mineola, NY: Dover; 2005

[18] Capolino GA. Progrès dans les machines électriques tournantes à courant alternatif: du principe de Galileo Ferraris à la technologie actuelle. Réunion Virtuelle du groupe IEEE France Section Life Members Affinity Group. Answer to a question on control methods by Giovanni F Crosta; Oct. 26th, 2021

Chapter 10

A New Approach to Solve Non-Fourier Heat Equation via Empirical Methods Combined with the Integral Transform Technique in Finite Domains

Cristian N. Mihăilescu, Mihai Oane, Natalia Mihăilescu, Carmen Ristoscu, Muhammad Arif Mahmood and Ion N. Mihăilescu

Abstract

This chapter deals with the validity/limits of the integral transform technique on finite domains. The integral transform technique based upon eigenvalues and eigenfunctions can serve as an appropriate tool for solving the Fourier heat equation, in the case of both laser and electron beam processing. The crux of the method consists in the fact that the solutions by mentioned technique demonstrate strong convergence after the 10 eigen-values iterations, only. Nevertheless, the method meets with difficulties to extend to the case of non-Fourier equations. A solution is however possible, but it is bulky with a weak convergence and requires the use of extra-boundary conditions. To surpass this difficulty, a new mix approach is proposed with this chapter resorting to experimental data, in order to support a more appropriate solution. The proposed method opens in our opinion a beneficial prospective for either laser or electron beam processing.

Keywords: non-Fourier equation, integral transforms technique, eigenfunctions and values, experimental data

1. Introduction

1.1 Mathematical background

The heat equation can be solved in a simpler mode *via* the Fourier heat equation, which involves the propagation of heat waves with infinite speed. This hypothesis is in particular valid for many applications, such as laser-metal interaction in the frame of two-temperature model [1, 2].

The solution of Fourier equations can be inferred using different mathematical techniques via Green function, integral, Laplace transform, or complex analysis. The

IntechOpen

predictions of the solutions given by the mentioned methods are of analytical or semianalytical nature and confirm the experimental data for certain situations such as laser–metal interaction.

One basically assumes that the heat waves propagation speed is inversely proportional to the square root of the relaxation time. A smaller relaxation time leads to higher heat speed waves, resulting in a good Fourier approximation. If one requires however a more accurate description of experimental data, one should introduce a more exact method to solve the non-Fourier equation involving a finite heat wave speed.

A mixed solution of the non-Fourier equation combines the theoretical method of finite integral transforms with information from experimental data. Thus, two additional boundary conditions can be imposed, which will lead to a semianalytical solution of the non-Fourier equation. The finite domains of the integral transform method for Fourier equations are eigenfunctions and values, which reach after 10 iterations a quite conform solution for the Fourier equation [3–6]. This method is applied to the non-Fourier equation, and the final form is obtained, with the support of experimental results.

A new heat transfer model was adopted in order to unify the thermal field distribution in both laser and electron beam processing. An analytical solution using non-Fourier heat equation has been developed corresponding to boundary conditions in the case of material processing. The model has been compared with the experimental data obtained using an in-house developed facility. A simplified and easy-to-use model via MATHEMATICA software stands for the novelty of the current work.

2. Non-Fourier equation

The non-Fourier equation is hyperbolic and can be written as:

$$\frac{\partial^2 T}{\partial r^2}+\frac{1}{r}\frac{\partial T}{\partial r}+\frac{\partial^2 T}{\partial z^2}+\frac{1}{r^2}\frac{\partial^2 T}{\partial \varphi^2}-\frac{1}{\gamma}\frac{\partial T}{\partial t}-\frac{\tau_0}{\gamma}\frac{\partial^2 T}{\partial t^2}=-\frac{P(r,\varphi,z,t)}{K}. \tag{1}$$

Here, T is target temperature, r is the radial coordinate, z is the spatial coordinate on the direction of laser beam propagation, t is time, τ_0 is relaxation time, and γ is thermal diffusivity. P stands for the source term, φ is the angular coordinate, while K is the target thermal conductivity. For a simple general solution, one assumes a cylindrical target with angular symmetry, under irradiation with a Gaussian laser beam with the center at $r = z = 0$. In this case, the temperature does not depend on φ:

$$\frac{\partial T}{\partial \varphi}=0.=>T=T(r,z,t). \tag{2}$$

The corresponding boundary conditions are:

$$K\frac{\partial T(r,z,t)}{\partial r}|_{r=b}+hT(b,z,t)=0. \tag{3}$$

Here $r = b$ is the cylinder radius, while h is the heat transfer coefficient. The boundary conditions for the z coordinate are:

$$K\frac{\partial T(r,z,t)}{\partial z}|_{z=0}-hT(r,0,t)=0, \tag{4}$$

and

$$K\frac{\partial T(r,z,t)}{\partial z}|_{z=a} + hT(r,a,t) = 0, \tag{5}$$

where a is the cylinder length. We will pass on the effective solution of the equation. The operator D_r was defined as:

$$D_r = \frac{\partial^2}{\partial r^2} + \frac{1}{r}\left(\frac{\partial}{\partial r}\right). \tag{6}$$

This applies to Eq. (1) with the boundary conditions:

$$\frac{\partial^2 K_r}{\partial r^2} + \frac{1}{r}\frac{\partial K_r}{\partial r} + \mu^2 K_r = 0, \tag{7}$$

and

$$K\frac{\partial K_r(r,z,t)}{\partial r}|_{r=b} + hK_r(b,z,t) = 0. \tag{8}$$

Eqs. (7) and (8) corroborate to Eq. (9):

$$\frac{h}{K}J_0(\mu_i b) - \mu_i J_1(\mu_i b) = 0. \tag{9}$$

One can deduce, based upon the theory of finite integral transforms, the eigenfunction $\tilde{K}_r(r,\mu_i)$ corresponding to the eigenvalue μ_i:

$$\tilde{K}_r(r,\mu_i) = J_O(\mu_i r)r\frac{1}{C_i}, \tag{10}$$

where the normalization constant is given by:

$$C_i(r) = \int_0^b r\,K_r^2(r,\mu_i)dr = \frac{b^2}{2\mu_i^2}\left(\frac{h^2}{k^2} + \mu_i^2\right)J_0^2(\mu_i b). \tag{11}$$

One defines:

$$\tilde{T}(\mu_i,z,t) = \frac{1}{C_i}\int_0^b T(r,z,t)r\,J_0(\mu_i\,r)dr, \tag{12}$$

and

$$\tilde{P}(\mu_i,z,t) = \frac{1}{C_i}\int_0^b P(r,z,t)\,r\,J_0(\mu_i r)dr. \tag{13}$$

Eq. (1) in this case converts to:

$$-\mu_i^2\cdot\tilde{T} + \frac{\partial^2\tilde{T}}{\partial z^2} - \frac{1}{\gamma}\frac{\partial\tilde{T}}{\partial t} - \frac{\tau_0}{\gamma}\frac{\partial^2\tilde{T}}{\partial t^2} = -\frac{\tilde{P}}{K}. \tag{14}$$

One obtains for z coordinate via similar mathematical calculation:

$$\frac{\partial^2 K_z}{\partial z^2} + \lambda^2 K_z = 0, \tag{15}$$

and

$$\left[K \frac{\partial K_Z}{\partial z} - h\, K_z \right]_{z=0} = 0, \tag{16}$$

as well as:

$$\left[K \frac{\partial K_Z}{\partial z} + h\, K_z \right]_{z=a} = 0. \tag{17}$$

The − and + signs for h in Eqs. (16) and (17) denote the target heat absorption and emission, respectively. One has:

$$K_z(z, \lambda) = cos\,(\lambda\, z) + \frac{h}{\lambda K} sin\,(\lambda\, z), \tag{18}$$

and

$$2 \cot\left(\lambda_j a\right) = \frac{\lambda_j k}{h} - \frac{h}{\lambda_j k}. \tag{19}$$

Here, λ_j denotes eigenvalues along the z-axis. From theory [7], it follows:

$$\tilde{K}_z\left(z, \lambda_j\right) = \frac{1}{C_j} K_z\left(z, \lambda_j\right), \tag{20}$$

and

$$C_j - \int_0^a K_z^2\left(z, \lambda_j\right) dz. \tag{21}$$

Note that Eqs. (12) and (13) discuss the eigenvalues along the r-axis, only. After introducing the eigenvalues along the z-axis, one can step ahead to generalize Eqs. (12) and (13) as:

$$\overline{T}\left(\mu_i, \lambda_j, t\right) = \frac{1}{C_i C_j} \int_0^a \int_0^b T(r, z, t) r K_r(\mu_i, r) K_z\left(\lambda_j, z\right) dr dz, \tag{22}$$

and

$$\overline{P}\left(\mu_i, \lambda_j, t\right) = \frac{1}{C_i C_j} \int_0^a \int_0^b P(r, z, t) r K_r(\mu_i, r) K_z\left(\lambda_j, z\right) dr dz. \tag{23}$$

Eq. (1) becomes now:

$$\left(\mu_i^2 + \lambda_j^2\right)\overline{T} + \frac{1}{\gamma}\frac{\partial \overline{T}}{\partial t} + \frac{\tau_0}{\gamma}\frac{\partial^2 \overline{T}}{\partial t^2} = \frac{\overline{P}}{K}. \tag{24}$$

We next applied the direct and inverse Laplace integral transform to solve Eq. (1) in relation to time. C[1] and C[2] stand for the normalizing coefficients with respect to the experimental data. The results are as follows:

$$T(r,z,t,\tau_0) = \sum_{i=1}^{10}\sum_{j=1}^{10}\left[\begin{array}{l}\dfrac{P(\mu_i,\lambda_j)}{\mu_i^2+\lambda_j^2} + C[1]\mathrm{e}^{\frac{\left(-\frac{1}{\gamma}-\sqrt{\frac{1}{\gamma^2}-4\mu_i^2\frac{\tau_0}{\gamma}-4\lambda_j^2\frac{\tau_0}{\gamma}}\right)\cdot t}{\frac{2\tau_0}{\gamma}}} + C[2] \\ \mathrm{e}^{\frac{\left(-\frac{1}{\gamma}+\sqrt{\frac{1}{\gamma^2}-4\mu_i^2\frac{\tau_0}{\gamma}-4\lambda_j^2\frac{\tau_0}{\gamma}}\right)t}{\frac{2\tau_0}{\gamma}}}\end{array}\right](K_r(\mu_i,r)K_z(\lambda_j,z)). \tag{25}$$

and

$$T(\mu_i,\lambda_j) = \frac{1}{C_iC_j}\int_0^a\int_0^b P(r,z,t)\, rK_r(\mu_i,r)K_z(\lambda_j,z)drdz. \tag{26}$$

We finally mention that for an intermediate point in the experimental curve, one has:

$$T(r,z,t) = T(\tau_0, C[1], C[2], r, z, t). \tag{27}$$

With the boundary conditions:

$$T(r,z,0,C[1],C[2],\tau_0) = 22^{\circ}\mathrm{C}. \tag{28}$$

$$T(r,z,\infty,C[1],C[2],\tau_0) = 22^{\circ}\mathrm{C}. \tag{29}$$

3. Two-temperature model in the non-Fourier version

The two-temperature model (TTM) is based upon two coupled equations:

$$AT_e\left(\frac{\partial T_e}{\partial t}\right) + \frac{K\tau_0}{\gamma}\left(\frac{\partial^2 T_e}{\partial t^2}\right) = K\left(\frac{\partial^2 T_e}{\partial x^2} + \frac{\partial^2 T_e}{\partial y^2} + \frac{\partial^2 T_e}{\partial z^2}\right) - G(T_e - T_i) + P_a\left(\vec{r},t\right), \tag{30}$$

$$C_i\left(\frac{\partial T_i}{\partial t}\right) = G(T_e - T_i). \tag{31}$$

Here T_e and T_i stand for the electron and phonon temperatures, respectively. G is the coupling factor between electrons and phonons. $P_a\left(\vec{r},t\right)$ is the heat source, which is induced via laser-metal interaction. The interaction could be considered of either classical or steady-state quantum mechanical type. A is the electron heat capacity, and K is the thermal conductivity of the metal. According to Ref. [8], G can be determined from:

$$G = \frac{\pi^2 mNv^2}{6\tau T_i}\left(\frac{T_e}{T_i}\right)^4 \times \int_0^{T_e/T_d}\left[x^4/(e^x-1)\right]dx, \tag{32}$$

where m is the electron mass, N is the conduction electron density, v is the velocity of sound in the metal, τ is the electron–phonon collision time, and T_D is the Debye temperature. The exact data for each metal (Cu, Ag, Al, or Fe) are available from text books and current literature (e.g., [8–10]). As for all metals, $C_i >> A$, one may assume, in a first approximation, that:

$$K\left(\frac{\partial^2 T_e}{\partial x^2}+\frac{\partial^2 T_e}{\partial y^2}+\frac{\partial^2 T_e}{\partial z^2}\right)-C_i\frac{\partial T_i}{\partial t}-\frac{K\tau_0}{\gamma}\left(\frac{\partial^2 T_e}{\partial t^2}\right)=-P_a\left(\vec{r},t\right). \tag{33}$$

According to the Nolte model [2], one has:

$$T_i=\kappa T_e, \tag{34}$$

where

$$\kappa=\frac{\tau_L}{\tau_L+\tau_i}. \tag{35}$$

Here, τ_i is the lattice cooling time while τ_L is the pulse duration time. Consequently, one has:

$$\frac{\partial T_i}{\partial t}=\kappa\frac{\partial T_e}{\partial t}. \tag{36}$$

Eq. (3) can be rewritten as:

$$K\left(\frac{\partial^2 T_e}{\partial x^2}+\frac{\partial^2 T_e}{\partial y^2}+\frac{\partial^2 T_e}{\partial z^2}\right)-C_i\kappa\frac{\partial T_e}{\partial t}-\frac{K\tau_0}{\gamma}\left(\frac{\partial^2 T_e}{\partial t^2}\right)=-P_a\left(\vec{r},t\right). \tag{37}$$

It follows that:

$$\left(\frac{\partial^2 T_e}{\partial x^2}+\frac{\partial^2 T_e}{\partial y^2}+\frac{\partial^2 T_e}{\partial z^2}\right)-\frac{1}{\gamma}\frac{\partial T_e}{\partial t}-\frac{\tau_0}{\gamma}\left(\frac{\partial^2 T_e}{\partial t^2}\right)=-\frac{-P_a\left(\vec{r},t\right)}{K} \tag{38}$$

with

$$\gamma=\frac{K}{C_i\kappa}. \tag{39}$$

Under the most general form, the heat source reads as:

$$P_a=\sum\nolimits_{m,n} I_{mn}(y,z)\left(\alpha_{mn}e^{-\alpha_{mn}x}\right)(1-r_{Smn})+r_{Smn}\delta(x)+qc)(H(t)-H(t-t_0)). \tag{40}$$

Here, $I_{mn}(y,z)$ stands for the laser transverse mode {m,n} while α_{mn} is the linear absorption coefficient, and r_{Smn} is the surface absorption coefficient. The quantum corrections (qc) are steady state for the respective mode. H stands for the step function, and t_0 for the exposure time. One model explains the continuous laser beam irradiation, while the other one, more realistic in our opinion, illustrates the laser beam in pulse form. The equivalence between the two models requires therefore that the intensity *versus* time plot should cover the same area.

In order to make a comparison with experiments, one needs besides analytical description, concrete numerical values. The next step is therefore to estimate the eigenvalues, numerically. For this purpose, Eq. (7) can be solved using the integral transform technique, and eigenfunctions and eigenvalues could be calculated. One has three differential equations as follows (K represents the eigenfunctions, while λ, μ, and ξ are the eigenvalues) [5]:

$$\frac{\partial^2 K_x}{\partial x^2} + \lambda_i^2 K_x = 0, \tag{41}$$

$$\frac{\partial^2 K_y}{\partial y^2} + \mu_j^2 K_y = 0, \tag{42}$$

$$\frac{\partial^2 K_z}{\partial z^2} + \xi_k^2 K_z = 0. \tag{43}$$

The final solutions could be achieved on the basis of Eqs. (41)–(43):

$$K_x = cos\,(\lambda_i x) + \frac{h}{K\lambda_i} sin\,(\lambda_i x), \tag{44}$$

$$K_y = cos\left(\mu_j y\right) + \frac{h}{K\mu_j} sin\left(\mu_j y\right), \tag{45}$$

$$K_z = cos\,(\xi_k z) + \frac{h}{K\xi_k} sin\,(\xi_k z). \tag{46}$$

The boundary conditions are:

$$\left[\frac{\partial K_x}{\partial x} - \frac{hK_x}{K}\right]_{x=0} = 0; \left[\frac{\partial K_x}{\partial x} + \frac{hK_x}{K}\right]_{x=a} = 0, \tag{47}$$

$$\left[\frac{\partial K_y}{\partial y} + \frac{hK_y}{K}\right]_{y=0} = 0; \left[\frac{\partial K_y}{\partial y} + \frac{hK_y}{K}\right]_{y=b} = 0, \tag{48}$$

$$\left[\frac{\partial K_z}{\partial_z} + \frac{hK_z}{K}\right]_{z=0} = 0; \left[\frac{\partial K_z}{\partial z} + \frac{hK_z}{K}\right]_{z=c} = 0. \tag{49}$$

Here, a, b, and c are the metal sample dimensions. As for the boundary conditions, the eigenvalues (h is the heat transfer coefficient) can be inferred from Eqs. (47) to (49), as:

$$2\,cot\,(\lambda_i a) = \frac{\lambda_i K}{h} - \frac{h}{K\lambda_i}, \tag{50}$$

$$2\,cot\left(\mu_j b\right) = \frac{\mu_j K}{h} - \frac{h}{K\mu_j}, \tag{51}$$

$$2\,cot\,(\xi_k c) = \frac{\xi_k K}{h} - \frac{h}{K\xi_k}. \tag{52}$$

The solution is obtained via integral transform technique as:

$$T(x,y,z,t,C[1],C[2],\tau_0)$$

$$= \sum_{i=1}^{10}\sum_{j=1}^{10}\sum_{k=1}^{10}\left[\begin{array}{l} \dfrac{P(\mu_i,\lambda_j,\xi_k)}{\mu_i^2+\lambda_j^2+\xi_k^2} + C[1]e^{\frac{\left(-\frac{1}{\gamma}-\sqrt{\frac{1}{\gamma^2}-4\mu_i^2\frac{\tau_0}{\gamma}-4\lambda_j^2\frac{\tau_0}{\gamma}-4\xi_k^2\frac{\tau_0}{\gamma}}\right)\cdot t}{\frac{2\tau_0}{\gamma}}} + C[2] \\ e^{\frac{\left(-\frac{1}{\gamma}+\sqrt{\frac{1}{\gamma^2}-4\mu_i^2\frac{\tau_0}{\gamma}-4\lambda_j^2\frac{\tau_0}{\gamma}-4\xi_k^2\frac{\tau_0}{\gamma}}\right)\cdot t}{\frac{2\tau_0}{\gamma}}} \end{array}\right] \quad (53)$$

$$\times\left(K_x(\mu_i,x)K_y(\lambda_j,y)K_z(\xi_k,z)\right)$$

The advantage of Eq. (53) in our model is related to a quick converging series. Thus, after 10 iterations, the solution's accuracy reaches already 10^{-2} K in the case of thermal distribution [11].

4. Experimental details

The experimental setup is operated by a Nd:YAG pulsed laser source (λ = 355 nm) (Surelite II from Continuum), generating pulses of 6 ns duration with (130 $\pm$ 0.6) mJ energy at a frequency repetition rate of 10 Hz. The laser beam had a spatial top-hat distribution. The laser beam was focused onto the metallic target surface by a lens with 240 mm focal lengths. An Al bulk target of (10 $\times$ 10 $\times$ 5) mm^3 was used in experiments. The laser fluence was set at ~7.5 J/cm^2 to surpass the ablation threshold but also to avoid the excessive plasma formation. A crater of 18 μm depth was dig into the sample after the application of 1000 subsequent laser pulses, as checked up by a Vernier Caliper instrument. During the multipulse laser irradiation, the thermal distribution was monitored on the sample back side via a FLUKA thermocouple connected to a computer having Lab view software, while the sample was irradiated at the top. All experiments were performed on an in-house developed equipment at Laser Department, National Institute for Laser, Plasma and Radiation Physics (INFLPR), Magurele, Romania.

5. Results and discussion

Experiments and simulations were carried out during the heating of a metallic target. The boundary conditions were described by Eq. (28). In all figures, the experimental data are plotted with dots while the simulations are represented by a continuous line. Relaxation time,τ_0, was assumed 0.5 ps (**Figure 1**), 1 ns (**Figure 2**), and 1 μs (**Figure 3**), respectively. For simulation, a heat transfer coefficient = 3 $\times$ 10^{-7} W mm^{-2} K^{-1} was selected. As known [12–14], for a very low heat transfer coefficient, the eigenvalues are positive very small numbers, resulting in a linear thermal distribution curve, as visible in **Figures 1–3**.

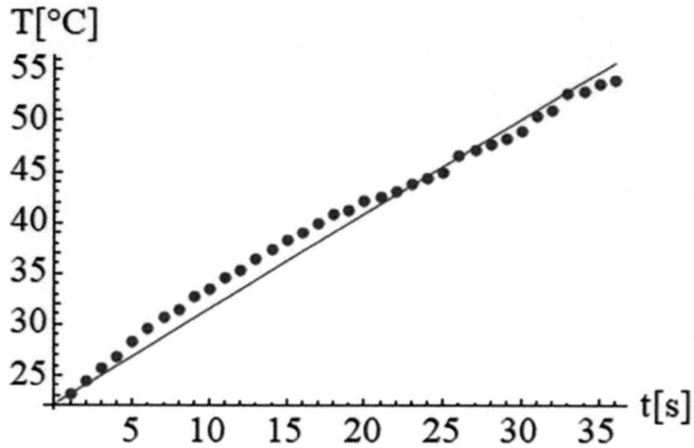

Figure 1.
Time evolution of temperature for a relaxation time of 0.5 ps: experiments (dotted line) vs. simulation (continuous line).

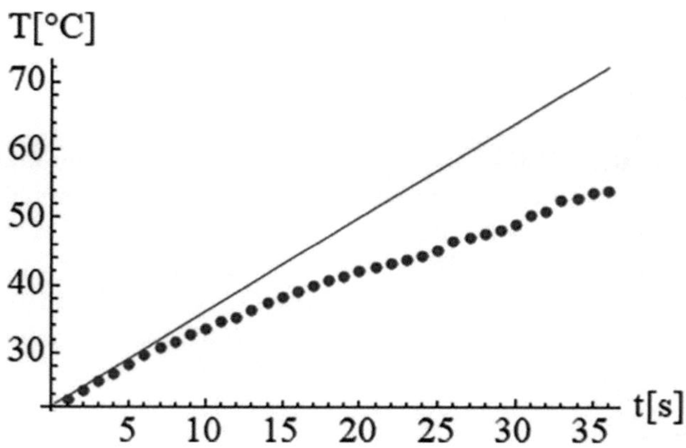

Figure 2.
Time evolution of the temperature for a relaxation time of 1 ns: experiments (dotted line) vs. simulation (continuous line).

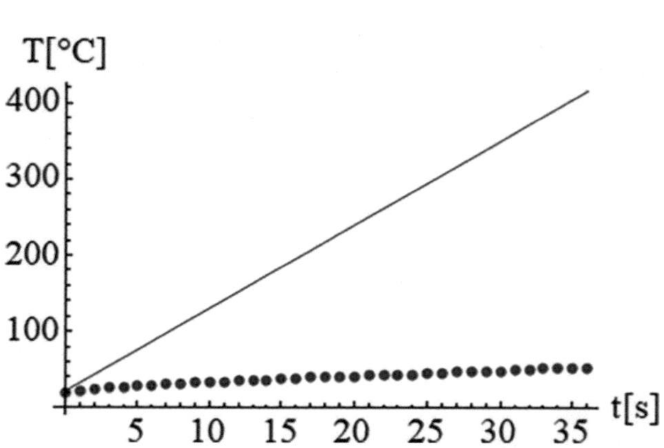

Figure 3.
Time evolution of the temperature for a relaxation time of 1 μs: experiments (dotted line) vs. simulation (continuous line).

The best agreement between theory and experiment was achieved for a relaxation time, τ_0, of 0.5 ps, as visible from **Figure 1**. We note that this is in accordance with available literature on subject [15].

6. Conclusions and outlook

The two-temperature model was generalized to the case of the non-Fourier approach via the electron-phonon relaxation time. Boundary conditions, Eq. (28) for heating and Eq. (29) for cooling, were considered to this purpose. The obtained solutions prove useful for experimental data analysis. The mathematical method belongs to the eigenvalues and functions family, while details on software are available from Ref. [6].

The exact nature of the metallic target (in our case aluminum) could be detected from the electron-phonon relaxation time using integral transform technique mix via acquired experimental data. The method can be extended to any experimental sample (metal) with the high accuracy.

Acknowledgements

CNM, MO, NM, and CR acknowledge for financial support by Romanian Ministry of Research, Innovation and Digitalization, under Romanian National NUCLEU Program LAPLAS VI–contract no. 16N/2019. CNM, NM, CR, and INM thank for the financial support from a grant of the Romanian Ministry of Education and Research, CNCS-UEFISCDI, project number ID code RO-NO-2019-0498 and UEFISCDI under the TE_196/2021 and PED_306/2020. M.A.M. received financial support from the European Union's Horizon 2020 (H2020) research and innovation program under the Marie Skłodowska–Curie grant agreement no. 764935.

Author details

Cristian N. Mihăilescu[1], Mihai Oane[1], Natalia Mihăilescu[1*], Carmen Ristoscu[1*], Muhammad Arif Mahmood[1,2] and Ion N. Mihăilescu[1]

1 National Institute for Laser, Plasma and Radiation Physics, Măgurele, Ilfov, Romania

2 Mechanical Engineering Program, Texas A&M University at Qatar, Doha, Qatar

*Address all correspondence to: natalia.serban@inflpr.ro and carmen.ristoscu@inflpr.ro

References

[1] Schoenlein RW, Lin WZ, Fujimoto JG, Eesley GL. Femtosecond studies of nonequilibrium electronic processes in metals. Physical Review Letters. 1987;**58**(16):1680-1683. DOI: 10.1103/PhysRevLett.58.1680

[2] Nolte S, Momma C, Jacobs H, Tunnermann A, Chichkov BN, Wellegehaussen B, et al. Ablation of metals by ultrashort laser pulses. Journal Optical Society of America B. 1997;**14**(10): 2716-2722. DOI: 10.1364/JOSAB.14.002716

[3] Oane M, Peled A, Medianu R. Notes on Laser Processing. Germany: Lambert Academic Publishing; 2013. ISBN 978-3-659-487-48739-2

[4] Oane M, Ticoş D, Ticoş CM. Charged particle beams processing versus laser processing. Germany: Scholars' Press; 2015. ISBN 978-3-639-66753-0

[5] Oane M, Sporea D. Temperature profiles modelling in IR optical components during high power laser irradiation. Infrared Physics & Technology. 2001;**42**(1):31-40. DOI: 10.1016/S1350-4495(00)00065-7

[6] Oane M, Mahmood MA, Popescu A. A state-of-the-art review on integral transform technique in laser-material interaction: Fourier and non-Fourier heat equations. Materials. 2021;**14**(16): 4733. DOI: 10.3390/ma14164733

[7] Koshlyakov NS, Smirnov MM, Gliner EB. Differential Equation of Mathematical Physics. Amsterdam: North-Holland; 1964

[8] American Institute of Physics Handbook. 3rd ed. New York: McGraw-Hill; 1972

[9] Du G, Chen F, Yang Q, Jinhai S, Hou X. Ultrafast temperature evolution in Au film under femtosecond laser pulses irradiation. Optics Communications. 2010;**283**:1869-1872. DOI: 10.1016/J.OPTCOM.2009.12.038

[10] Damian V, Oane M, Buca A. The Fourier approach of the two temperature model for laser beam-metal interaction: Experiment versus theory. Lasers in Engineering. 2016;**33**(1-3):181-186

[11] Visan T, Sporea D, Dumitru G. Computing method for evaluating the absorption coefficient of infrared optical elements. Infrared Physics & Technology. 1998;**39**(5):335-346

[12] Oane M, Vutova K, Mihailescu IN, Donchev V, Florescu G, Munteanu L, et al. The study of vacuum influence on spatial-temporal dependence of thermal distributions during laser-optical components interaction. Vacuum. 2012; **86**:1440-1442

[13] Oane M, Toader D, Iacob N, Ticos CM. Thermal phenomena induced in a small tungsten sample during irradiation with a few MeV electron beam: Experiment versus simulations. Nuclear Instruments and Methods in Physics Research B. 2014;**337**:17-20

[14] Oane M, Toader D, Iacob N, Ticos CM. Thermal phenomena induced in a small graphite sample during irradiation with a few MeV electron beam: Experiment versus theoretical simulations. Nuclear Instruments and Methods in Physics Research B. 2014;**318**:232-236

[15] Girardeau-Montaut JP, Afif M, Girardeau-Montaut C, Moustaizis SD, Papadogiannis N. Aluminium electron-phonon relaxation-time measurement from subpicosecond nonlinear single-photon photoelectric emission at 248 nm. Applied Physics A. 1996;**62**:3-6. DOI: 10.1007/BF01568079

Chapter 11

Advanced Methods for Solving Nonlinear Eigenvalue Problems of Generalized Phase Optimization

Mykhaylo Andriychuk

Abstract

In the process of solving the problems of generalized phase optimization the necessity to apply an eigenvalue approach often appears. The practical statement of the optimization problems consists of using the amplitude characteristics of functions that are sought. The usual way of optimization is deriving the Euler equation of the functional, which is used as criterion of optimization. As a rule, such equation is an integral one. It is worth pointing out that the integral equations of the generalized phase optimization are nonlinear ones. The characteristic property of such equations is non-uniqueness of solutions and their branching or bifurcation. The determination of branching solutions leads to the investigation of the corresponding homogeneous equations and the respective eigenvalue problem. This problem is nonlinear because of specificity of the statement of the optimization problem. The study of the above problem allows us to determine a set of points, in which the respective eigenvalues are equal to unity that determines the branching points of solutions. The data of calculations testify to the ability of the approach proposed to determine the solutions of nonlinear equations numerically with not large computations.

Keywords: nonlinear optimization, variational approach, radiation characteristic, nonlinear eigenvalue problem, bifurcation of solutions, computational modeling

1. Introduction

The nonlinear eigenvalue approach is used in this chapter for the study of the properties to solutions of the generalized phase problem related to the synthesis of radiation systems through the incomplete data. Such incompleteness is considered here in the example of an indeterminate phase characteristic of function, which characterizes the radiation of the plane antenna arrays.

The problems with an indeterminate phase of the wave field arise in various applications and are widely described in the literature. The most well-known of these is the so-called phase problem (see, for example, [1–4]). It consists in restoring the phase distribution (argument) of the Fourier transform of a finite function by its amplitude (module) given (measured) along the entire real axis. This problem belongs to the classical problems of recovery (identification) and requires the conditions of existence of a unique solution.

In this chapter, another class of inverse problems is considered, and it can be termed optimization (design) problems. In sense of the Fourier transform, this can be, for example, the problem of finding such a finite complex function, the modulus of its Fourier transform satisfies a certain requirement (e.g., is close to a given positive function). As a rule, such requirements are formulated in the variational form, as the minimization of certain functionals. Obviously, such a formulation does not require a uniform solution. On the contrary, the existence of many solutions is often desirable because the above allows many degrees of freedom to determine an appropriate solution. The characteristic applications of such phase optimization problems include the theory of power transmission lines, field converters, antennas and resonators. The first works dealing with nonlinear inverse problems of such type appeared in the second part of the last century (see, for instance [5–10]).

In mathematical terms, problems of this type are reduced to the nonlinear integral equations of Hammerstein type [11–13]. They contain a linear kernel and a nonlinear multiplier that depends on a complex unknown function as an integrand. As a rule, the argument (phase) of this function appears there separately from the module. Similar equations are found in the literature in the context of the mentioned phase problem [14, 15]. They have different solutions, and the study of their structure and process of branching or bifurcation is an interesting mathematical problem [16].

Due to their nonlinearity, the problems under consideration require the development and application of special analytical and numerical methods for their solving. Along with the iterative methods that simulate the physical processes of field formation, the various modifications of Newton's method could be the most promising in this direction [17]. One such modification, which uses solving the nonlinear eigenvalue problems and searching for the zero curves of respective determinants, is proposed in this Chapter. It allows simultaneously with the finding of the branch of solutions to detect the presence of branching points on it and to determine them approximately, provided by this the initial approximations for more accurate calculation.

The nonlinear eigenvalue problems arise in pure and applied mathematics, as well as in the different areas of science that investigate the nonlinear phenomena [18, 19]. A variety of analytical-numerical methods have been elaborated till now for solving the nonlinear problems in acoustics, electrodynamics, fluid dynamics and other areas of applied science [20, 21]. The methods, developed until that time, were focused mainly on solving one-dimensional problems. The difficulties of analytical and computational nature appear if to apply them to a multidimensional problem. The method of implicit function is one of effective tools that been applied for solving the two- and three-dimensional nonlinear eigenvalue problem in the last two decades [22–24]. The extension of this method, which leads to solving the Cauchy problem (21) and (22), we apply in Section 3 to solve the nonlinear two-dimensional eigenvalue problem.

2. The operators of direct electrodynamics problem

In the physical relation, the radiation system represents the plane array with the rectangular or hexagonal placement of radiators. Firstly, we consider the array with the rectangular ordering of separate elements (**Figure 1a**).

Consider a plane rectangular array consisting of $N_2 \cdot M_2 = (2N + 1) \cdot (2M + 1)$ identical elements (radiators), which are located in the xOy plane of the Cartesian coordinate system equidistantly for each of the coordinates. Since the radiators are identical, it is possible to formulate the synthesis problem not for the whole three-

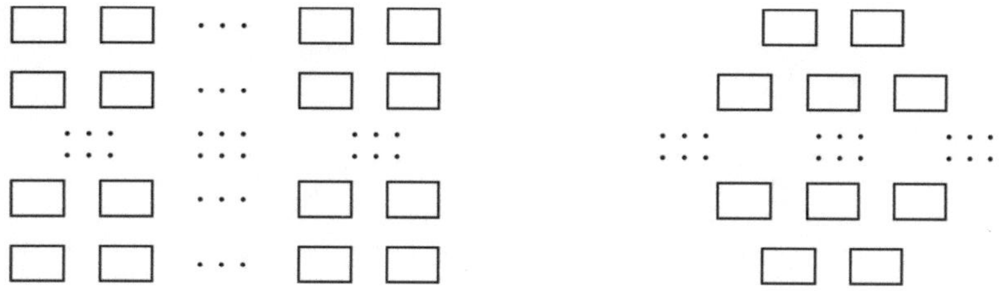

Figure 1.
The placement of elements for the considered arrays. a) rectangular. b) hexagonal.

dimensional vector directivity pattern (DP), but only for some complex scalar function $f(x_1, x_2)$ that is termed as the array multiplier. This function for a rectangular equidistant array has the form [25]:

$$f(x_1, x_2) = A\mathbf{I} \equiv \sum_{n=-N}^{N} \sum_{m=-M}^{M} I_{nm} e^{i(c_1 n x_1 + c_2 m x_2)} \tag{1}$$

where $\mathbf{I} = \{I_{nm}, -N \le n \le N, -M \le m \le M\}$ is a set of excitations (currents) in the array's elements, $x_1 = \sin\theta\cos\varphi/\sin\alpha_1$, $x_2 = \sin\theta\sin\varphi/\sin\alpha_2$ are the generalized angular coordinates, $c_1 = kd_1\sin\alpha_1$, $c_2 = kd_2\sin\alpha_2$, $k = 2\pi/\lambda$ is wave number, d_1 and d_2 are the distances between radiators along the Ox axis and Oy axis respectively, α_1 and α_2 are the angular coordinates, within which the desired power DP $P(x_1, x_2)$ is not equal to zero ($P(x_1, x_2) \equiv 0$ outside these angles). The function $f(x_1, x_2)$ possesses $2\pi/c_1-$periodicity with respect to x_1 and $2\pi/c_2-$ periodicity with respect to x_2. Let us denote the region of change of coordinates x_1 and x_2 on one period as $\Omega = \{(x_1, x_2) : |x_1| \le \pi/c_1, |x_2| \le \pi/c_2\}$. Below, the function $f(x_1, x_2)$ is termed as the DP of array.

A similar formula can be derived for the array with the hexagonal placement of separate elements (**Figure 1b**)

$$f(x_1, x_2) = A\mathbf{I} \equiv \sum_{m=-M_2}^{M_2} \sum_{n=-N_1(m)}^{N_1(m)} I_{nm} e^{i(c_1 n x_1 + c_2 m x_2)} \tag{2}$$

where $M = 2M_2 + 1$ is quantity of the linear subarrays, then $N = 2N_1(m) + 1$ is the number of elements in the $m-$th subarray.

Eqs. (1) and (2) for DP $f(x_1, x_2)$ represent the result of a linear operator A, which acts on a complex-valued space $H_I = C^{N_2 \times M_2}$ (rectangular case) or $H_I = C^{N_0 \times M}$ (hexagonal case) to the space of complex functions of two variables defined in the domain Ω. The value N_0 determines the number of elements in the central linear subarray in the hexagonal case.

Assume that the desired power DP $P(s_1, s_2)$ is not equal to zero in some regions $\overline{G} \subseteq \Omega$, and it is equal to zero outside. The optimization problem is formulated as the minimization problem of the functional

$$\sigma_\alpha(\mathbf{I}) = \left\| P - |AI|^2 \right\|_f^2 + \alpha \|\mathbf{I}\|_I^2 \tag{3}$$

where $\|\cdot\|_f$ and $\|\cdot\|_I$ determine the norms in the space of DPs and space of currents respectively, which are defined by the inner products

$$\|f\|_f^2 = (f_1, f_2)_f = \iint_{\Omega} f_1(x_1, x_2)\bar{f}_2(x_1, x_2)dx_1dx_2 \quad (4)$$

$$\|\mathbf{I}\|_I^2 = (\mathbf{I}_1, \mathbf{I}_2)_I = \frac{4\pi^2}{c_1c_2}\sum_{n=-N}^{N}\sum_{m=-M}^{M} I_{1nm}\bar{I}_{2nm} \quad (5)$$

here the values $\bar{f}_2(x_1, x_2)$ and $\bar{I}_{2nm}$ are conjugated to $f_2(x_1, x_2)$ and I_{2nm}.

The nonlinear integral equation for the complex vector $\mathbf{I}$of currents in space $H_\mathbf{I}$, which is derived using the necessary condition of the minimum of functional (3), has the form [26].

$$\alpha\mathbf{I} + 2A^*(|A\mathbf{I}|A\mathbf{I}) - 2A^*(PA\mathbf{I}) = 0 \quad (6)$$

Here A^* is the operator adjoint to A, its form is defined by equality $(A\mathbf{I}, f)_f = (\mathbf{I}, A^*f)_I$. Using the inner products (4), (5) and Eq. (1) we obtain.

$$(A^*f)_{nm} = \frac{c_1c_2}{4\pi^2}\iint_{\Omega} f(x_1, x_2)e^{-i(c_1nx_1+c_2mx_2)}dx_1dx_2, n = -N, -N+1\ldots, \quad (7)$$
$$N-1, N, m = -M, -M+1, \ldots, M-1, M.$$

If to act by operator A on both the parts of (6), we get a nonlinear integral equation of Hammerstein type for the function f

$$\alpha f + 2AA^*(|f|f) - 2AA^*(Pf) = 0 \quad (8)$$

The kernel of the AA^* operator for the rectangular array is defined as

$$K(c_1, c_2, x_1, x_1', x_2, x_2') = K_1(c_1, x_1, x_1')K_2(c_2, x_2, x_2'), \quad (9)$$

where

$$K_1(x_1, x_1', c_1) = \frac{c_1}{\pi}\frac{\sin(N_2c_1(x_1 - x_1')/2)}{\sin(c_1(x_1 - x_1')/2)} \quad (10)$$

$$K_2(x_2, x_2', c_2) = \frac{c_2}{\pi}\frac{\sin(M_2c_2(x_2 - x_2'))/2)}{\sin(c_2(x_2 - x_2'))/2)} \quad (11)$$

The kernel of the AA^* operator for the hexagonal array is more complicated because we can not to present it in the form of two multipliers

$$K(c_1, c_2, x_1, x_1', x_2, x_{12}') = \frac{\sin[c_1(N_1(0) - 1/2)(x_1 - x_1')]}{\sin(1/2c_1(x_1 - x_1'))} +$$
$$+2\sum_{m=1}^{M_2}\cos mc_2(x_2 - x_2')\begin{cases}\dfrac{\sin[c_1(N_1(m) - 1/2)(x_1 - x_1')]}{\sin(1/2c_1(x_1 - x_1'))}, & N_1(m) \text{ is odd,} \\ 2\displaystyle\sum_{n=1}^{N_1(m)}\cos[c_1(n - 1/2)(x_1 - x_1')], & N_1(m) \text{ is even.}\end{cases} \quad (12)$$

The kernels (9) and (12) of the integral Eq. (8) are real and degenerate. Since Eqs. (6) and (8) are nonlinear ones, both may have a non-unique solution. The number of solutions and their properties is studied according to the method proposed in [16, 27]. In the practical applications, the solution of Eqs. (6) and (8) is performed by the method of successive approximations. The convergence of the method depends on the parameter α, desired DP $P(x_1, x_2)$, as well as the parameters $c_1$and $c_2$contained in the kernels (9) and (12).

3. Search for the bifurcation curves

We should use the linear integral equation to define the bifurcation curves according to [16]. Based on this equation, we pass to the respective eigenvalue problems, solutions of which allow us to find the characteristic values of parameters c_1 and c_2 in the kernel of equation, at which the bifurcation appears.

3.1 Description of procedure

The linear equation

$$\alpha f = 2AA^*(Pf) \tag{13}$$

is used to study the properties of Eq. (8).

In contrast to a similar equation for the amplitude DP synthesis problem [25], Eq. (8) does not have a trivial nonzero initial solution f_0 for all parameters c_1 and c_2; the trivial solution f_0 is zero for it, so in contrast to the problem of synthesis by amplitude DP, we are not talking about the branching of solutions, but about their bifurcation.

The problem of finding bifurcation curves is reduced to the corresponding eigenvalue problem. The equation for eigenfunctions and corresponding eigenvalues, which refers to (13), is

$$g(x_1, x_2) = 2\lambda\alpha^{-1}\iint\limits_{\Omega} g(x_1', x_2')K_1(c_1, x_1, x_1')K_2(c_2, x_2, x_2')dx_1'dx_2' \tag{14}$$

As stated by the branching theory of solutions of the nonlinear equations [16], the bifurcation points can be those values of $c_1$and c_2 at which Eq. (14) has nonzero solutions.

Using the properties of the degeneracy of the kernel AA^*, we reduce Eq. (14) to the equivalent system of the linear algebraic equations (SLAE). The coefficients of matrix of this equation depend on the parameters c_1 and c_2 analytically. To this end, the equations for eigenfunctions corresponding to (13) are written as

$$g(x_1, x_2) = \sum_{n=-N}^{N}\sum_{m=-M}^{M} x_{nm}e^{i(c_1 n x_1 + c_2 m x_2)} \tag{15}$$

where

$$x_{nm} = \frac{c_1 c_2}{4\pi^2}\iint\limits_{\Omega} P(x_1', x_2')g(x_1', x_2')e^{-i(c_1 n x_1' + c_2 m x_2')}dx_1'dx_2' \tag{16}$$

Multiplying both the parts of (15) on $P(x_1', x_2')e^{-i(c_1 k x_1' + c_2 l x_2')}$ at $k = -N, -N+1, \dots, N-1, N, \;\; l = -M, -M+1, \dots, M-1, M$ and integrating over the domain Ω, we obtain a system of linear algebraic equations to determine the quantities x_{nm}

$$x_{kl} = \sum_{n=-N}^{N} \sum_{m=-M}^{M} a_{nm}^{(kl)}(c_1, c_2) x_{nm}, \;\; k = -N, -N+1, \dots, N-1, N, \\ l = -M, -M+1, \dots, M-1, M, \tag{17}$$

where

$$a_{nm}^{(kl)} = \frac{c_1 c_2}{4\pi^2} \iint_{\Omega} P(x_1, x_2) e^{-i[(c_1(n-k)x_1 + c_2(m-l)x_2)} dx_1 dx_2 \tag{18}$$

and matrix of the coefficients $a_{nm}^{(kl)}$ is self-adjoint and Hermitian.

Thus, we obtained a two-parameter nonlinear spectral problem corresponding to a homogeneous SLAE (17). This problem can be given as

$$(E_M - A_M(c_1, c_2))\mathbf{x} = 0 \tag{19}$$

where A_M is the matrix of coefficients $a_{nm}^{(kl)}$, E_M is a unit matrix of dimension $N_2 \cdot M_2$. For the system (19), the equality

$$\Psi(c_1, c_2) = \det[E_M - A_M(c_1, c_2)] = 0 \tag{20}$$

must be met to have a non-zero solution.

One can easy to make sure that the function $\Psi(c_1, c_2)$ is real. Moreover, since $A_M(c_1, c_2)$ is the Hermitian matrix, then $E_M - A_M(c_1, c_2)$ is Hermitian too. The determinant of the Hermitian matrix is a real number [28]. Thus, $\Psi(c_1, c_2)$ is a real function of real arguments $c_1$and c_2.

Consequently, the problem to find the eigenvalues of Eq. (14) or to determine the solution of the equivalent SLAE (19) is reduced to finding zeros of function $\Psi(c_1, c_2)$.

If to consider the equation $\Psi(c_1, c_2) = 0$ as a problem of determining an implicit function $c_2 = c_2(c_1)$ in the vicinity of some point c_1, we get Cauchy problem [29].

$$\frac{dc_2}{dc_1} = \frac{\Psi_{c_1}{}'(c_1, c_2)}{\Psi_{c_2}{}'(c_1, c_2)} \tag{21}$$

$$c_2\left(c_1^{(0)}\right) = c_2^{(0)} \tag{22}$$

To retrieve the initial conditions (22) we pass to an auxiliary one-dimensional nonlinear spectral problem if to substitute c_2 by $c_2 = \gamma c_1$ in Eq. (20) with some real parameter γ. As a result, we get the one-dimensional eigenvalue problem

$$(E_M - A_M(c_1, \gamma c_1))\tilde{\mathbf{x}} \equiv \left(E_M - \tilde{A}_M(c_1)\right)\tilde{\mathbf{x}} = 0 \tag{23}$$

Eq. (20), which corresponds to Eq. (23), is

$$\Psi(c_1, \gamma c_1) = \det\left[E_M - \tilde{A}_M(c_1)\right] = 0 \tag{24}$$

Let $c_1^{(0)}$ be the solution of the Eq. (24), then $\left(c_1^{(0)}, c_2^{(0)}\right) = \left(c_1^{(0)}, \gamma c_1^{(0)}\right)$ is the point that corresponds to eigenvalue $\lambda_0 \approx 1$ of Eq. (15). By solving Eqs. (21) and (22) in a small vicinity of point $\left(c_1^{(0)}, c_2^{(0)}\right)$, we find the spectral curve of the matrix-function $A_M(c_1, c_2)$, which is the curve $c_2(c_1)$ defining a set of the bifurcation points.

The eigenfunctions of Eq. (14) are defined as the eigenvectors of matrix $A_M(c_1, c_2)$ using the resulting solution of the Cauchy problem with the sought solutions $\Psi(c_1, c_2)$. In this procedure, a four-dimensional matrix $A_M(c_1, c_2)$ is reduced to a two-dimensional one by the relevant renouncement of its elements.

3.2 Defining the area of nonzero solutions

Due to the peculiarity of the problem statement according to desired power DP $P(x_1, x_2)$, Eq. (8) has zero solution at arbitrary values of the parameters c_1, c_2, α. From an engineering point of view, this is a significant drawback, but for some desired DPs $P(x_1, x_2)$ it is possible to fix an area of parameters c_1, c_2, α at which a nonzero solution exists. At the small c_1 and c_2, the kernel (9) is given approximately in the form

$$K\left(c_1, c_2, x_1, x_1', x_2, x_2'\right) \approx \frac{M_2 N_2 c_1 c_2}{\pi^2} \tag{25}$$

Assuming that $f(x_1, x_2)$ is constant, the integral Eq. (8) can be rewritten as (usually for small c_1 and c_2 $f(x_1, x_2) \approx \text{const}$).

$$\frac{\pi^2 \alpha}{2M_2 N_2 c_1 c_2} = \int_{-1}^{1}\int_{-1}^{1} P(x_1, x_2) dx_1 dx_2 - 4|f(x_1, x_2)|^2 \tag{26}$$

The area of integration Ω in Eq. (8) is reduced in the last formula to the area $[-1, 1] \times [-1, 1]$ because of definition of both the arguments x_1, x_2 and parameters c_1, c_2.

Taking into account that $|f(x_1, x_2)|^2$ is positive, we get the following relationship between the function $P(x_1, x_2)$ and the parameters c_1, c_2, and α:

$$\int_{-1}^{1}\int_{-1}^{1} P(x_1, x_2) dx_1 dx_2 - \frac{\pi^2 \alpha}{2M_2 N_2 c_1 c_2} > 0 \tag{27}$$

Finally, considering the case $P(x_1, x_2) \equiv 1$, we obtain:

$$c_1 c_2 > \frac{\pi^2 \alpha}{8 M_2 N_2} \tag{28}$$

In fact; inequality (28) determines the area of parameters c_1, c_2, α, where nonzero solutions exist. In **Figure 2**, the dependence curves $c_2 = c_2(c_1)$ for three different values M_2 and N_2 are shown. The results are given for array with the number of elements $M_2 = N_2 = 3$ (curve 1), $M_2 = N_2 = 5$ (curve 2) and $M_2 = N_2 = 11$ (curve 3). The area of values c_1 and c_2, where the existence of zero solutions is possible, according to the estimate (28) is located below and to the left of the presented curves.

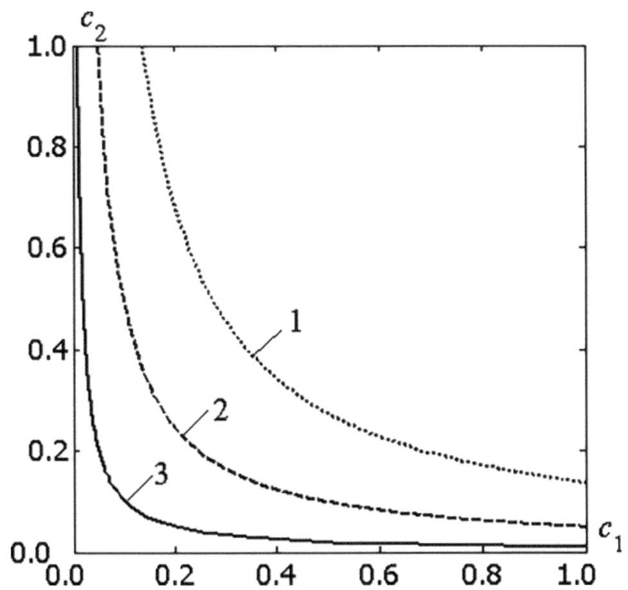

Figure 2.
The curves $c_2 = c_2(c_1)$ at the different M_2 and N_2.

As can be seen, the area of zero values decreases significantly with increasing $N_2$and M_2. The obtained results testify that the zero solutions of Eq. (8) for a given constant power DP can exist either at a small value c_1c_2 corresponding to low frequencies (at a given size of array), or at the values of c_1 that significantly exceeding c_2 and vice versa. The last case corresponds to arrays with a large difference in distance of elements along the coordinate axes. Such arrays are usually rarely used in practice.

3.3 Determination of bifurcation lines

3.3.1 The case of rectangular array

The finding of bifurcation lines of the nonlinear Eq. (8) was performed for the array containing $N_2 * M_2 = 11 * 11 = 121$ radiators for the desired power DP $P(x_1, x_2) = 1$ at $\Lambda_c = \{(c_1, c_2), 0 < c_1, c_2 \leq 2\}$ for the different values of the parameter α in (3).

The search for bifurcation lines can be performed directly by investigating the properties of the determinant (20) as a function of the parameters c_1 and c_2. In addition, the function (20) depends on the parameter α; so the set of its eigenvalues also depends on this parameter, i.e. the set of spectral curves that separate the areas of zero and nonzero solutions.

The behavior of the corresponding curves when changing the parameter α is shown in **Figure 3**. The behavior of the determinant (24) depending on the parameters $c_1$and c_2 at $\alpha = 0.5$ is given in **Figure 3a**; and in **Figure 3b–d**, the intersection of this function with a plane $\Psi(c_1, \gamma c_1) = 0$ is illustrated at the different α. This results in a set of curves that correspond to a set of spectral lines separating the area of zero and nonzero solutions. At a fixed size of array, the area where zero solutions can exist expands if the parameter α increases, this area is located below the left of the first curve.

The curves marked by number 1 correspond to the solutions with constant (zero or even) phase DP; curves with number 2 correspond to the solutions with phase DP that

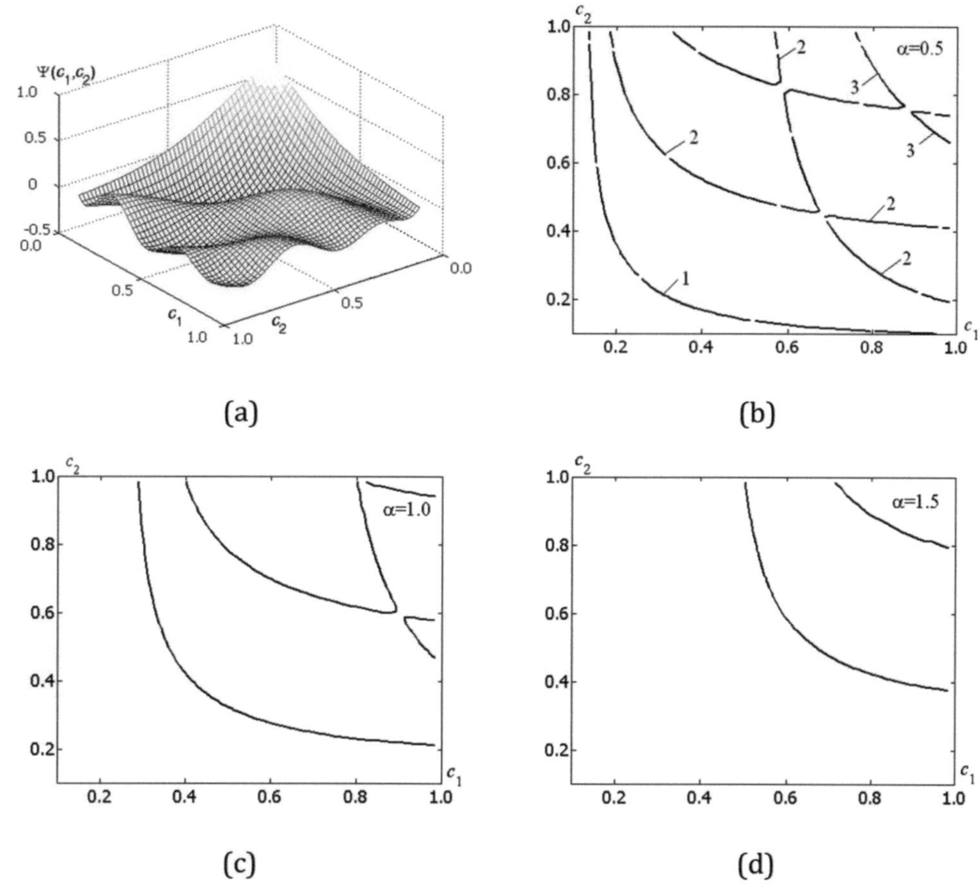

Figure 3.
The spectral curves of Eq. (19) at the different α.

is even with respect to the Ox_1 axis, and odd with respect to the Ox_2 axis, and curves numbered by 3 correspond to the solutions with a phase DP odd with respect to two coordinate axes. The proposed procedure is quite approximate, it does not allow to separate the curves that correspond to different types of solutions and thus identify the areas where there is a nonzero solution for the synthesized power DP with the specified phase property.

The method of implicit function proposed in [23] and developed for plane array in [30] is devoid of this drawback.

At the first step of this method, a series of one-dimensional eigenvalue problems is solved, by this the different values of parameter γ are prescribed by the relation $c_2 = \gamma c_1$and a one-dimensional problem is solved with respect to c_1. In **Figures 4** and **5**, the first four eigenvaluesof the problem at $\gamma = 1.0$ and $\gamma = 0.2$ are shown. The values $\left(c_1^{(i)}, c_2^{(i)} = c_1^{(i)}\right)$, $i = 1, 2, 3, 4$, at which $\lambda_i = 1$, are the bifurcation points in the plane c_1Oc_2. By this, the set of points (c_1, c_2)at which the eigenvalue $\lambda^{(i)} = 1$ is determined approximately from the graphical data.

The next step is to refine the values $\left(c_1^{(i)}, c_2^{(i)}\right)$ by solving the transcendental Eq. (20), and the point $\left(c_1^{(i)}, c_2^{(i)}\right)$, which is considered as the initial approximation.

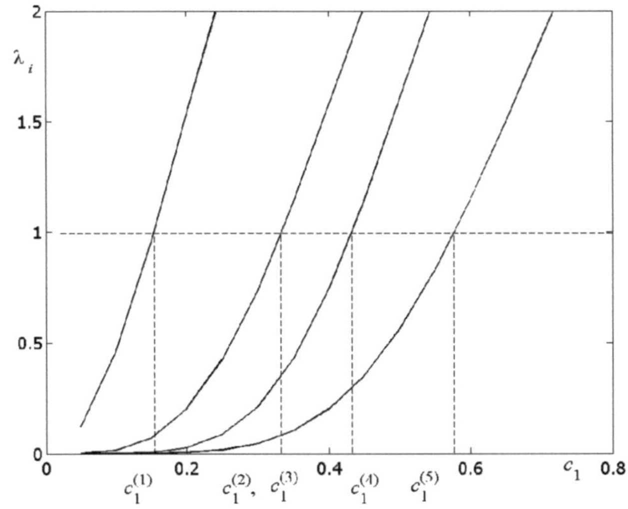

Figure 4.
The first eigenvalues at the ray $c_1 = c_2$, $\alpha = 0.5$.

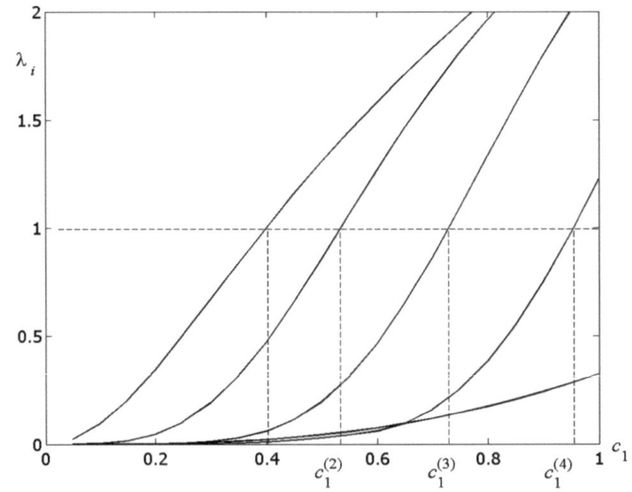

Figure 5.
The first eigenvalues at the ray $c_2 = 0.2c_1$, $\alpha = 0.5$.

In the final step, the bifurcation curve in the plane (c_1, c_2) is determined by solving Eqs. (21) and (22), after specification of the values $\left(c_1^{(i)}, c_2^{(i)}\right)$. In **Figure 6**, the bifurcation curves $\left(c_1^{(1)}, c_2^{(1)}\right)$–$\left(c_1^{(4)}, c_2^{(4)}\right)$ that correspond to the first four eigenvalues are shown. The curve with number 1 corresponds to the solution with the zero (even) phase of the created DP. This curve corresponds to that is marked by 1 in **Figure 3b**.

There are no nonzero solutions with such a phase property for the values c_1 and c_2 above and to the right of this curve. Curves 2 and 2′ correspond to solutions in which the phase DP is symmetric about one axis and asymmetric about the other axis (obviously, for a plane array, there are two such curves and they are antisymmetrical). Curve 2 corresponds that is marked by 2 in **Figure 3b**. The curve with number 3

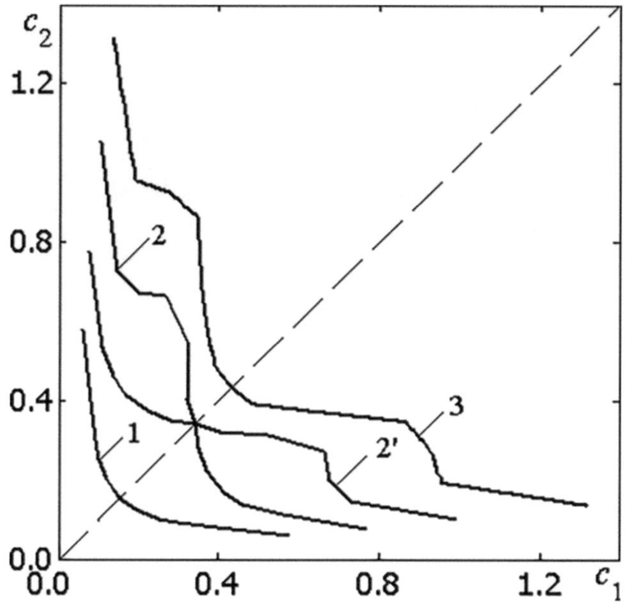

Figure 6.
The bifurcation curves corresponding to set $\left(c_1^{(i)}, c_2^{(i)}\right), i = 1, \ldots, 4, N_2 = M_2 = 11.$

corresponds to a solution with a phase DP antisymmetrical (odd) with respect to both the axes. The location of the areas of zero and non-zero solutions is the same as in **Figure 3b**. It should be noted that the problem of refining the roots of Eq. (20) is the most time-consuming in computational relation because refining the roots of this equation requires a series of computational experiments with different values $\left(c_{1,0}^{(i)}, c_{2,0}^{(i)}\right)$ of initial parameters close to approximate values.

3.3.2 The case of hexagonal array

Firstly, we consider the procedure of determination of bifurcation curves by finding zero lines of determinant (24). The results, similar to those are presented in **Figure 3a** and **b** for the rectangular array, are shown in **Figures 7** and **8**. One can see that the behavior of function $\Psi(c_1, c_2)$ is more complex than in the case of rectangular array. The obtained graphs testify that the solutions with other different behavior of phase $\arg(f(x_1, x_2))$ of the DP appear additionally. One such solution is marked by number 4. Other solutions appear when parameters c_1 and c_2 increase at the fixed α.

Search of the bifurcation curves is carried out similarly to the case of rectangular array. The numerical results are presented for the array with $N_{tot} = 61$ elements for the desired power DP $N_0(x_1, x_2) = 1$ at $\Lambda_c = \{(c_1, c_2), 0 < c_1, c_2 \le 2.0\}$ for the different values of α in the functional (3). At the first step, the one-dimensional eigenvalue problems were solved at the different values of parameter γ. In **Figure 9**, the first four eigenvalues are shown at $\gamma = 1.0$, and in **Figure 10**, they are shown at $\gamma = 0.2$. Similar to the case of rectangular array, the points, in which $\lambda_i = 1$ are moved to right and the distance between them increases at $\gamma = 0.2$. The values $\left(c_1^{(i)}, c_2^{(i)} = \gamma c_1^{(i)}\right)$, where

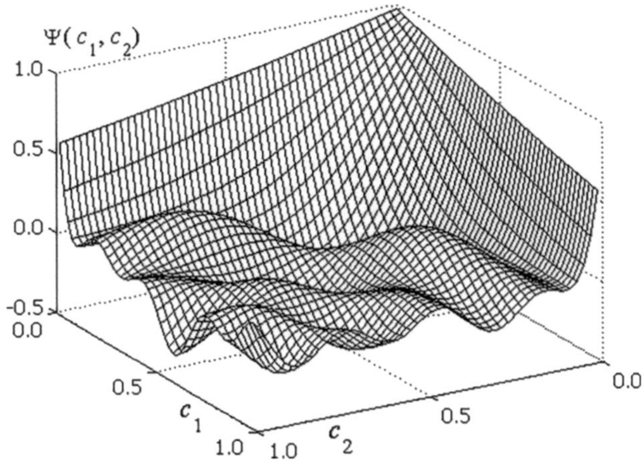

Figure 7.
The surface of determinant (24) values for the hexagonal array, $\alpha = 0.5$.

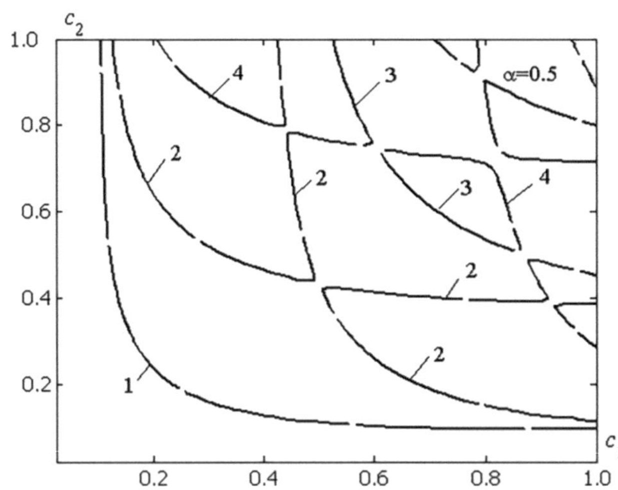

Figure 8.
Zero lines of determinant (24) for the hexagonal array, $\alpha = 0.5$.

$i = 1, 2, 3, 4$, are the bifurcation points in the plane (c_1, c_2). The points $\left(c_1^{(i)}, c_2^{(i)}\right)$, for which the eigenvalues $\lambda^{(i)} = 1$ are determined approximately in this step.

The specification of values $\left(c_1^{(i)}, c_2^{(i)}\right)$ by solving Eq. (20) is carried out in the next step, and the points $\left(c_1^{(i)}, c_2^{(i)}\right)$ of the graph data from the **Figures 9** and **10** are used as initial approximations. The usual numerical half-division method is used for this goal.

The bifurcation points $\left(c_1^{(i)}, c_2^{(i)}\right)$, $i = 1, 2, 3, 4$ for the first four eigenvalues in the rays, $c_2 = \gamma c_1$ are shown in **Figure 11**. The respective curves of bifurcations, which are obtained by solving Eqs. (21) and (22), are shown in **Figure 12**. As in the case of a rectangular array, to obtain the necessary data, we should carry out precise computations.

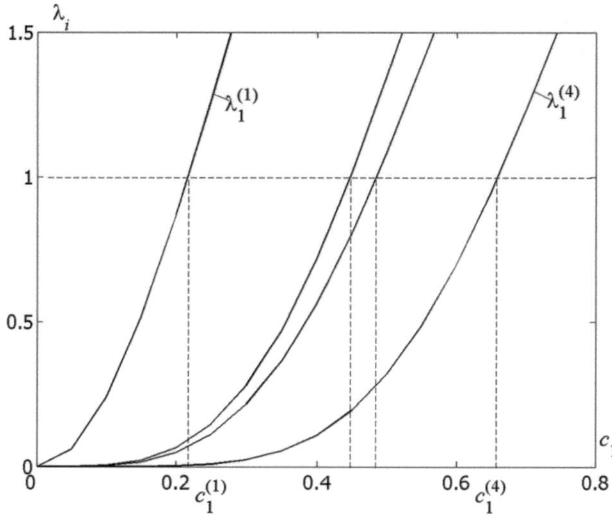

Figure 9.
The first eigenvalues for the ray $c_1 = c_2$ at $\alpha = 0.5$.

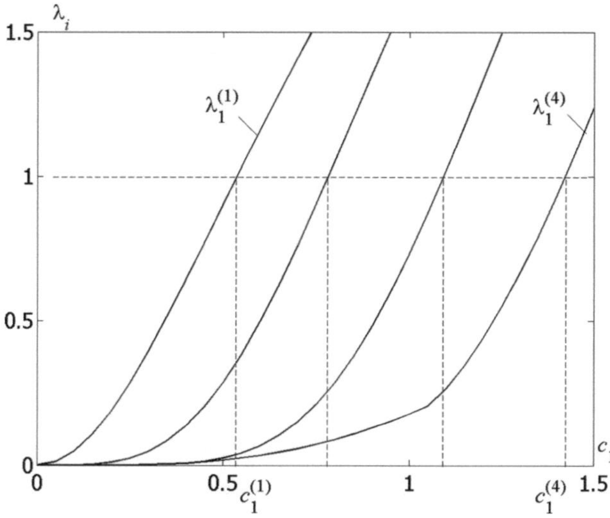

Figure 10.
The first eigenvalues for the ray $c_2 = 0.2c_1$, $\alpha = 0.5$.

4. The engineering applications

The results presented in this Section demonstrate how the knowledge about the point of bifurcation obtained as the solutions of the nonlinear eigenvalue problems allows us to understand better the process of bifurcation and how to get the solutions, which are the most optimal in sense of the used criterion of optimization.

4.1 The method of successive approximations

The properties of solutions to Eq. (8) obtained by using the method of successive approximation are related directly with the properties of phase characteristic of the

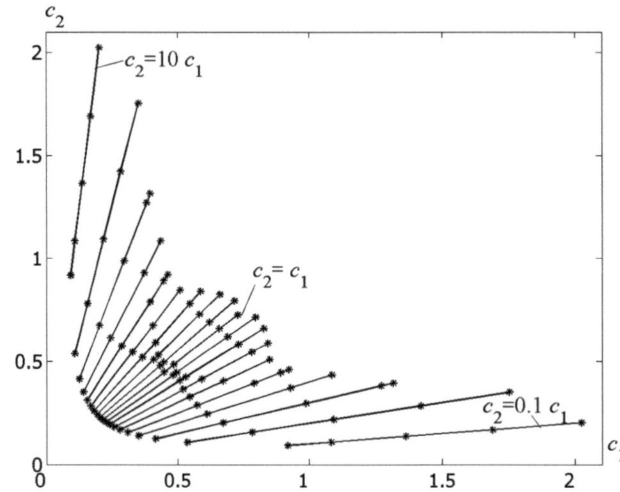

Figure 11.
The bifurcation points at the rays $c_2 = \gamma c_1$.

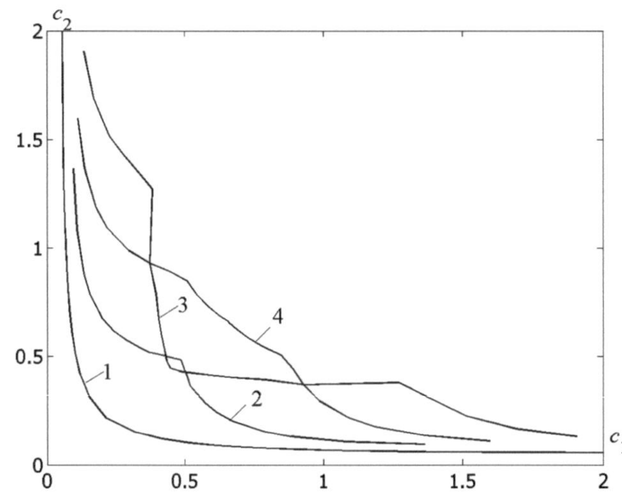

Figure 12.
The bifurcation curves in the plane (c_1, c_2).

eigenfunctions, which are determined at solving the eigenvalue problem. Prescribing the initial approximation f_0 for the iterative process for solving Eq. (8) with the specified property of the phase $\arg f_0$, we could receive the solution of Eq. (8) with the same phase property in the wide range of characteristic parameters c_1 and c_2. This is important for the engineering design of arrays having the fixed phase characteristics of radiation in the defined range of frequencies.

The method of successive approximations

$$f_{n+1} - \beta f_n + (1-\beta)B(f_n) = 0, n = 0, 1, 2, \ldots \tag{29}$$

is used for solving Eq. (8) with a set of specific physical parameters of array. In the last formula,

$$B(f) = \frac{2}{\alpha}\left[AA^*\left[(P \cdot f) - \left(|f|^2 \cdot f\right)\right]\right] \tag{30}$$

Parameter $\beta \in [0, 1]$ in (29) is used to accelerate the convergence of iterative process. To substantiate the condition of convergence of the iterative process (29), we apply Theorem 2.6.2 [26] (p. 133), which states that the operator AA^* be contraction one. This requirement is met when the inequality

$$\alpha > 2\left\|AA^*\left[f\left(N_0 - |f|^2\right)\right]\right\| \tag{31}$$

met. The results of numerical calculations show that condition (31) is overestimated and for some values of the problem parameters the iterative process (29) converges for values αthat do not satisfy the estimate (31).

4.2 The case of rectangular array

In **Figure 13**, the dependence of the convergence of the iterative process (29) on the value of parameter α for a desired power DP $P(x_1, x_2) = 1$ at the fixed values $\beta = 0.1$, $c_1 = c_2 = 2.0$, the number of radiators $M_2 \cdot N_2 = 11 \cdot 11 = 121$ is shown. The required accuracy $\varepsilon = 10^{-3}$.

The results of solving the optimizing problem for this desired power DP at $\alpha = 0.5$ are shown in **Figure 14**. The approximation quality to a desired DP P significantly depends on the parameters c_1 and c_2 at both fixed N_2 and M_2. The mean-square deviation (MSD) (the first term in (3)) is equal to 0.0847 for $c_1 = c_2 = 1.0$, and it is equal to 0.0075 for $c_1 = c_2 = 3.14$.

The synthesized DP $|f|$for larger c_1, c_2 has not only a more optimal mean-square approximation, but it is also closer to the shape of the desired DP P. The optimal amplitudes $|I_{nm}|$ of currents in the array's elements are close to constant at such parameters c_1 and c_2.

When solving the optimizing problem for desired power DP of a more complex form, the quality of the approximation significantly depends on both the parameter α

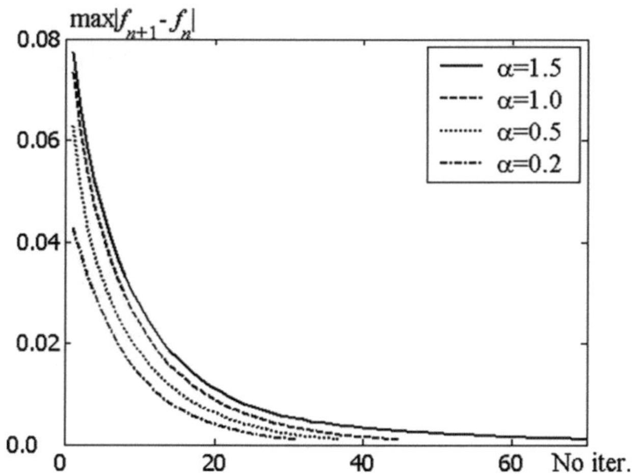

Figure 13.
The character of convergence of iterative process (29) at the different α, $M_2 = N_2 = 11$.

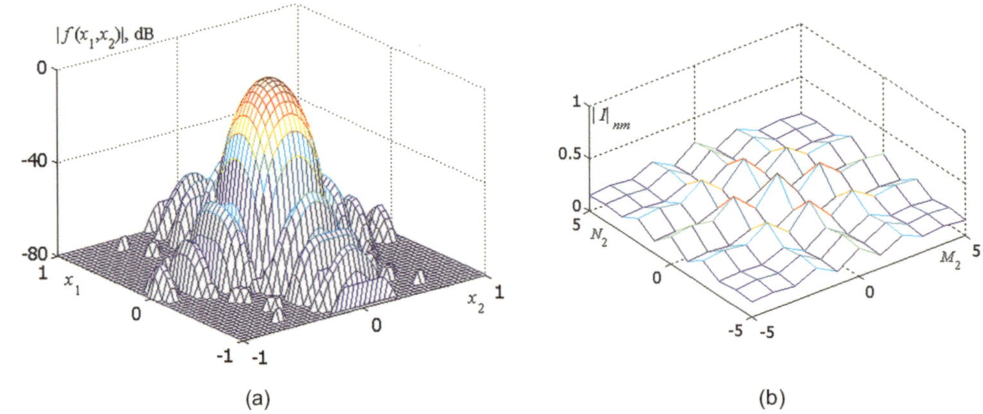

Figure 14.
The created power RP $|f|$ (a) and optimal distribution of currents $|I_{nm}|$ (b) at $c_1 = c_2 = 1.0$.

and the type of initial approximation for the phase of a given DP. The results are shown for the desired power DP

$$P(x_1, x_2) = |\sin(\pi x_1) \cdot \sin(\pi x_2)|, \quad -1 \le x_1 \le 1, \ -1 \le x_2 \le 1 \tag{32}$$

at $c_1 = c_2 = 3.14$ and $\alpha = 0.2$ in **Figure 15**. Despite the fact that the shape of desired DP $P(x_1, x_2)$ is more complex that in the previous example, decrease of α (from 0.5 to 0.2) and simultaneous increase of c_1 and c_2 (from 1.0 to 3.14) allows us to get the amplitude $|f(x_1, x_2)|$ of created DP, which is very close to the $P(x_1, x_2)$. The optimal distribution of currents' amplitudes $|I_{nm}|$ (**Figure 15b**) approaches the shape of the created DP.

We have used an additional optimization parameter β in Eq. (29), which, as shown by the results of numerical calculations, accelerates the convergence of iterative process significantly. In **Figure 16**, the results of the study of the influence of this parameter on the rate of convergence at a fixed value of the parameter $\alpha = 0.5$ are shown. The results are given for the desired power DP $P(x_1, x_2) = 1$ at $c_1 = c_2 = 2.0$. In order to achieve the accuracy 10^{-3} of calculations, one needs 157 iterations at $\beta = 0.01$. If parameter β increases to a certain value, the number of iterations decreases significantly, so at $\beta = 0.05$, $\beta = 0.10$, $\beta = 0.15$, $\beta = 0.20$, and $\beta = 0.25$ one requires

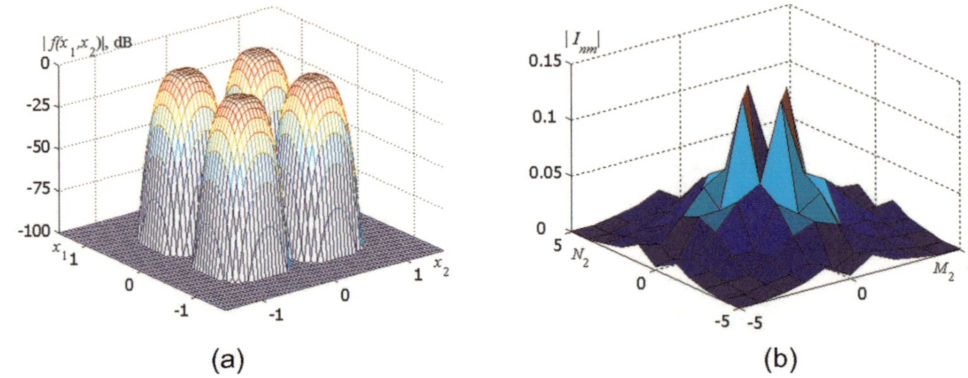

Figure 15.
The created power RP $|f|^2$ at $c_1 = c_2 = 3.14$.

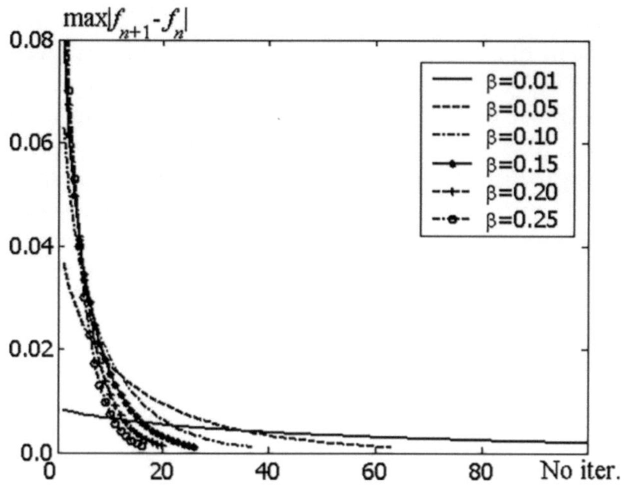

Figure 16.
The convergence of iterative process (29) for the different β.

62 iterations, 37 iterations, 28 iterations, 20 iterations, and 16 iterations, respectively. At the subsequent increase, the number of required iterations begins to increase and already at $\beta = 0.30$ the iterative process begins to diverge. Numerical calculations show that the limit valueof β, at which the iterative process (29) begins to diverge, significantly depends on the value of the parameter α. So, if this parameter decreases, the threshold value of β increases. The dependence of the convergence on the array's parameters $(c_1, c_2, M_2, N_2, d_1, d_2)$ is not so significant.

The values of the functional (3) for the created DP with different phases are shown in **Figure 17**. The solid curve corresponds to the phase DP even with respect to two axes, the dotted curve corresponds to phase DP even with respect to one axis and odd with respect to the other, the dashed curve corresponds to the phase DP odd with respect to both the axes.

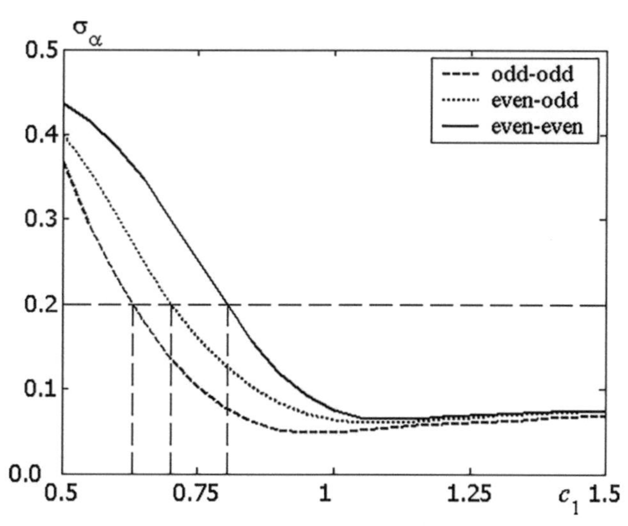

Figure 17.
The values of functional (3) versus the phase of created DP.

One can see that the values of functional at the fixed c (frequency) significantly depend on the phase of the created DP. The value $\sigma = 0.2$ is achieved for the "even-even" solution at $c = 0.79$, for the "even-odd" solution at $c = 0.71$ and for the "odd-odd" solution at $c = 0.625$. That is, within the used criterion, the latter type of solution is 21% better than the first one. From this fact, it follows that at a fixed distance between the radiators for the desired DP $P(x_1, x_2) = 1$, the number of array's elements can be reduced by 21% with the same value of MSD. A similar situation is observed for the characteristics of DP at $\sigma = 0.1$, i.e. "odd-odd" solution is better on 19.4% than "even-even".

4.3 The case of hexagonal array

The results of solution of the optimization problem for two given power DPs $P_1(x_1, x_2) \equiv 1$ and

$$P_2(x_1, x_2) = \begin{cases} 2\sqrt{x_1^2 + x_2^2}\sqrt{1 - x_1^2 - x_2^2}, x_1^2 + x_2^2 \leq 1, \\ 0, x_1^2 + x_2^2 > 1, \end{cases} \quad (33)$$

in the form of body of rotation are shown in **Figures 18** and **19** at $\alpha = 0.5$.

As previously, the optimization problem consists of solving Eq. (8) by the method of successive approximation (29). The MSD (the value of the first term in (3)) for the first desired DP is equal to 0.3774, and it is equal to 0.2218 for the second desired DP.

Similar to the case of rectangular array, the approximation quality to the desired DP P depends on both the parameters c_1, c_2, and α. The characteristic of MSD of DPs for α at the different c_1 on the ray $c_2 = 1.118c_1$ is shown in **Figures 20** and **21**. The chosen relation between c_1 and c_2 provides the regularity of the array's geometry, and as the numerical computations have shown, gives the ability to get the close characteristics of radiation in the planes $x_1$and x_2.

The largest MSD for the P_1 is achieved at $\alpha = 1.0$ for $c_1 = 0.5$, and it is equal to 1.96443, it diminishes almost linearly if parameter α decreases. The largest MSD for DP P_2 is equal to 1.43685. One should note that the value of MSD diminishes if α

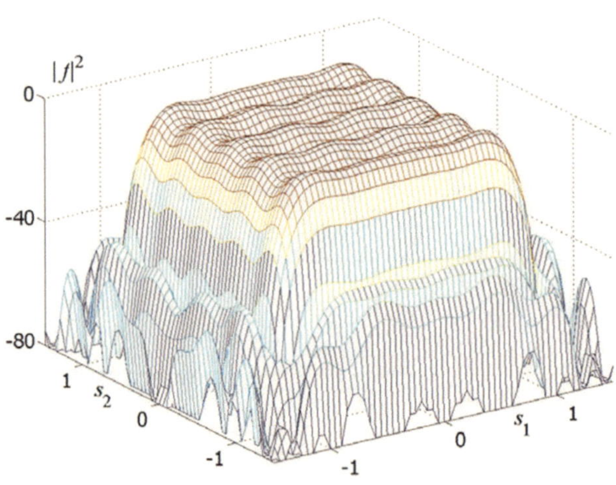

Figure 18.
The amplitude of created DP $|f|^2$ for P_1 at $c_1 = 2.0, c_2 = 2.236$.

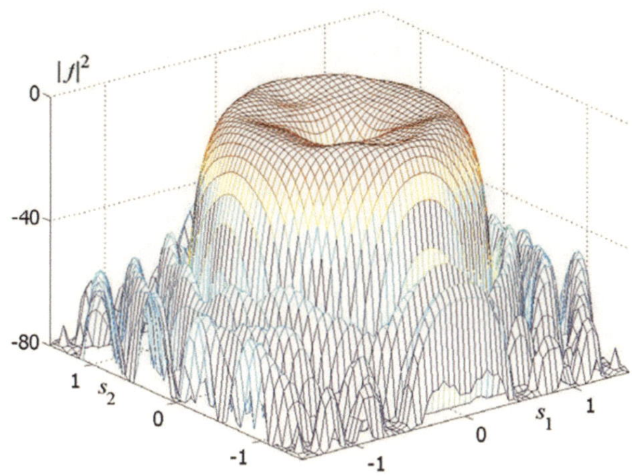

Figure 19.
The amplitude $|f|^2$ *of created DP for* P_2 *at* $c_1 = 2.0, c_2 = 2.236$.

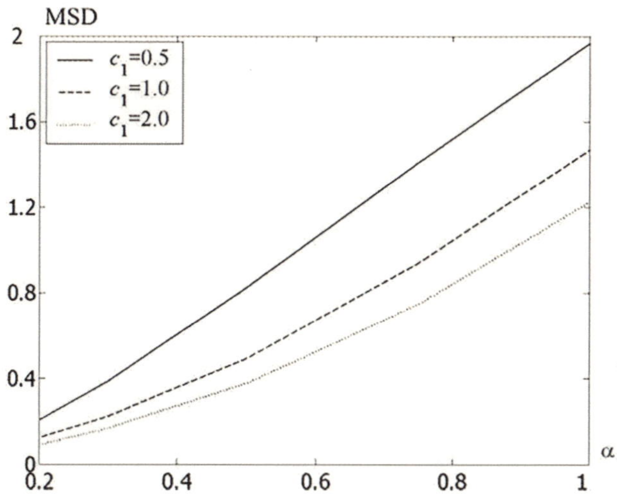

Figure 20.
The MSD versus the parameter α *for the DP* P_1.

decreases, but the norm $\|\mathbf{I}\|_{H_I}$ of current growth that is unacceptable from the engineering point of view.

The approximation quality to the desired DP depends also on the type of initial data, which are prescribed for the iterative process (29). The dependence of the values of functional (3) on the parity of phase of the initial approximation f_0 for the DP P_1 is shown in **Figure 22**. The results are shown for four types of initial approximation: odd with respect to both the axes (dotted curve), even with respect to the Ox_1 axis and odd with respect to the Ox_2 axis (dashed curve), odd with respect to the Ox_1 axis and even with respect to the Ox_2 axis (dash-dot curve), and even with respect to both the axis (solid curve). The initial approximation f_0, corresponding to the even phase with respect to both the axis, is optimal for this DP, moreover the values of σ_α for the small values of parameters c_1 and c_2 differ significantly, but starting from $c_1 = 0.8$ this difference does not exceed 10%.

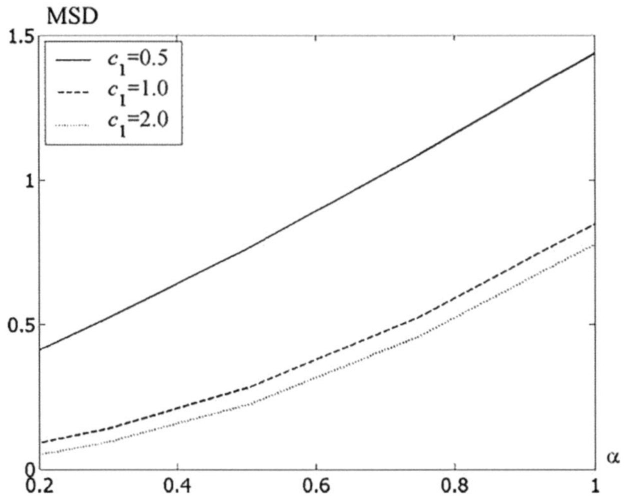

Figure 21.
The MSD versus parameter α for DP P_2.

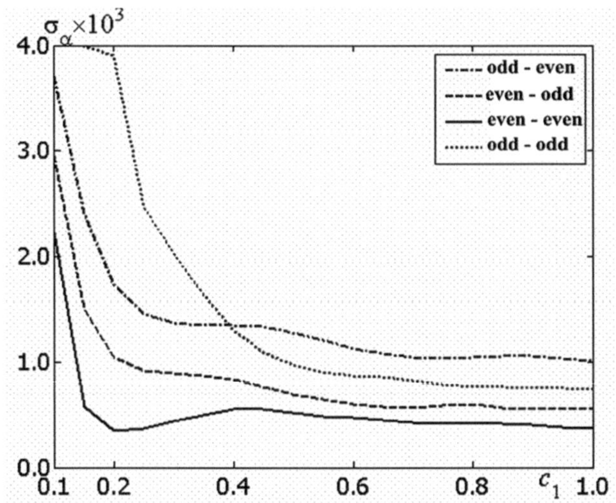

Figure 22.
The values of σ_α versus the initial approximation of initial approximation for the iterative process (29).

The dependence of convergence of the iterative process (29) on the parameter α at the fixed $\beta = 0.1$ is shown in **Figure 23**. As in the case of a rectangular array, the iterative process converges most slowly at $\alpha = 1.5$, one needs 67 iterations to achieve the accuracy that is equal to 10^{-3}. The number of iterations decreases if α diminishes. For example, one needs 30 iterations to achieve the same accuracy at $\alpha = 0.2$.

The dependence of convergence on the parameter β is studied too. The necessary number of iterations that needs to achieve the accuracy 10^{-3}at $\beta = 0.01$, $\beta = 0.05$, $\beta = 0.10$, $\beta = 0.15$, $\beta = 0.20$, and $\beta = 0.30$ (curves 1–6 respectively) is shown in **Figure 24**. It is substantiated that the iterative process converges most slowly at $\beta = 0.01$, it is necessary 152 iterations to achieve the prescribed accuracy. The most optimal among the considered β is $\beta = 0.20$, one needs 20 iterations only to achieve

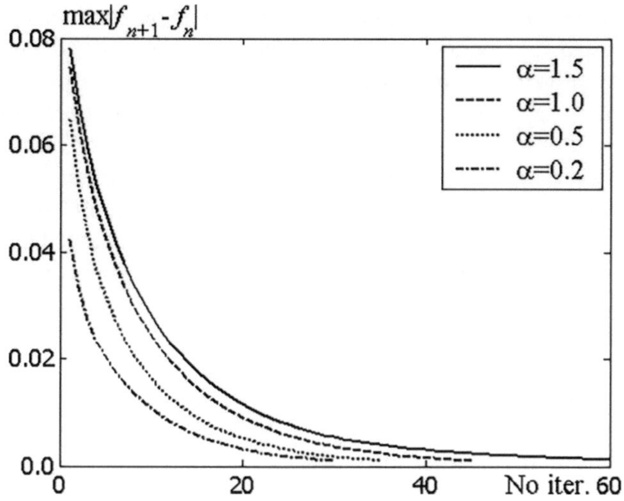

Figure 23.
The convergence of iterative process (29) versus α.

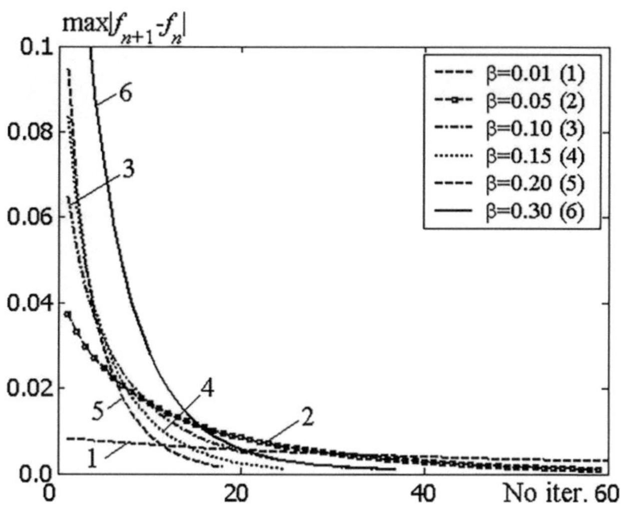

Figure 24.
The convergence of iterative process (29) versus number of iteration, $\alpha = 0.5$.

the above accuracy. The iterative process becomes slow at the subsequent growth of β. For example, one needs 37 iterations to achieve this accuracy; the iterative process (29) becomes convergent at $\beta > 0.40$. This testifies that in the process of computations one should to limit by non-large values of β ($\beta \le 0.20$) that guarantees the convergence and considerably grows its speed on the contrast with small β ($\beta \le 0.01$).

More information about the problem under investigation one can find in [30–34].

5. Conclusions

The problem of finding the solutions to the nonlinear integral equations and their properties is reduced to nonlinear two-dimensional eigenvalue problems that lead to the subsequent application of an implicit function method for solving the Cauchy problem for the linear differential equation. The area of non-zero solutions to the above equations is determined by involving the solving transcendental equation, which is got by equating to zero of determinant related to the eigenvalue problem. The results of solving the nonlinear eigenvalue problems are applied subsequently for specification of the bifurcation points and obtaining the bifurcation curves. The approach does not depend on the form of operator determining the radiation properties of physical system (plane rectangular and hexagonal arrays). The obtained results are the constructive basis on which a series of practical engineering problems of optimization was solved numerically.

Conflict of interest

The author declares no conflict of interest.

Author details

Mykhaylo Andriychuk[1,2]

1 Pidstryhach Institute for Applied Problems of Mechanics and Mathematics. NASU, Lviv, Ukraine

2 Lviv Polytechnic National University, Lviv, Ukraine

*Address all correspondence to: andr@iapmm.lviv.ua

References

[1] Burge RE, Fiddy MA, Greenaway AH, Ross G. The phase problem. Proceedings of the Royal Society of London. 1976; **350**(1661):A350191-A350212. DOI: 10.1098/rspa.1976.0103

[2] Zenkova CY, Gorsky MP, Ryabiy PA, Angelskaya AO. Additional approaches to solving the phase problem in optics. Applied Optics. 2016;**55**:B78-B84. DOI: 10.1364/AO.55.000B78

[3] Taylor G. The phase problem. Acta Crystallographica Section D. 2003; **59**(11):1881-1890. DOI: 10.1107/S0907444903017815

[4] Hauptman HA, Langs DA. The phase problem in neutron crystallography. Acta Crystallographica Section A. 2003; **59**(3):250-254. DOI: 10.1107/S010876730300521X

[5] Dumber AS. On the theory antenna beam scanning. Journal of Applied Physics. 1958;**13**(5):31-39

[6] Tartakovskyi LB, Tikhonova VK. Synthesis of a linear radiator with a given distribution of amplitudes. Radiotechnica & Electronica. 1959;**4**(12): 2016-2019 (In Russian)

[7] Chony YI. To the synthesis of an antenna system for a given amplitude radiation pattern. Izvestiya Vysshikh Uchebnykh Zavedenii. Radioelektronika. 1968;**11**(2):1325-1327 (In Russian)

[8] Sklyanin EK. The method of the inverse scattering problem and the quantum nonlinear Schrödinger equation. Doklady Akademii Nauk SSSR. 1979;**244**(6):1337-1341 (In Russian)

[9] Ramm AG. Multidimensional Inverse Scattering Problems. New York: Longman Scientific & Wiley; 1992. p. 385

[10] Ikehata M. Reconstruction of an obstacle from the scattering amplitude at a fixed frequency. Inverse Problems. 1998;**14**:949-954

[11] Colton D, Kirsch A. A simple method for solving inverse scattering problems in the resonance region. Inverse Problems. 1996;**12**:383-393

[12] Precup R. Methods in Nonlinear Integral Equations. Alphen aan den Rijn, Netherlands: Kluwer; 2002

[13] Masujima M. Applied Mathematical Methods in Theoretical Physics. Weinheim, Germany: Wiley-VCH; 2005

[14] Bauschke HH, Combettes PL, Luke DR. Phase retrieval, error reduction algorithm, and Fienup variants: A view from convex optimization. Journal of the Optical Society of America. 2002;**19**(7): 1334-1345

[15] Fienup JR. Phase retrieval algorithms: A comparison. Applied Optics. 1982;**21**:2758-2769

[16] Weinberg MM, Trenogin VA. Branching Theory of Solutions to Nonlinear Equations. Moscow: Nauka; 1969. p. 528 (In Russian)

[17] Deuflhard P. Newton Methods for Nonlinear Problems.Affine Invariance and Adaptive Algorithms.Springer Series Computational Mathematics. Berlin, Heidelberg: Springer-Verlag; 2004. p. 35: xii+424

[18] Güttel S, Tisseur F. The nonlinear eigenvalue problem. ActaNumerica. 2017;**26**:1-94. DOI: 10.1017/S0962492917000034

[19] Voss H. Nonlinear eigenvalue proble ms. In: Hogben L, editor. Handbook of

Linear Algebra. 2nd ed. Boca Raton, FL: Chapman and Hall/CRC; 2014. p. 1904

[20] Ruhe A. Algorithms for the nonlinea r eigenvalue problem. SIAM Journal on Numerical Analysis. 1973;**10**(4):674-689. DOI: 10.1137/0710059

[21] Mehrmann V, Vos H. Nonlinear eigenvalue problems: A challenge for modern eigenvalue methods. GAMM-Mitteilungen. 2004;**27**(2):121-152. DOI: 10.1002/gamm.201490007

[22] Jittorntrum K. An implicit function theorem. Journal of Optimization Theory and Applications. 1978;**25**(4):575-577. DOI: 10.1007/BF00933522

[23] Savenko PA, Protsakh LP. Implicit function method in solving a two-dimensional nonlinear spectral problem. Russian Mathematics. 2007;**51**(11): 41-44. DOI: 10.3103/s1066369x07110060

[24] Savenko P, Tkach M. Numerical approximation of real finite nonnegative function by the modulus of discrete Fourier transformation. Applied Mathematics. 2010;**1**:41-51

[25] Andriychuk MI, Voitovich NN, Savenko PA, Tkachuk VP. Antenna Synthesis According to the Amplitude Radiation Pattern. Numerical Methods and Algorithms. Kyiv: NaukovaDumka; 1993. p. 256 (In Russian)

[26] Savenko PO. Nonlinear Problems of the Radiation System Synthesis. Theory and Methods of Solution. Lviv: IAPMM of NASU; 2002. p. 320 (In Ukrainian)

[27] Kravchenko VF, Protsakh LP, Savenko PA, Tkach MD. Mathematical peculiarities of plane equidistant array synthesis by given amplitude radiation pattern. Antennas. 2010; **3**(154):34-48

[28] Gantmacher FR. The Theory of Matrices.Volume One. New York: The Chelsea Publishing Company; 1959. p. 276

[29] Savenko P. Computational methods in the theory of synthesis of radio and acoustic radiating systems. Applied Mathematics. 2013;**4**:523-549. DOI: 10.4236/am.2013.43078

[30] Andriychuk MI, Savenko PO, Tkach MD. Synthesis of plane equidistant array according to power radiation pattern. In: Proceedings of XVIIth International Seminar/Workshop on Direct and Inverse Problems of Electromagnetic and Acoustic Wave Theory (DIPED-2012); 24-27 September 2012. Tbilisi, New York: IEEE; 2012. pp. 68-74

[31] Andriychuk MI, Voitovich NN. Antenna synthesis according to power radiation pattern with condition of norm equality. In: Proceeding of 2013 XVIIIth International Seminar/Workshop on Direct and Inverse Problems of Electromagnetic and Acoustic Wave Theory (DIPED); 23–26, September 2013. Lviv, New York: IEEE; 2013. pp. 137-140

[32] Andriychuk MI, Kravchenko VF, Savenko PO, Tkach MD. The plane radiation system synthesis according to the given power radiation pattern. FizicheskiyeProblemyPriborostroeniya. 2013;**2**(3):40-55 (In Russian)

[33] Andriychuk M, Savenko P, Tkach M. Non-linear synthesis problems for plane radiating systems according to the prescribed power directivity pattern. Open Journal of Antennas and Propagation. 2013;**1**:23-34. DOI: 10.4236/ojapr.2013.12006

[34] Andriychuk MI. Antenna Synthesis through the Characteristics of Desired Amplitude. Newcastle, UK: Cambridge Scholars Publishing; 2019. p. xvi+150

Chapter 12

Using Matrix Differential Equations for Solving Systems of Linear Algebraic Equations

Ioan R. Ciric

Abstract

Various ordinary differential equations of the first order have recently been used by the author for the solution of general, large linear systems of algebraic equations. Exact solutions were derived in terms of a new kind of infinite series of matrices which are truncated and applied repeatedly to approximate the solution. In these matrix series, each new term is obtained from the preceding one by multiplication with a matrix which becomes better and better conditioned tending to the identity matrix. Obviously, this helps the numerical computations. For a more efficient computation of approximate solutions of the algebraic systems, we consider new differential equations which are solved by simple techniques of numerical integration. The solution procedure allows to easily control and monitor the magnitude of the residual vector at each step of integration. A related iterative method is also proposed. The solution methods are flexible, permitting various intervening parameters to be changed whenever necessary in order to increase their efficiency. Efficient computation of a rough approximation of the solution, applicable even to poorly conditioned systems, is also performed based on the alternate application of two different types of minimization of associated functionals. A smaller amount of computation is needed to obtain an approximate solution of large linear systems as compared to existing methods.

Keywords: matrix equations, large linear algebraic systems, solution by numerical integration

1. Introduction

Exact analytic expressions in the form of infinite series of matrices for the solution of linear systems of algebraic equations were derived in [1] by integrating associated ordinary differential equations. These differential equations were obtained using a quadratic functional related to the system of algebraic equations and describe the orthogonal trajectories of the family of hypersurfaces representing the functional. More convergent matrix series were presented in [2] which can be applied to approximate the solution of the system of equations. Solution of linear systems based on the numerical integration of differential equations has originally been formulated in [3].

In Section 2 of the present book chapter, we use recently derived highly convergent series formulae for matrix exponentials [3] in order to construct improved iterative methods for solving approximately large systems of algebraic equations. In Section 3, we use novel functionals that allow to formulate differential equations which lead to a substantial increase in the efficiency of the solution process [4]. Independently of the starting point, the paths of integration of these equations converge all to the solution point of the system considered. At each step of the numerical solution, one passes, in fact, from one path to a slightly different one due to computation errors. The procedure does not require to find accurately an entire path but only the solution point which is common to all the paths. This is why we apply the simple Euler method [5] to integrate the differential equations. The computation errors are now determined by the magnitude of the second derivative of the position vector with respect to the parameter defining the location along the path. A related iterative method [6] is also described. In Section 4, two different kinds of minimization of the system functionals are applied alternately for quick computation of a rough solution of large linear systems [7].

2. Matrix series formulae for the solution of linear algebraic systems

Consider a system of equations written in matrix form as

$$Ax - b = 0 \tag{1}$$

where $A \in R^{n\times n}$ is an arbitrary nonsingular matrix, $b \in R^n$ is a given n-dimensional vector and $x = (x_1, x_2, \dots, x_n)^T$ is the unknown n-dimensional vector, with T indicating the transpose. Assume that x is a continuous function of the real variable v over a certain interval and associate to (1) the vector differential equation of the first order

$$\frac{dx}{dv} = f(v)(Ax - b) \tag{2}$$

where $f(v)$ is a continuous function to be chosen.

2.1 Exact analytic expressions for the solution of (2)

Imposing the condition $x(v_o) = x_o$, x_o being a chosen position vector, (2) has a unique solution over a specified interval [4], namely,

$$x(v) = x_o + e^{-Ag(v_o)} \sum_{k=0}^{\infty} \frac{A^k}{(k+1)!} \left[(g(v))^{k+1} - (g(v_o))^{k+1} \right] (Ax_o - b) \tag{3}$$

where $g(v) \equiv \int f(v) dv$ is a primitive of $f(v)$, i.e., $f(v) = dg/dv$. If $f(v)$ is taken to be $f(v) = 1/v$, then $g(v) = \ln v$. Choosing $v_o = 1$ gives $g(v_o) = 0$ and (3) becomes now

$$x(v) = x_o + (\ln v) \sum_{k=0}^{\infty} \frac{(A \ln v)^k}{(k+1)!} (Ax_o - b) \tag{4}$$

Over the interval $v \in (0, 2)$, $x(v)$ can also be expressed in the form [3]

$$x(v) = x_o - (1-v)\left[I + \sum_{k=1}^{\infty} \frac{(1-v)^k}{k+1}(I-A)\left(I-\frac{A}{2}\right)\cdots\left(I-\frac{A}{k}\right)\right](Ax_o - b) \tag{5}$$

I denoting the identity matrix of order n. The solution of (1) is theoretically obtained for $v = 0$, the series being in general extremely low convergent. Since the rate of convergence of the series in (5) is very small for v very close to the value corresponding to the solution of (1), i.e., $v = 0$, this formula should be applied repeatedly for a v not too close to zero, $v \in (0,1)$, until a satisfactory small distance to the solution $x(0)$ is reached.

2.2 Highly convergent series iteration formulae

Practical formulae for an approximate solution of (1) can be derived by using matrix series that are much more convergent than the one in (5). This can be done by writing (4) in the form

$$x(v) = x_o + A^{-1}\left(e^{A\ln v} - I\right)(Ax_o - b) \tag{6}$$

and by expressing the matrix exponential in terms of series of superior convergence given recently in ref. [3]. Very close to the solution $v = 0$, say for $v = e^{-N}, N \gg 1$, we have $e^{A\ln v} = e^{-NA}$ and with [3]

$$e^{-NA} - I = -10^{-q}c_q NA\left[I + \sum_{k=1}^{\infty}\frac{(-1)^k 10^{-qk}}{k+1}(I + c_q NA)\right.$$
$$\left.\times\left(I + \frac{c_q NA}{2}\right)\cdots\left(I + \frac{c_q NA}{k}\right)\right] \tag{7}$$

where $q > 0$ and $c_q \equiv 1/\ln(1 + 10^{-q})$, we get

$$x\left(e^{-N}\right) = x_o - 10^{-q}c_q N\left[I + \sum_{k=1}^{\infty}\frac{(-1)^k 10^{-qk}}{k+1}(I + c_q NA)\right.$$
$$\left.\times\left(I + \frac{c_q NA}{2}\right)\cdots\left(I + \frac{c_q NA}{k}\right)\right](Ax_o - b) \tag{8}$$

In order to perform numerical computations, the number of terms in the series has to be appropriately chosen. To have a small c_q in (8) one has to take a small q. For q = 0.5, e.g., $c_q = 3.639409$. One may start with N = 15 which would require to retain about 50 terms in this alternating series to get a rough approximation of $x(e^{-15})$. The computation is repeated with the new x_o taken to be the preceding $x\left(e^{-N}\right)$ until an acceptable accuracy of the solution of (1) is reached.

Much more convergent formulae are constructed if we apply in (6) the expansion [3]

$$e^{-NA} = (1+10^{-q})^{-p}\left\{I + p!\sum_{k=1}^{p}\frac{10^{-qk}}{k!(p-k)!}\left(I-\frac{c_qNA}{p}\right)\left(I-\frac{c_qNA}{p-1}\right)\cdots\left(I-\frac{c_qNA}{p-k+1}\right)\right.$$

$$-10^{-(p+1)q}c_qNA(I-c_qNA)\left(I-\frac{c_qNA}{2}\right)\cdots\left(I-\frac{c_qNA}{p}\right)\left[\frac{1}{p+1}I\right.$$

$$\left.\left.+p!\sum_{k=1}^{\infty}\frac{(-1)^k10^{-qk}}{(k+1)(k+2)\cdots(k+p+1)}(I+c_qNA)\left(I+\frac{c_qNA}{2}\right)\cdots\left(I+\frac{c_qNA}{k}\right)\right]\right\}, \tag{9}$$

$$p = 1, 2, \ldots$$

The multiplication with A^{-1} from (6) is avoided by arranging the first summation in (9) in terms of powers of A and by taking into account that

$$p!\sum_{k=1}^{p}\frac{10^{-qk}}{k!(p-k)!} = (1+10^{-q})^p - 1 \tag{10}$$

It is obvious that using (9) leads to computations with a much smaller number of terms retained in the series for a given N. With the same amount of computation we obtain an $x\left(e^{-N}\right)$ which is much closer to the exact solution $x(0)$ of the system of equations (1).

3. Methods of numerical integration

In this Section, we use special kinds of functionals that lead to the construction of differential equations which allow a substantial increase in the efficiency of the solution of large linear algebraic systems.

3.1 Vector differential equations and their application to the solution of (1)

Consider a functional of the form

$$F(x) = \|Ax-b\|^{\alpha} \tag{11}$$

associated to (1) where α is a positive real number to be chosen. A real variable $v, v > 0, v \in (v_o, v_S)$, is now defined by

$$F(x(v)) = F(x(v_o))\,h(v) \tag{12}$$

where $h(v)$ is an appropriately selected real function with definite first and second derivatives in $(v_o, v_S), v_o$ corresponding to a starting point $x(v_o) \equiv x^{(0)}$, with $h(v_o) = 1$, and v_S corresponding to the solution vector of (1) $x(v_S) \equiv x_S$, with $h(v_S) = 0$. Denoting $F\left(x^{(0)}\right) \equiv F_o$, we have

$$F_o\frac{dh}{dv} = \alpha\|Ax-b\|^{\alpha-2}\left(A^T(Ax-b)\right)^T\frac{dx}{dv} \tag{13}$$

Thus,

$$\frac{dx}{dv} = \frac{F_o}{\alpha}\frac{dh}{dv}\frac{A^T(Ax-b)}{\|Ax-b\|^{\alpha-2}\left\|A^T(Ax-b)\right\|^2} \tag{14}$$

which is the differential equation to be integrated from $v = v_o$ to $v = v_S$. The second derivative of x is

$$\begin{aligned}\frac{d^2x}{dv^2} &= \frac{F_o}{\alpha}\frac{d^2h}{dv^2}\frac{A^T(Ax-b)}{\|Ax-b\|^{\alpha-2}\left\|A^T(Ax-b)\right\|^2} \\ &+\left(\frac{F_o}{\alpha}\right)^2\left(\frac{dh}{dv}\right)^2\frac{1}{\|Ax-b\|^{2(\alpha-1)}\left\|A^T(Ax-b)\right\|^2} \\ &\times\left\{\frac{\|Ax-b\|^2}{\left\|A^T(Ax-b)\right\|^2}\left[A^TA-2\frac{\left\|AA^T(Ax-b)\right\|^2}{\left\|A^T(Ax-b)\right\|^2}I\right]+(2-\alpha)I\right\}A^T(Ax-b)\end{aligned} \tag{15}$$

From (11) and (12) we get a useful relationship,

$$\|Ax-b\| = (F_o\,h(v))^{1/\alpha} \tag{16}$$

that allows to simply monitor the magnitude of the residual vector of (1) during the computation process.

As explained in the Introduction we apply the Euler method for the solution of (14) and compute successively

$$x^{(i+1)} = x^{(i)} + \eta\left(\frac{dx}{dv}\right)_{x=x^{(i)}}, \quad i = 0,1,2,\dots \tag{17}$$

where η is the step size. In the absence of any hint about a good starting point $x^{(0)}$ corresponding to $v = v_o$, we have used the point along the normal from the origin $x = 0$ to the surface $F(x) = const$ in (11) which is the closest to the solution point x_S [8], i.e.,

$$x^{(0)} = \frac{\|b\|^2}{\left\|A^Tb\right\|^2}A^Tb \tag{18}$$

The function $h(v)$ and the parameter α are chosen such that the first and the second derivatives of x in (14) and (15) remain finite when $v \to v_S$, i.e., when the residual $\|Ax-b\| \to 0$, while the errors evaluated with the second derivative are kept reasonably small along the interval of integration $v \in (v_o, v_S)$. For each $h(v)$, α is determined by imposing the condition that the second derivative in (15) tends to zero as $\|Ax-b\| \to 0$. To decide on the value of α for a given $h(v)$, we require to have a good rate of decrease of $\|Ax-b\|$ at the beginning of the computation process, for instance to have (see (11) and (12))

$$\frac{\left\|Ax^{(1)}-b\right\|}{\left\|Ax^{(0)}-b\right\|} = (h(\eta))^{1/\alpha} \cong 0.8 \tag{19}$$

when $\eta = 0.1$. Finally, to solve (1), we determine the actual step size to be used for the numerical solution such that the errors at the beginning of the computation process are small, say

$$\frac{\eta^2}{2}\frac{1}{\|x^{(1)}\|}\left\|\frac{d^2x}{dv^2}\right\|_{x=x^{(0)}} < 0.01 \tag{20}$$

If the magnitude of the residual vector, which is computed at each step, does not decrease anymore the computation is continued with a new cycle of integration by applying (17) with $x^{(0)}$ replaced by a new starting point, i.e.,

$$x^{(new)} = x^{(last)} - \frac{\|Ax^{(last)} - b\|^2}{\left\|A^T\left(Ax^{(last)} - b\right)\right\|^2} \times A^T\left(Ax^{(last)} - b\right) \tag{21}$$

where $x^{(last)}$ is the position vector from the preceding step. $x^{(new)}$ is the closest point to x_S along the normal to the surface $F(x) = const$ taken at $x^{(last)}$ [8]. Subsequent cycles of integration are performed in the same way until a satisfactory approximate solution of (1) is obtained. If the difference between $x^{(new)}$ and $x^{(last)}$ is insignificant we find the point along the normal to $F(x) = const$, taken at $x^{(new)}$, where F has a minimum and, then, apply (21) again. It should be noted that as one approaches the solution point the direction of the gradient $A^T(Ax - b)$ tends to become more and more perpendicular to the direction of $x_S - x$ and, thus, the residual $\|Ax - b\|$ and the relative error $\|x - x_S\|/\|x_S\|$ will not decrease any more as expected. This is why the computation has to be continued by opening a new cycle of integration. For systems with higher condition numbers, this happens more quickly.

Numerical experiments obtained using $h(v) = 1 - v$ with $\alpha = 0.45$ and also using $h(v) = (1 - v)^2$ with $\alpha = 0.9$ shows that only two up to five integration cycles with a step size of 0.1 are needed in order to get an accuracy of about 1% for the solution of systems with condition numbers up to 100.

3.2 A related iterative method

The basic idea of this method is to find, starting from a point $x^{(0)}$, a point $x^{(1)}$ along the gradient of a functional (11) associated with the general system (1), such that the magnitude of the new residual vector is an established fraction of its initial value, $\|Ax^{(1)} - b\| = \tau\|Ax^{(0)} - b\|, \tau < 1$. Instead of performing an integration as in Section 3.1, one proceeds iteratively, i.e.,

$$\|Ax^{(i+1)} - b\| = \tau\|Ax^{(i)} - b\|, \quad i = 0, 1, 2, \ldots \tag{22}$$

for each iteration the starting point being the point found in the preceding iteration, with τ maintained as tight as possible at the same value. To do this, we impose that $F(x)$ in (11) varies from $x^{(i)}$ to $x^{(i+1)}$ in the same way for each iteration, namely,

$$F(x) = F\left(x^{(i)}\right) h(v) \tag{23}$$

where $h(v)$ is now a real function of a real variable, monotone decreasing in the interval $v \in [v_o, v_o + \eta], v_o \geq 0, \eta > 0$, with definite first and second derivatives, and with $h(v_o) = 1$ such that

$$F\left(x^{(i+1)}\right) = F\left(x^{(i)}\right) h(v_o + \eta), \quad i = 0, 1, 2, \ldots \tag{24}$$

Then, from (11),

$$\frac{F\left(x^{(i+1)}\right)}{F\left(x^{(i)}\right)} = \frac{\left\|Ax^{(i+1)} - b\right\|^{\alpha}}{\left\|Ax^{(i)} - b\right\|^{\alpha}} = \tau^{\alpha} \tag{25}$$

and

$$\tau = (h(v_o + \eta))^{1/\alpha} \tag{26}$$

$x^{(i+1)}$ is computed as

$$x^{(i+1)} = x^{(i)} + \eta \left(\frac{dx}{dv}\right)_{v=v_o} \tag{27}$$

i.e., with (14) where F_o is replaced with $F\left(x^{(i)}\right)$,

$$x^{(i+1)} = x^{(i)} + \frac{\eta}{\alpha}\left(\frac{dh}{dv}\right)_{v=v_o} \frac{\left\|Ax^{(i)} - b\right\|^2}{\left\|A^T\left(Ax^{(i)} - b\right)\right\|^2} \times A^T\left(Ax^{(i)} - b\right), \quad i = 0, 1, 2, \ldots \tag{28}$$

in which, taking into account (21), one has to have $\left|(\eta/\alpha)(dh/dv)_{v=v_o}\right| < 1$. This expression corresponds to the first step in the numerical integration by Euler's method of the differential Eq. (14), starting from $x^{(i)}$ with a step η, or to the first two terms of the Taylor series expansion of $x(v_o + \eta)$.

The starting value $x^{(0)}$ and the function $h(v)$ are chosen in the same way as in Section 3.1. For selected ratios η/α the iteration cycle continues as long as the residual $\left\|Ax^{(i)} - b\right\|$ decreases at a proper rate. Theoretically, to make $\left\|Ax^{(i)} - b\right\| = \varepsilon\left\|Ax^{(0)} - b\right\|$ with $\varepsilon \ll 1$ one would need to conduct $(\ln \varepsilon)/\ln \tau$ iterations. Since computation errors are introduced at each iteration (as in the Euler method), the initially chosen value of τ cannot be maintained the same as the iterative process continues. An approximate solution of (1) is obtained at the end of the iteration cycle as before, applying (21). Subsequent iteration cycles are performed with the starting point in each cycle being the point determined in the preceding cycle.

Numerical results were generated using, as in the method presented in Section 3.1, $h(v) = 1 - v$ but with $\eta/\alpha = 0.2$ and $h(v) = (1 - v)^2$ with $\eta/\alpha = 0.1$. Now, in both cases, $v_o = 0$, $(\eta/\alpha)(dh/dv)_{v=0} = -0.2$ in (28), and $\tau \cong 0.8$. A substantial increase in τ, approaching $\tau = 1$, or oscillations of its value during the first few iterations show that the computation errors evaluated from (15) (with F_o replaced by $\|Ax - b\|^{\alpha}$) are too big. In such a situation, one has to decrease the factor $(\eta/\alpha)(dh/dv)_{v=0}$ in (28), i.e., to decrease η/α for a given $h(v)$, which leads to an increase of τ and a decrease of $(1/2)\eta^2\|d^2x/dv^2\|_{v=v_o}$. For accuracy of 1% for the solution of (1), one needs a number of three to six short iteration cycles, with about eight iterations per cycle, for systems with condition numbers below 100. Of course, as in the previous method, an increased number of iteration cycles is required to reach the same accuracy for the solution of poorly conditioned systems.

Remarks. One can easily see that $x^{(i+1)}$ in (28) can be expressed in the form

$$x^{(i+1)} = x^{(0)} + \sum_{\ell=0}^{i} x_d^{(\ell)}, \quad i = 0, 1, 2, \ldots \tag{29}$$

where

$$x_d^{(\ell)} \equiv x^{(\ell+1)} - x^{(\ell)} = \frac{\eta}{\alpha}\left(\frac{dh}{dv}\right)_{v=v_o} \frac{\left\|Ax^{(\ell)} - b\right\|^2}{\left\|A^T\left(Ax^{(\ell)} - b\right)\right\|^2} \times A^T\left(Ax^{(\ell)} - b\right) \tag{30}$$

with the solution of (1) given by the infinite series

$$x_S = x^{(0)} + x_d^{(0)} + x_d^{(1)} + x_d^{(2)} + \cdots \tag{31}$$

$x_d^{(\ell)}$ in (30) can also be written as

$$x_d^{(\ell)} = \frac{\eta}{\alpha}\left(\frac{dh}{dv}\right)_{v=v_o} \frac{\left\|b^{(\ell)}\right\|^2}{\left\|A^T b^{(\ell)}\right\|^2} A^T b^{(\ell)} \tag{32}$$

where

$$\begin{aligned} b^{(\ell)} &= Ax_d^{(\ell-1)} - b^{(\ell-1)}, \quad \ell = 1, 2, \ldots; \\ b^{(0)} &= Ax^{(0)} - b \end{aligned} \tag{33}$$

This shows that the difference $x_d^{(\ell)}$ is a "rough approximation" to the solution of a system (1) whose right-hand side is, at each iteration, just the residual vector of the system in the preceding iteration, which decreases in magnitude from one iteration to the next. Thus, the method presented in this Section represents a practical, concrete implementation of the well-known idea of successive approximations/perturbations [9].

To search for a possible increase in efficiency, more general functionals of the form

$$F(x) = \mathcal{F}(\|Ax - b\|) \tag{34}$$

could be tested, where $\mathcal{F}$ and its first derivative are finite and continuous at all the points within the interval of integration. Now the corresponding (14) is

$$\frac{dx}{dv} = F_o \frac{dh}{dv} \left(\frac{d\mathcal{F}}{d\|Ax-b\|} \right)^{-1} \frac{\|Ax-b\|}{\left\|A^T(Ax-b)\right\|^2} \times A^T(Ax-b) \tag{35}$$

Note. The paths of integration in the methods presented can be looked at as being the field lines of a Poissonian electrostatic field in a homogeneous region bounded by a surface of constant potential $F(x) = \|Ax_o - b\|^\alpha$ and with a zero potential at the solution point, $F(x_S) = 0$. In the particular case of $\alpha = 2$ the ratio of the volume density of charge within the region to the material permittivity is constant, namely, $-2\sum_{i=1}^{n}\sum_{k=1}^{n} a_{ik}^2$ where a_{ik} are the entries of A. By altering this electrostatic field one could eventually make quicker the approach to the solution point along the integration path.

4. Method of alternate minimizations

This simple method is based on the property of a functional of the form (34) associated with a general system of equations (1) to allow not only to minimize the value of the functional but also the distance to the solution point of (1). Using only minimizations along ordinary gradients of the functional is not efficient unless the system is very well-conditioned.

For computing efficiently an approximate solution of general, large linear systems of algebraic equations, we propose in this Section the alternate application of minimizations of a functional and of the distance to the solution point, along the direction of the gradient of the functional.

Consider the functional in (34) where $\mathcal{F}$ is a real function defined for all x from a chosen starting point $x^{(s)}$ to the solution point x_S of (1), monotone decreasing with $\|Ax-b\|$, $\mathcal{F}(0) = F(x_S)$. The gradient of $F(x)$ is

$$\nabla F(x) = \frac{d\mathcal{F}(\|Ax-b\|)}{d\|Ax-b\|} \frac{A^T(Ax-b)}{\|Ax-b\|} \tag{36}$$

i.e., in the direction of the vector $A^T(Ax-b)$.

The minimum of $F(x)$ along a straight line through $x^{(s)}$ in the direction defined by an n-dimensional vector d is found from the condition that $x - x^{(s)} = \lambda d$, where λ is a scalar to be determined, and $A^T(Ax-b)$ are perpendicular. This gives λ and x for the minimum of $F(x)$, namely,

$$x^{(d)}_{\min F} = x^{(s)} - \frac{\left[(Ad)^T\left(Ax^{(s)} - b\right)\right]}{\|Ad\|^2} d \tag{37}$$

The point at which $F(x)$ is minimum along the normal taken at $x^{(s)}$ is determined by replacing d with $A^T\left(Ax^{(s)} - b\right)$ in (37),

$$x_{\min F} = x^{(s)} - \frac{\left\| A^T \left(Ax^{(s)} - b \right) \right\|^2}{\left\| AA^T \left(Ax^{(s)} - b \right) \right\|^2} A^T \left(Ax^{(s)} - b \right) \tag{38}$$

The minimum distance between the solution point x_S and a point x along the direction of the gradient of $F(x)$ taken at $x^{(s)}$ is at

$$x = x^{(s)} + \mu A^T (Ax - b) \tag{39}$$

where the scalar μ is determined by requiring the distance $\|x - x_S\|$ to be minimum. This gives μ and x for this minimum, namely,

$$x_{\min D} = x^{(s)} - \frac{\left\| Ax^{(s)} - b \right\|^2}{\left\| A^T \left(Ax^{(s)} - b \right) \right\|^2} A^T \left(Ax^{(s)} - b \right) \tag{40}$$

which depends only on the residual vector and on $A^T(Ax - b)$ at the point $x^{(s)}$, independently of the form of $\mathcal{F}$.

As already mentioned, repeated minimizations using only (38) are not efficient for solving a general system (1). The same is true when using only (40). To obtain an approximate solution of (1), we apply the formulae (40) and (38) alternately, the starting point $x^{(s)}$ being each time the point determined in the preceding minimization. As in Section 3, when there is no indication about a convenient first starting point $x^{(s)}$, one can use the origin $x^{(s)} = 0$. Only a few iterations, up to ten, are needed for a solution with a relative error $\|\tilde{x} - x_S\| / \|x_S\|$ of about 1% for systems with condition numbers of up to 100.

The procedure is surprisingly efficient for a rough solution even for very ill-conditioned systems. For example, for the system $Hx = b$ where H is the Hilbert matrix of order eight and whose solution is $x_S = [1, 1, \ldots, 1]^T$ [10], a solution of 6% accuracy is obtained with a starting point $x^{(s)} = 0$ by performing only seven alternate minimizations (40), (38), with no equilibration or regularization preoperated on the system. By comparison [10], for the Gauss elimination method, the accuracy is only 40.6%, for the Gauss elimination with equilibration 9.15%, and for the Householder method with equilibration 5.6%.

Experimental results show that, as the new point x given by (40), (38) becomes closer to the solution point x_S, the direction $A^T(Ax - b)$ of the gradient in (36), i.e., of the correction terms in (40) and (38), becomes more and more orthogonal to the direction of $x - x_S$. This causes the magnitude $\|Ax - b\|$ of the residual vector of (1) and the relative difference $\|x - x_S\| / \|x_S\|$ to not significantly decrease anymore. To progress with the computation one can intercalate minimizations of $F(x)$ along directions that are different from that of $A^T(Ax - b)$. By numerical testing, it is observed that $x - x^{(0)}$, where $x^{(0)}$ is the original starting point, has at this stage, in most cases, a significant component along the direction of $x - x_S$. Thus, one can use such a minimization direction to try to improve the solution accuracy.

5. Conclusions

Highly convergent iteration formulae for solving general, large linear systems of algebraic equations are derived from exact analytic solutions of particular differential equations based on new, accurate series representations of the matrix exponential recently published in ref. [3]. Specialized differential equations which make it possible to monitor and control the computation errors and the decrease of the magnitude of the residual vector $\|Ax - b\|$ of (1) at each stage of the computation procedure are constructed and integrated numerically to approximate the solution of these systems. Two methods of the solution have been presented in this book chapter and the simplest Euler method is applied for the numerical integration of the vector differential equations. In the first method cycles of integration are used, each cycle starting from a convenient value of the unknown and continuing until the rate of decrease of $\|Ax - b\|$ becomes too small. The second method is an iterative method where a fixed rate of decrease of $\|Ax - b\|$ is imposed at the beginning of the iteration cycles. In this method, only the first step in the Euler method is computed at each iteration and each cycle of iteration is conducted until there is no significant change in the magnitude of the residual vector. These two methods are highly efficient for large systems with condition numbers below 100 since only up to six cycles with less than ten steps per cycle are necessary to get a solution accuracy of 1%, at each step within a cycle having to compute two matrix-vector multiplications. The method in Section 3.2 seems to be more efficient for systems with bigger condition numbers. The number of cycles of integration/iteration increases with the condition number and preconditioning should be done for ill-conditioned systems before attempting to apply the methods presented in this work.

The iterative method of alternate minimizations presented in Section 4 is intended for computing quickly a rough approximation of the solution of linear systems of equations. In this method, preequilibration or preregularization/preconditioning are not required to obtain useful results even for systems with poorly conditioned matrices.

The present Book Chapter has been intended to constitute a review of the work done so far on the subject matter. It describes the proposed new methods for an approximate solution of large linear algebraic systems using appropriately chosen matrix functionals and shows the procedures for constructing the concrete solution algorithms. At this stage, the results presented are only validated by preliminary numerical experiments which indicate the efficiency of the proposed procedures for deriving approximate/rough solutions of large systems. More numerical experiments involving systems with higher condition numbers, as well as theoretical results, will be presented in future work. It is my hope that other researchers will be attracted to this new area and rigorous theoretical results will also be established (theorems, etc.).

Author details

Ioan R. Ciric
The University of Manitoba, Winnipeg, Canada

*Address all correspondence to: ioan.ciric@umanitoba.ca

References

[1] Ciric IR. Series solutions of vector differential equations related to linear systems of algebraic equations. In: Proceedings of the 12th International Symposium on Antenna Technology and Applied Electromagnetics and Canadian Radio Sciences Conference (ANTEM/URSI). Montréal, Canada: IEEE; 2006. pp. 317-321. ISBN: 978-0-9638425-1-7

[2] Ciric IR. Rapidly convergent matrix series for solving linear systems of equations. In: Proceedings of the 17th International Symposium on Antenna Technology and Applied Electromagnetics (ANTEM). Montréal, Canada: IEEE; 2016. DOI: 10.1109/ANTEM.2016.755022/978-1-4673-8478-0

[3] Ciric IR. New matrix series formulae for matrix exponentials and for the solution of linear systems of algebraic equations. In: Shah K, Okutmustur B, editors. Functional Calculus. London: IntechOpen; 2020. DOI: 10.5772/Intechopen.2022.77599/978-1-83880-007-9

[4] Ciric IR. Approximate solution of large linear algebraic systems using differential equations. In: Proceedings of the 19th International Symposium on Antenna Technology and Applied Electromagnetics (ANTEM). Winnipeg, Canada: IEEE; 2021. ISBN: 978-1-6654-0335-1

[5] Burden RL, Faires JD. Numerical Analysis. 5th ed. Boston: PWS-KENT; 1993. ISBN: 0-534-93219-3

[6] Ciric IR. Using selected rates of residual vector decrease for solving iteratively large linear systems. In: Proceedings of the 19th International Symposium on Antenna Technology and Applied Electromagnetics (ANTEM). Winnipeg, Canada: IEEE; 2021. ISBN: 978-1-6654-0335-1

[7] Ciric IR. Alternate minimizations for an approximate solution of general systems of linear algebraic equations. In: Proceedings of the 19th International Symposium on Antenna Technology and Applied Electromagnetics (ANTEM). Winnipeg, Canada: IEEE; 2021. ISBN: 978-1-6654-0335-1

[8] Ciric IR. A geometric property of the functional associated with general systems of algebraic equations. In: Proceedings of the 12th International Symposium on Antenna Technology and Applied Electromagnetics and Canadian Radio Sciences Conference (ANTEM/URSI). Montréal, Canada: IEEE; 2006. pp. 311-315. ISBN: 978-0-9638425-1-7

[9] Lanczos C. Applied Analysis. New York: Dover Publications, Inc.; 1988. ISBN: 0-486-65656-X (Paperback)

[10] Hämmerlin G, Hoffmann K-H. Numerical Mathematics. New York: Springer-Verlag; 1991